Environmental Dispute Resolution

ENVIRONMENT, DEVELOPMENT, AND PUBLIC POLICY

A series of volumes under the general editorship of
Lawrence Susskind, *Massachusetts Institute of Technology, Cambridge, Massachusetts*

ENVIRONMENTAL POLICY AND PLANNING
Series Editor:
Lawrence Susskind, *Massachusetts Institute of Technology, Cambridge, Massachusetts*

CAN REGULATION WORK?
Paul A. Sabatier and Daniel A. Mazmanian

PATERNALISM, CONFLICT, AND COPRODUCTION
Learning from Citizen Action and Citizen Participation in Western Europe
Lawrence Susskind and Michael Elliott

BEYOND THE NEIGHBOHOOD UNIT
Residential Environments and Public Policy
Tridib Banerjee and William C. Baer

RESOLVING DEVELOPMENT DISPUTES THROUGH NEGOTIATIONS
Timothy J. Sullivan

ENVIRONMENTAL DISPUTE RESOLUTION
Lawrence S. Bacow and Michael Wheeler

Other subseries:

CITIES AND DEVELOPMENT
Series Editor:
Lloyd Rodwin, *Massachusetts Institute of Technology, Cambridge, Massachusetts*

PUBLIC POLICY AND SOCIAL SERVICES
Series Editor:
Gary Marx, *Massachusetts Institute of Technology, Cambridge, Massachusetts*

ENVIRONMENTAL DISPUTE RESOLUTION

LAWRENCE S. BACOW
Massachusetts Institute of Technology
Cambridge, Massachusetts

AND

MICHAEL WHEELER
New England School of Law
Boston, Massachusetts
and Massachusetts Institute of Technology
Cambridge, Massachusetts

PLENUM PRESS • NEW YORK AND LONDON

Library of Congress Cataloging in Publication Data

Bacow, Lawrence S.

Environmental dispute resolution.

(Environment, development, and public policy. Environmental policy and planning)
Bibliography: p.
Includes index.
1. Environmental mediation—United States. I. Wheeler, Michael, 1943–
II. Title. III. Series.
KF3775.B32   1984   344.73′046′0269   84-16066
ISBN 0-306-41594-1   347.304460269

©1984 Plenum Press, New York
A Division of Plenum Publishing Corporation
233 Spring Street, New York, N.Y. 10013

All rights reserved

No part of this book may be reproduced, stored in a retrieval system, or transmitted in any form or by any means, electronic, mechanical, photocopying, microfilming, recording, or otherwise, without written permission from the Publisher

Printed in the United States of America

To our children

Preface

This book has its origins in an M.I.T. research project that was funded by the U.S. Environmental Protection Agency (EPA). Our immediate objective was to prepare a set of case studies that examined bargaining and negotiation as they occurred between government, environmental advocates, and regulatees throughout the traditional regulatory process. The project was part of a larger effort by the EPA to make environmental regulation more efficient and less litigious. The principal investigator for the research effort was Lawrence Susskind of the Department of Urban Studies and Planning. Eight case studies were prepared under the joint supervision of Susskind and the authors of this book.

Studying the negotiating behavior of parties as we worked our way through an environmental dispute proved enlightening. We observed missed opportunities for settlement, negotiating tactics that backfired, and strategies that appeared to be grounded more in intuition than in thoughtful analysis. At the same time, however, we were struck by how often the parties ultimately managed to muddle through. People negotiated not out of some idealistic commitment to consensus but because they thought it better served their own interests. When some negotiations reached an impasse, people improvised mediation. These disputants succeeded in spite of legal and institutional barriers, even though few of them had a sophisticated understanding of negotiation.

It soon became clear that the case studies we were developing had a powerful teaching potential. The studies provided documented examples of opportunities and obstacles to negotiation in a variety of regulatory contexts, among them permitting, enforcement, grant making, and rulemaking. Our pedagogical goal was twofold: first, to prepare materials that would help environmentalists, developers, and regulators negotiate more effectively and intelligently; and second, to identify for legislators, planners, and managers ways in which laws could be amended and procedures revised to encourage nonadversarial dispute resolution.

Whether one is locked into a particular dispute or is concerned with broader policy, a sophisticated understanding of the dynamics of bargaining and negotiation is essential. This book represents an attempt to teach a structured, analytic approach to the major issues likely to be encountered when people work their way through environmental controversies. Our mode of analysis draws heavily from decision theory. Other scholars might view this topic from a different perspective—a psychological approach, to name one. A variety of perspectives can be valuable, but we are convinced that everyone can benefit from sharper analytic negotiating skills.

We have taken the original case studies prepared for the EPA, edited them heavily for teaching purposes, and supplemented them with essays, notes, questions, problems, additional readings, and descriptions of still other cases. (For those who are interested, a full version of the cases is available in a book entitled *Resolving Environmental Regulatory Disputes* by Lawrence Susskind, Lawrence Bacow, and Michael Wheeler [New York: Schenkman, 1984].) Although this book should be of interest to anyone who may someday be involved in an environmental controversy, it is organized as a self-contained text for a one-semester graduate-level course in environmental dispute resolution. A course by that name has been offered by the New England School of Law and M.I.T. each of the past four years. The materials have been revised substantially in that time. The book is designed to be accessible to students from a variety of backgrounds—law, planning, management, public administration, and engineering. In fact, our course has usually drawn a mix of students from such schools and has thus served as a forum for examination of different perspectives on environmental problems. Our materials assume no prior training or exposure either to bargaining, negotiation, or environmental policy. Draft portions of the book have also been used in courses at the University of California, the University of Colorado, the University of Hawaii, Harvard Law School, and the Harvard Graduate School of Education.

Readers will quickly recognize that we believe that, in many instances, face-to-face negotiation of environmental disputes is more likely than litigation to produce a fair and efficient outcome that serves the interests of all sides. Yet we do not view negotiation as a panacea. Through questions and notes, we have tried to highlight the pitfalls and shortcomings of negotiation as well as the advantages. More fundamentally, we believe it is important to understand the interplay of negotiation and litigation; in many instances, both paths are followed.

The organization of the book reflects the order in which negotiation issues arise in practice. The first chapter examines the nature of environmental conflict, its sources, its costs, and the frustrating characteristics of litigation that often give rise to a search for alternatives. The second chapter on dispute resolution theory introduces the analytic approach to negotiation that is developed in the remainder of the book; in this chapter, we develop a vocabulary for analyzing negotiation problems. Chapter 3 focuses on the first problem that confronts a prospective negotiator: What are my incentives to negotiate and what are the incentives of the other parties? This matter of incentives first arises at the start of bargaining but remains relevant throughout a dispute as the parties continually assess and reassess the factors that keep them at the bargaining table. Chapter 4 analyzes one of the strongest incentives for negotiation: the prospect of mutually beneficial gains through joint problem solving. Chapter 5 explores the problems inherent in resolving disputes that appear to be highly technical; that is, disputes in which data, modeling, and differing expert opinions lie at the core of the problem. The next two chapters look at how bargaining and negotiation change when more than two parties are at the table and at how issues of compliance affect the negotiation process. Chapters 8, 9, and 10 are devoted to mediation. They examine the circumstances under which the presence of a nonpartisan facilitator may help to achieve agreement; they also raise a number of important questions about the ethical responsibilities of the mediator in an environmental dispute. Chapter 11 analyzes multiparty negotiation as it occurs at the policy development stage. Chapter 12 adopts a systemic perspective and reviews attempts to institutionalize negotiation through reforms in a traditional dispute resolution process. The book concludes with a look at a number of themes that are touched on but not directly addressed in earlier chapters.

Many people made important contributions to the writing of this book. Foremost among them was our colleague and friend, Larry Susskind. He took responsibility for organizing and administering the original research project; he proposed the idea of a casebook; he patiently reviewed drafts and offered advice based on his use of the teaching materials; and he gently prodded us to keep the project on track. We are indebted to him for both his advice and his enthusiasm.

The original versions of the cases that appear in this book were researched and written by a talented group of students and postdoctoral fellows. The original research group consisted of Heidi Burgess, David Gilmore, Stephen Hill, Diane Hoffman, Alexander Jaegerman, Jennifer Knapp-Stump, Mary Lucci, Douglas Smith, and Timothy Sullivan. The original author of each case is identified where his or her case appears in the book. We are grateful to Julia Wondolleck for her permission to use portions of her Grayrocks Dam case. We are also indebted to our students for their comments on earlier drafts of this book. Our colleague from the Harvard Negotiation Project, David Kuechle of the Harvard Graduate School of Education, contributed greatly through additional research and updating of the Brown Paper case that appears in Chapter 3. Thomas Schelling of Harvard University generously shared with us some of his superb teaching materials on bargaining.

We also wish to thank Henry Beal, formerly of the EPA, who helped initiate the research project that gave rise to this book. The Department of Urban

Studies and Planning at M.I.T., the New England School of Law, and the Program on Negotiation at Harvard Law School also contributed institutional support to this project. Finally, we wish to thank Heather Worrel and Audrey Latimer who displayed great patience and skill in helping to prepare the manuscript for publication.

Contents

CHAPTER 1

The Nature of Environmental Conflict 1

    Introduction 1
    The Sources of Environmental Conflict 5
    The Storm King Litigation 10
    Litigating Environmental Disputes 12
    Study Questions 16
    The Scope of Judicial Review 17
    Negotiation as an Alternative 18

CHAPTER 2

Dispute Resolution Theory 21

    Introduction 21
    Negotiation Analysis 22
    Incentives to Negotiate 26
    Obstacles to Consensus 28
    Problem 1 30
    Zero-sum and Nonzero-sum Disputes 33
    Problem 2 34
    Problem 3 35
    Bargaining Strength 38
    Problem 4 40
    Conclusion 41

Chapter 3

Incentives to Negotiate 42

    Introduction 42
    Incentives to Settle a Lawsuit 43
    Incentives to Negotiate in Other Contexts 44
    Case Study: Grayrocks Dam 46
    Study Questions 51
    Inducements and Obstacles 52
    Problem 54

Chapter 4

Joint Problem Solving 56

    Introduction 56
    Case Study: Brown Paper 56
    Study Questions 70
    Problem 1 71
    Problem 2 71
    Problem 3 72
    The Search for Solutions 73

Chapter 5

Data Negotiation 76

    Introduction 76
    The Holston River Case 77
    Study Questions 89
    Judicial Review of Technical Decisions 91
    Courts and Technical Issues 96
    The Elusive Nature of Facts in Environmental Cases 99

Chapter 6

Two-Party Versus Multiparty Negotiations 104

    Introduction 104
    Case Study: The West Side Highway 105

CONTENTS   xiii

    West Side Highway Study Questions........................... 108
    Negotiation Participants: Representation 109
    Study Questions on Sullivan's Proposal......................... 113
    Case Study: The Snoqualmie Dam Dispute 113
    Questions on Snoqualmie...................................... 115
    Multiparty Negotiation and Coalitions.......................... 115
    Problem 1 117
    Problem 2 118
    Problems of Cost Sharing..................................... 119
    Study Questions.. 122
    Cross-References.. 122
    Study Questions.. 123
    Reading References... 124

CHAPTER 7

Prospects for Compliance....................................... 126

    Introduction 126
    Case Study: Jackson, Wyoming—201 Grants
       for Municipal Wastewater Treatment 127
    General Study Questions 143
    Hypothetical Problem... 145
    The Compliance Issue 145
    Cross-Reference: A Look Ahead................................ 154

CHAPTER 8

Mediation Techniques 156

    Introduction 156
    Case Study: Mediation and the Brayton Point Coal Conversion 158
    Study Questions.. 184
    Cross-Reference... 185
    Mediation Study Questions 186
    Mediation Skills.. 187

CHAPTER 9

Mediating Large Disputes...................................... 195

Introduction 195
Case Study: The Foothills Water Treatment Project 196
General Study Questions 240
Mediation Study Questions 241
Foothills Epilogue.. 243
The Role of the Press 244

CHAPTER 10

Mediation Ethics... 248

Introduction 248
The Concept of Accountability 248
Problems of Implementing Accountability........................ 253
Study Questions... 262
Related Ethical Issues... 262
Study Questions: Incentives and Ethics......................... 269
The Mediator With Clout 270
Conclusion 275

CHAPTER 11

Negotiated Rulemaking.. 279

Introduction 279
Case Study: Water Treatment Rulemaking 280
Case Study Questions.. 303
Competition or Cooperation?................................... 304
Study Questions on Negotiations and the Rulemaking Process 305
Reform.. 305
Study Questions... 316

CHAPTER 12

Institutionalizing Negotiation............................... 323

Introduction 323
Removing Procedural Obstacles 324
Conflict Anticipation and Its Kin.............................. 326
Mandatory Negotiation 328

Study Questions on Colstrip 336
Cross Reference Problem 336
Encouraging Negotiation through Incentives 339
Study Questions on the Siting Act 343

CHAPTER 13

Epilogue.. 347

Introduction 347
Small-Scale Disputes 347
Problem 1 350
Problem 2 354
Politics and Alternative Dispute Resolution 359
Study Questions 362
Conclusion: Negotiation and Public Policymaking 362

Bibliography 365

Index 369

1

THE NATURE OF ENVIRONMENTAL CONFLICT

INTRODUCTION

Not that long ago, most people regarded economic development as the cornerstone of social progress. Industrial expansion, municipal growth, highway construction, energy development, and mineral extraction were promoted both as ends in themselves and as the means by which the lives of all members of society would constantly improve.

In the last two decades, however, public attitudes about development have changed significantly. People have come to realize that few of the benefits of development come without significant costs. Many of the same industries that produce goods and provide jobs also pollute the air and water. Likewise, unregulated urban growth may increase the housing stock but cause urban sprawl and overload municipal services. The massive federal highway program expanded the nation's transportation system, but sometimes it cut into fragile landscapes. In turn, energy development and mineral extraction have jeopardized other natural resources. Increasing concern about such impacts has been at the heart of the environmental protection movement.

The negative consequences of economic development have always been with us. What has changed, however, has been people's attitudes about them. The causes for this emerging environmental consciousness still are not completely clear. One plausible explanation is that the costs of development, though substantial, often are long term and cumulative. Hence, it may take years, even generations, before people can fully appreciate all the impacts of industrial pollution. Moreover, the consequences of certain kinds of pollution can be subtle and hard to detect. People failed to oppose the dumping of toxic wastes 20 or 30 years ago, not because they were unconcerned with their health, but because they had no way of knowing about the long-term hazards.

A second explanation for the protectionist movement is that people's values have fundamentally changed. Social historians might well remind us that the recent wave of environmentalism did not take shape in an empty sea; instead, it grew in a turbulent period (the 1960s and 1970s) of political ferment and change. Thus, some may construe environmentalism as one reaction to excessive materialism. Others may relate it to a more basic antiestablishment impulse. (For a summary of the historical forces that underlie the environmental movement, see Scott Mernitz, *Mediation of Environmental Disputes*. New York: Praeger, 1980, pp. 1–22.)

A third explanation for the shift in public opinion focuses on political and legal institutions. The passage of new legislation and the creation of court precedents have not only empowered opponents of development in specific cases, but they have strengthened the environmental movement as a whole. The success that conservationists achieved in lobbying for stricter water quality regulations in the late 1960s, for example, contributed to the political momentum that led to enactment of clean air legislation in the early 1970s.

In short, the recent environmental movement has many roots; the three that are suggested here simply begin the list. As distinct as these three might appear at first, they share an important element: they are all grounded in conflict. To the extent that environmentalism is related to growing awareness of the long-term impacts of development, for example, some people will suffer such costs more acutely than do others. The priorities of those who live next to a plastics factory are bound to be different from those who work in it or buy its products.

Likewise, even though environmentalism may reflect changed social values, the attitudes of particular individuals and groups may still be in sharp conflict. Public opinion polls that show increased general concern with protecting natural resources also reveal that views are sharply divided on specific issues such as balancing the need for national energy independence against protection of the wilderness. Finally, to the extent that legal and political institutions have fostered environmentalism, they have always operated in arenas of contention. Environmental statutes have been enacted only after intense lobbying and heated debate. Landmark court decisions have been the products of a system premised on adversary proceedings.

Whatever its roots, environmental conflict has been manifested repeatedly in a wide range of cases and in a variety of familiar forms. There have been demonstrations at nuclear power plant sites, court injunctions to prevent federal funding for highways, objections raised in regulatory hearings to the granting of discharge permits, and legislative logrolling over tightening or relaxing current statutory standards.

Conflicts over specific developments or activities often generate significant costs of their own. When environmental disputes go to court, for example, litigants on all sides—industry, government, and citizen groups—can incur substantial legal fees. Moreover, delays in the courtroom or in protracted regulatory hearings may impose still greater costs on developers and, ultimately, their consumers. Likewise, to the extent that there is uncertainty about how the dispute will be resolved, affected parties may have to engage in expensive contingency planning. Opportunities that were open at one point may be lost long before the dispute can be settled. In short, when industry, government, and citizen groups get locked into contentious battles, they all inevitably must consume resources that could be used elsewhere productively.

The human energy and economic resources that are lost through this sort of friction may be regarded as transaction costs, that is, costs that are related to the methods of resolving conflict. However, costs can be reflected in ends as well as means. Indeed, the frequent inadequacy of conventional dispute resolution processes, most notably lawsuits, to produce efficient and equitable settlements may account for the most significant cost of environmental conflict.

There is no reason to believe that the fundamental conflict over environmental issues will soon diminish, let alone disappear. Development will continue to benefit some people and harm others. People undoubtedly will continue to hold disparate values and priorities. An analysis of 1,800 reported cases of environmental conflict between 1970 and 1977 revealed the following current trends:

> Environmental conflict is spreading geographically, but once it emerges in any particular region, it remains. Environmental conflict is spreading to encompass a wider range of industrial facilities . . . Environmental conflict is more and more focused on new projects moving into an area rather than on problems in existing facilities. The frequency of environmental conflict is steadily rising with an increasing percentage of heavy industrial projects encountering community opposition. (Gladwin, "Environmental Conflict," 2 *EIA Review* 48–49, 1978)

Although environmental conflict is almost certain to remain with us, there is still reason to hope that it can be managed better. Although the costs of conflict may not be eliminated, they likely can be reduced. Even if perfection will always be out of reach, the quality of decisions in environmental cases surely can be enhanced.

For the most part, the environmental debate has centered around specific substantive questions: Should work on the Tellicoe Dam in Tennessee be halted because of the apparent risk to the snail darter? Should industrial plants in the Midwest be required to burn low sulfur fuel in order to prevent acid rain in New England? Should the federal government permit western cities to build dams and reservoirs on public lands?

Although there has been no end of articles, documentaries, and studies on the scientific and economic dimensions of substantive issues, there has been surprisingly little attention to the variety of methods that can be used to address them. The following questions are germane. What is lost (and what is gained) when law-trained judges must resolve highly complex and often controversial scientific questions? In what way do current administrative procedures breed disputes instead of preventing them? Is it possible to revise our procedures so as to promote more equitable and efficient environmental policymaking?

Such questions necessarily invite examination of the competence of courts to deal with broad social issues. It is always open season for criticism of lawyers and judges, but what are the alternatives to environmental litigation? Why is it often so difficult to settle these cases out of court? To what extent do collective bargaining and mediation of international disputes offer useful lessons?

This book is about the process of environmental dispute resolution. Extensive case studies in the chapters that follow will describe specific problems of air and water pollution and of land use and energy development. They have been chosen, however, to illustrate processes and procedures, not substantive law or technology. Moreover, although the cases generally involve large state and federal problems, the manner in which they were resolved should be instructive for the handling of local disputes among neighbors. Likewise, the cases describe natural resource issues, but they are also relevant to community confrontations over human rights and resources.

The book has two principal goals. One is to help teach practitioners whose work involves them in fights over the development and preservation of limited resources to represent their clients and constituents more effectively. These practitioners—lawyers, managers, planners, consultants, and government officials—need both the technical skills and knowledge of their particular disciplines and a broader capacity to analyze and employ competing modes of dispute resolution (among them litigation, negotiation, mediation, arbitration, and fact-finding).

The second goal of the book is to suggest a perspective for policymakers. If environmental conflict is inevitably with us, can we manage it better by revising our legal procedures and implementing new processes? There is so much at stake that even modest improvements in current approaches might yield substantial social gains.

This first chapter introduces themes that will recur throughout the book. Far more issues are raised at this juncture than are resolved. Moreover, the authors acknowledge their own skepticism about the appropriateness of the judicial resolution of complex environmental problems; this skepticism is undoubtedly reflected in the materials chosen for the book and in its organization. Thus cautioned, readers should be better situated to make their own critical judgments. Indeed, as a starting point, consider more fully the impacts of environmental conflict: (1) What other costs can be cataloged? and (2) are there not also some countervailing benefits? For more extensive consideration of conflict resolution generally, see Morton Deutsch, *The Resolution of Conflict* (New Haven: Yale University Press, 1973, pp. 3–19); Paul Wehr, *Conflict Regulation* (Boul-

der: Westview Press, 1979, *pp. 1–24); and* Kenneth Boulding, *Conflict and Defense* (New York: Harper, 1962, pp. 305–328).

The Sources of Environmental Conflict

Most environmental disputes arise because people have different views over what constitutes good policy for the environment. A utility may propose to build a power-generating dam, but farmers and conservationists fight it because of its effect on irrigation and wildlife downstream. The government may license a new regional landfill that is opposed by neighboring residents who fear the noise it will generate. By adopting a new regulation that requires municipalities to improve their wastewater treatment facilities significantly, the Environmental Protection Agency (EPA) may unwittingly invite opposition both from the affected cities who claim the regulations are needlessly stringent and costly and from environmental groups who argue that the regulations are not strict enough to protect water quality. In all such cases the essence of the dispute is a question of policy: Should the dam be built? Is the landfill located in the right place? Is the EPA regulation cost-effective? People inevitably disagree about what constitutes proper policy in such cases. If environmental disputes are to be resolved, it is essential to understand the more fundamental conflicts that underlie them.

Environmental conflict has many sources. People often take opposing positions because they have quite different stakes in the outcome. For example, fishermen tend to oppose dams because they are harmful to fishing, whereas farmers support them if they will provide more water for agriculture. A simple assessment of the distributional consequences—who wins and who loses—can provide important insights into the politics of environmental controversies. Furthermore, such an assessment can be the first step in creating solutions to environmental disputes. Where the gains generated by a project will exceed losses, people who will feel the negative impacts may drop their opposition if there is some appropriate compensation. For example, a developer who wants to use some valuable open space may be able to assuage residents by dedicating other land to recreational uses. (The use of compensation is discussed throughout the book, particularly in Chapter 3.) If a project will produce more losses than gains, it is of course inefficient, and it should be abandoned.

In many environmental controversies, however, it is not at all obvious who will win and who will lose. Environmental policymaking often involves considerable uncertainty. For example, the construction of a power plant on a shoreline may—or may not—threaten the coastal ecosystem, depending upon whether pollution control devices turn out to be effective. If the technology is new or the geology unique, no one can be absolutely sure of all the consequences. Likewise, building the power plant may potentially help the local economy by providing jobs and tax revenues, or it may ultimately prove detrimental if construction imposes excessive demands on municipal services. Whether the project proves to be a boon or bane may finally depend on general economic conditions and trends that defy accurate prediction.

The following simple example illustrates how uncertainty about physical and economic impacts is central to many environmental disputes; it also introduces a mode of analysis for making decisions when the consequences are not completely clear. Imagine a community that must decide how to dispose of its municipal solid waste. It is weighing two alternatives: constructing a sanitary landfill or building a new incinerator. Both options present possible environmental risks. Landfills, if not adequately constructed and maintained, will leach pollutants and contaminate groundwater. Incinerators, if not properly built, will pollute the air.

The choice facing the community is represented graphically in Figure 1. The upper branch of the decision tree represents the option of building the landfill; the lower branch represents the alternative of constructing an incinerator. The community is well aware of the options, of course. It is the consequences that are of concern. Either the landfill or the incinerator may prove to be clean or dirty environmentally. These consequences may be grafted onto the branches of the decision tree, but because they represent matters of chance rather than choice, they are symbolized by the circular "chance node" rather than the boxlike decision point. (In practice, of course, there would be a range of possible outcomes; the extent to which either facility might pollute would be a matter of degree. Decision theory can accommodate this complexity, but for the sake of clarity the options here have been simplified to be *clean* or *dirty*. For an exposition of this theory, see Howard Raiffa's *Decision Analysis*. Reading, Mass.: Addison-Wesley, 1968.)

In some instances, even though the outcomes are uncertain, there may be pretty good estimates of their probability. Experience may show, for example, that a certain control technology fails to be effective 2% of the time. In other cases, however, the probabilities themselves may be less clear. In Figure 2, the chance that the landfill will leach pollutants is set at the probability p; the chance that it will not pollute thus must be $1-p$. It would be an odd coincidence, indeed,

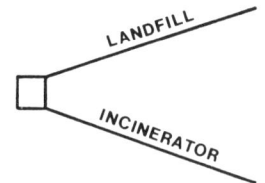

FIGURE 1. A simple decision tree.

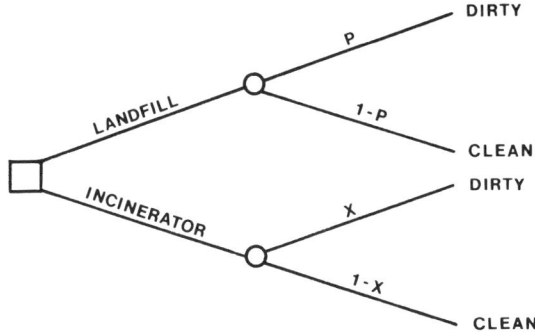

FIGURE 2. Decision making under uncertainty.

if the risk of pollution by the incinerator was identical. Therefore, the chance of its failure is represented by x in the lower portion of the diagram; $1-x$ represents the chance that the system will not pollute.

So that the significance of uncertainty is highlighted, assume for the moment that in all other respects the facilities are equivalent: construction and maintenance costs are the same, and the facilities dispose of solid waste equally well. Which option should the community select?

The answer, of course, is that it depends. All other things being equal, the community should prefer the alternative that poses the lesser threat to the environment. That threat, however, is a function of two factors: the probability of system failure (p or x in the previously mentioned figure) and the magnitude of resulting harm. People often make radically different probability estimates. Some may have great confidence that a facility will operate safely, whereas others genuinely fear that it will fail. Sometimes such differences may rest on access to different data. In environmental disputes, it is not unusual for both project proponents and project opponents to have only partial information about impacts. Even when people are looking at the same data (or anecdotal information), they may come to entirely different conclusions about their meanings.

Some environmental disputes are thus generated by different assessments of probabilities. They are battles, one might say, between optimists and pessimists. There still may be conflict, however, even where there is no real disagreement about the likelihood of future impacts. In the solid waste disposal example, it is conceivable (though perhaps unlikely) that citizens might agree that the chances of the landfill failing are in the order of one percent. Still, there might be sharp disagreement over the extent of harm should there be such a failure. Some might regard the possible contamination of local water as a costly inconvenience but one that could be rectified by diverting water from other sources. Other people might regard such a possibility as a disaster, threatening as it might human health—a priceless commodity. Although the possibility of system failure might be remote, such people would view the consequences as so threatening that they would vigorously oppose the landfill. To the extent that environmental disputes are triggered by different assessments of impacts, they are really conflicts over values. People who value clean water more than clean air will tend to favor the incinerator over the landfill, and vice versa.

Differences in opinion over the possible impacts of a project can also spark debate over the proper scope of the dispute. For example, when community officials consider the air pollution problems that might be generated by the incinerator, may they consider only the possible impacts on local residents or must they also take into account the interests of people who live elsewhere in the air shed? In like manner, what are the responsibilities of such officials when the interests of current voters may be at odds with the interests of future generations? Although there are philosophical and economic justifications for defining disputes broadly so that all the impacts on every affected party are taken into account, existing political systems are based on geographic boundaries that may artificially limit the perceived impacts (good as well as bad) of a project.

An environmental dispute thus may be based on different estimates of probabilities and impacts; it may be compounded by differences over the scope of the dispute in terms of both place and time. Even where there is complete accord in this regard, however, there still may be conflict over what sort of risks are tolerable. Suppose, for example, that everyone in the community agrees about the probability of failure and resulting impacts for both the incinerator and the landfill. In the case of the incinerator, it is clear that x is very high, virtually 100%—incinerators are predictably dirty—but the potential damage is relatively low, say $25 thousand. By contrast, in the case of the landfill, the probability of system failure p is quite low, 1%, but should anything go wrong it would cause $1 million in damage. Landfills are very expensive to clean up. Figure 3 represents the two options graphically.

What should the community do? Some residents might argue for the landfill because it is extremely likely to be environmentally benign. Others might favor the incinerator, which although almost certain to pollute, is not expected to do much damage. An actuary might calculate the expected cost of each facility by multiplying its probability of failure by its particular harm. Thus, the expected cost of the incinerator would be $24,750 (99% × $25,000), whereas the expected cost of the landfill would be only $10,000 (1% × $1,000,000). If the community had to make such decisions repeatedly it would be ahead in the long run by minimizing expected cost. Yet, because such decisions are seldom repeated and because the consequences of a failure would be so enormous, people might rationally want to avoid such a risk no matter how remote. To take an example from another setting, one person may be comforted by the fact that there is only one chance in a thousand of a serious accident at a proposed nuclear power

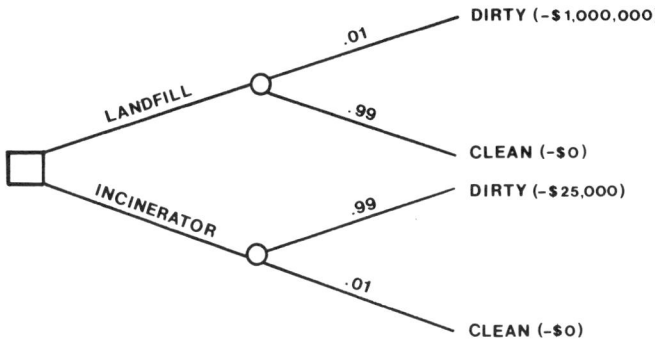

FIGURE 3. Comparing probabilities and impacts.

plant, whereas another may be terrified by the same statistic. In short, even in a hypothetical world where probabilities and impact were known by all, differing attitudes toward risk would breed conflict. In practice, of course, there are likely to be difference on each of these scores.

Most environmental disputes are similar structurally to our landfill–incinerator example. Conflict arises not only because of the distributional consequences of the project but because people assess probabilities, outcomes, and risks differently. In the Storm King case described in the section that follows, some nearby residents objected to a massive water project because it would detract from their scenic views. In their eyes, they had nothing to gain and much to lose; their opposition rested on distributional grounds. That dispute also involved the project's threat to fishing. Fishermen and conservationists thought it would be very harmful, whereas the Consolidated Edison Company of New York (Con Ed, the utility promoting the project) claimed it would have a negligible effect. These were differences of probability and impact assessment, but there were also disagreements in values. Con Ed officials apparently were willing to tolerate a modest decrease in the Hudson River fish population but the fishermen were not. To project supporters, the environmental risks were acceptable; to opponents they were not.

When an environmental dispute springs from different views of what the future holds and how it should be valued, it is tempting to appeal to experts—biologists, chemists, economists, and the like—to settle the question. The use of expertise is considered throughout the book. The Holston River case described in chapter 5 illustrates how technical information may be developed and, in a sense, negotiated. Sometimes experts can narrow the range of disagreement and focus attention on new solutions. It is a mistake, however, to assume that environmental disputes would disappear if there was sufficient technical information. Notwithstanding the continuing advances of science, there is still much that is unknown about the consequences of natural resource use. So long as new products and technologies develop, there will be some uncertainty about their long-term impacts. More fundamentally, even when there is no disagreement about the facts, there may be legitimate differences of priorities, values, and attitudes toward risk. A nuclear engineer may be well situated to compute the probability of a meltdown, but there is no reason why his or her opinion about what constitutes an acceptable risk should be given more weight than the view of any other citizen. It may be tempting to refer technical issues to seemingly neutral experts or "blue ribbon commissions," but experience shows that their reports often become ammunition for the contending parties.

The Storm King Litigation

In 1962, the Consolidated Edison Company of New York announced plans to build a pumped storage hydroelectric project at Storm King Mountain on the Hudson River in Cornwall, New York. The proposed facility would pump water from the Hudson to a reservoir located over 1,000 feet above the river. Electricity from Con Ed's conventional steam plants in New York City would provide electric power to force the water up the mountain during off-peak hours. In periods of peak power demand, the water would be released to generate additional electricity.

Con Ed's proposals raised a number of concerns among local residents, environmentalists, and fishermen. People feared that a project of such massive scale would ruin the scenic vistas and recreational opportunities in the area, threaten the local ecology, and jeopardize the fishing along the Hudson. The groups banded together to form the Scenic Hudson Preservation Conference (SHPC) and intervened in the subsequent Federal Power Commission (FPC) hearings. SHPC claimed (1) that the new facility was unnecessary; (2) that Con Ed could better meet its power demands through the use of gas turbines; (3) that underground transmission lines would minimize scenic damage; and (4) that Con Ed had neglected to consider seriously the impact of the project on the Hudson River fisheries.

The first lawsuit in the Storm King case was filed in 1965. That year the United States Court of Appeals for the Second Circuit ruled that the Federal Power Commission (FPC) had erred in rejecting testimony offered by the intervenors concerning the likely impacts from the project and alternatives to mitigate those impacts. The court remanded the case to the FPC for another hearing to gather further evidence. *Scenic Hudson Preservation Conference v. FPC*, 354 F.2d 608 (2d Cir. 1965); cert. denied, 384 U.S. 941 (1966). After a rehearing, the FPC issued a new construction license to Con Ed and SHPC went back to court again. In the second case, which was decided in 1972, the Court of

Appeals rejected SHPC's claims that the FPC's decisions regarding Storm King were unsupported by substantial evidence and thus ruled in favor of Con Ed. *Scenic Hudson Preservation Conference v. FPC*, 453 F.2d 463 (2d Cir. 1971), cert. denied; 407 U.S. 926 (1972).

The second case, however, did not signal the end to the Storm King's legal battles. In 1974, SHPC and the Hudson River Fishermen's Association (HRFA) sought to have the FPC reopen licensing hearings on the Storm King plant, arguing that newly discovered evidence of tidal flow in the Hudson greatly increased the vulnerability of fish larvae to destruction at the project outflow. Although the court declined to revoke the project license as requested by the plaintiffs, it nonetheless ordered the FPC to hold a hearing on the Fishermen's association's motion for a suspension of the project's operations during spawning season. *Hudson River Fishermen's Ass'n v. FPC*, 498 F.2d 827 (2d Cir. 1974).

The last round in the Storm King legal battle was won by the SHPC, which obtained an injunction in 1974 barring Con Ed from dumping rock and other fill material from the project into the Hudson without first obtaining a permit under the Federal Water Pollution Control Act. *Scenic Hudson Preservation Conference v. Callaway*, 499 F.2d 127 (2d Cir. 1974). Indeed, the final chapter on the Storm King controversy was not concluded until 1981 when, after 16 years of court battles and a 20,000-page administrative record, the parties finally settled their differences with the benefit of an outside mediator.

What should we make of such protracted cases? The battle over Storm King has been hailed by some as a great victory for environmentalism and pilloried by others as an example of obstructionism by a wealthy elite. It is hard, however, to regard the Storm King dispute as anything but a major loss for society as a whole. Project supporters and their opponents struggled at great expense for well over a decade to produce an outcome that apparently could have been achieved far more cheaply and quickly. For all their involvement, the courts almost always focused on narrow legal questions (such as compliance with statutes setting out requirements for administrative hearings) rather than addressing the substantive issues that were the heart of the Storm King controversy. In the end, no participant could feel well served by the litigation process.

Although not many cases are as lengthy or as costly as was Storm King, environmental disputes that go on for years are not uncommon. Moreover, they are expensive no matter what the final outcome is. If objections are ultimately found to be frivolous and the project is ultimately approved, its costs may well have multiplied several times; in the interim, citizens and consumers must make do with other services and products. In other cases, it may require enormous effort to vindicate environmental concerns; expedited decision making would liberate money and labor to be used productively in other causes and activities.

The costs of litigation can be formidable in environmental disputes of much smaller scale. Neighbors and town boards who are battling over the proposed use of some open space may find it difficult to underwrite even modest legal expenses. The delays that occur in such disputes can be just as lengthy and disruptive as those that occur in the celebrated cases.

The same kinds of delays and expenses that occur in litigation also take place in the administrative process, and for much the same reasons. According to former Secretary of Labor John Dunlop,

> The regulatory process encourages conflict, rather than acting to reconcile opposing interests. Moreover, there is a sense that it is wrong for the regulatory agency to try to bring parties together and develop consensus. Reliance on public and highly formal proceedings makes the development of consensus extremely difficult, if not impossible. And unless this consensus can be developed, neither party has any stake in the promulgated rule. Thus both are free to complain that it is biased, stupid, or misguided. Moreover, each side is free to continue the controversy in the form of endless petitions for review, clarification, and litigation before the agency and the courts. Nothing is ever settled because true settlement can come only through agreement, consent, or acquiescence. (Dunlop, "The Limits of Legal Compulsion," 27 *Labor L. J.* 67, 70, 1976)

What is it about regulatory disputes of any scale that tend to make them so difficult? Why is it that a legal system that at least passably handles tort claims, contract disputes, and probate matters often appears ill-suited to the resolution of complicated environmental problems? Perhaps most important, why is it that parties to environmental disputes, often mindful of the shortcomings of litigation, nevertheless turn to it rather than negotiation, mediation, or some other form of dispute resolution? Litigation, as we shall see, is not without its virtues. There also can be obstacles that deter disputants from turning to alternatives.

Litigating Environmental Disputes

The supposed shortcomings of the litigation process have been trumpeted in popular and academic journals. Skepticism about the courts is hardly new, nor is it confined to nonlawyers. More than half a century ago, Judge Learned Hand stated, "As a litigant, I should dread a lawsuit beyond almost anything short of sickness and death." (Hand, "Deficiencies of Trials to Reach the Heart of the Matter," in 3 *Lectures on Legal Topics* 89, 105, 1926.) As the recent Storm King saga confirms, legal cases can drag on for years without ever getting to the substance of a dispute. Nevertheless, environmental advocates continue to resort to the courts. It is important to consider why.

For all its limitations, litigation offers empowerment. Small groups, even individuals, can take on giant corporations and powerful government agencies and sometimes win. Litigation is also attractive in that it usually forces action.

When one side brings a suit and makes allegations, the other side must respond. The process goes forward under highly structured rules that are well known to the parties in advance. Litigation not only provides for an aggrieved party to get a hearing, but also, to some extent at least, to set the agenda. Environmental activists sometimes prefer to put their claims before a judge rather than before an agency official or a legislative committee. Judges may be slow in rendering an order, and to some extent they may choose which issues to address. However, they are less able to duck controversial problems than are other policymakers. Also, judges who are appointed for life enjoy a political insularity that may make it easier for them to come to politically unpopular decisions.

Even if outright victory is unlikely, the mere filing of a lawsuit may give an environmental organization important leverage. A company that is reasonably confident of ultimately prevailing in court may nonetheless offer to mitigate environmental damage if its opponents will withdraw their suit and let the project go forward immediately. (If the claims of the opponents are utterly groundless, of course, they and their lawyers may be liable for harrassment.) In theory, at least, plaintiffs may face a dilemma in such instances: by accepting an attractive settlement offer in a particular case, they may forfeit the opportunity to establish a favorable legal precedent for other cases.

Environmental advocates may be drawn to litigation for other reasons. Lawsuits can be a means of educating the public and galvanizing opinion. By going to court to stop chemical dumping at a particular site, an environmental group may also hope to develop support for stricter legislative standards generally. Bringing a lawsuit may also help strengthen an organization by demonstrating its vigilance and dedication. Abstract commitments to environmental quality are made concrete when a case goes to court. The sense of immediacy may help attract new members. Political demonstrations may accomplish some of the same things, but bringing a lawsuit casts the issues in terms of legal rights.

Litigation can be attractive because it is cheap, relatively speaking, at least in the early stages. The costs of instituting a lawsuit are usually minimal. Subsequent stages—retaining expert witnesses, engaging in extensive discovery, and the like—can be extremely expensive, of course, but some of the advantages of litigation noted previously may be obtained at the outset. (For a description of the kinds of pretrial expenses that can be incurred in major cases, see E. Green, J. Marks, and R. Olson, "Settling Large Case Litigation: An Alternative Approach," 11 *Loyola L.A. L. Rev* 493, 497–501. 1977-78.) Even when cases continue all the way through trial and appeal, costs for the environmentalists are sometimes manageable if professionals are willing to donate their time to the cause.

The decision to go to court is made easier by the fact that it is not irrevocable. Environmental groups often pursue negotiation and mediation at the same time they are prosecuting a lawsuit. One of the themes that is explored in this book is the relationship between adversary and consensual processes. (For example, some environmental advocates believe that it is foolhardy to begin to negotiate without first bringing a suit, so that there is some sort of credible threat to the other side; other representatives contend that litigation tends to polarize the parties; hence, litigation may make any talk of compromise more difficult.)

In cases that fail to settle and therefore go all the way through trial, court procedures constrain the introduction of evidence, they limit the relevant arguments, and they define the way in which judges must view disputes. Judicial decision making is quite different from policymaking that occurs in other branches of government. The author of the article excerpted later assesses the role of the courts as arbiters of social policy. Following the selection are a series of questions that explore the application of his general observations to the field of environmental dispute resolution. As you read this material, keep two general questions in mind:

1. How do legal procedures likely affect the way in which concurrent processes, like negotiation, are carried out?
2. What specific aspects of judicial decision making might tend to make it more or less attractive to environmental disputants?

Horowitz, Donald. In *The Courts and Social Policy*. Washington, D.C.: Institute, 1977, pp. 34–35.

> 1. Adjudication is focused. The usual question before the judge is simply: Does one party have a legal right? Does another party have a duty? This should be contrasted with the question before a "planner," whether legislative or bureaucratic: What are the alternatives? These are quite different ways of casting problems for decision. For the judge, alternatives may be relevant, but they are relevant primarily to the subsequent issue of what "remedies" are appropriate to redress "wrongs" done to those who possess "rights." In other words, the initial focus on rights tends to defer the question of alternatives to a later stage of the inquiry and to consider it a purely technical question.
>
> As this suggests, the initial focus on rights is also a serious impediment to the analysis of costs, for, in principle at least, if rights exist they are not bounded by considerations of cost. If a person possesses a right, he possesses it whatever the cost. . . .
>
> The contrast between rights and alternatives suggests the much broader framework in which non-adjudicative policymakers function. . . . [Horowitz notes that a judge's power is largely coercive; while he has some flexibility to tailor remedies to the particular problem at hand, usually his options are limited to directing the parties to do or refrain from doing something.]
>
> Legislators and administrators, on the other hand, have a wider range of tools in their kit. They may resort to the same kinds of sanctions judges invoke, or they may use taxation, incentives and subsidies of various kinds,

interventions in the marketplace, the establishment of new organizations or the takeover of old ones, or a number of other ways of seeking to attain their goals. The judiciary, having no budget (save for administrative expenses), no power to tax or to create new institutions, has much less ability to experiment or to adjust its techniques to the problems it confronts. . . .

2. Adjudication is piecemeal. The lawsuit is the supreme example of incremental decisionmaking. As such, it shares the advantages and the defects of the species. The outcome of litigation may give the illusion of a decisive victory, but the victory is often on a very limited point. The judge's power to decide extends, in principle, only to those issues that are before him. Related issues, not raised by the instant dispute, must generally await later litigation. . . .

Piecemeal decisions also isolate artificially what in the real world is merged. It is a truism that everything is related to everything else, and of course this cliche proves too much, because no institution can or should attempt to deal with everything simultaneously. But the litigation setting creates the danger of doing too little at one time and thus magnifies the possibility of unanticipated consequences that a more comprehensive view might perceive and attempt to limit or control. . . . Piecemeal decisions result in the seriatim consideration of policy priorities. The judge cannot frame his issue in terms of more health care versus less prison reform, though (depending on whether and how executives and legislators respond to his decision) this may be the exact result of a decision that purports to make choices in one of these areas or the other. Again, the focus on rights obscures the ultimate nature of the social policy choices being made, and so does the judges' lack of budgetary authority or responsibility. . . .

3. Courts must act when litigants call. The passivity of the judicial process is one of its most prominent characteristics. Judges sit to hear disputes brought to them by parties; they do not initiate action. This makes the sequencing of judicially ordered change dependent on the capricious timing of litigants rather than the planning of a public body. It also makes it difficulty to ascertain the extent to which the situation of the litigants faithfully represents or illustrates the dimensions of the problem they bring to court. . . . As a matter of litigation strategy, plaintiff's lawyers are likely to bring not the most representative case but the most extreme case of discrimination, of fraud, of violation of statute, of abuse of discretion, and so on. . . .

[Similarly, there is] no assurance that litigants constitute a random sample of the class of cases that might be affected by a decree. Because courts respond only to the cases that come their way, they make law from what may be very special situations. Courts see the top of the iceberg as well as the bottom of the barrel. The law they make may be law for the worst case or for the best, but it is not necessarily law for the mean or modal case.

The unrepresentative character of litigants raises another problem. Unlike legislation, litigation is not a finely tuned device for registering intensities of preference. Bargaining and compromise—at least bargaining and compromise beyond the confines of the individual case—are more difficult because of the adversary setting and the limited number of interested participants. Dependent as it is on an uncompromisingly partisan presentation, the adversary process is not conducive to the ordering of preferences. It compels litigants to argue favorable positions with a vigor that may be out of proportion to their actual preferences and that may therefore mislead the judge; in any case, their preferences may have little support in the wider social group the litigants ostensibly represent. In ascertaining the configuration and intensity of public preferences, the judge is, for the most part, left to roam at large.

This problem is naturally exacerbated by the deliberately imposed isolation of judges from their communities. The prohibition on judges discussing pending cases with individuals or groups interested in the outcome is obviously designed to insure the independence and impartiality of the judiciary. But what fosters the detachment of judges is necessarily at odds with their sensitivity to social forces.

4. Fact-finding in adjudication is ill-adapted to the ascertainment of social facts. [Horowitz discusses the difficulties courts face in ascertaining the validity of hypotheses governing social relationships. For example, does pornography stimulate the commission of sex crimes, or does it provide a cathartic release for those who might otherwise commit such offenses? He argues that the structure of litigation renders the courts ill suited to decide such issues. The analogy in environmental disputes is to the resolution of highly technical matters. This problem is considered in depth in chapter Five and in the Holston River case.]

5. Adjudication makes no provision for policy review. . . . Judges base their decisions on antecedent facts, on behavior that antedates the litigation. Consequential facts—those that relate to the impact of a decision on behavior—are equally important but much neglected.

This, of course, is a result of the focus on rights and duties rather than alternatives. Litigation is geared to rectifying the injustices of the past and present rather than to planning for some change to occur in the future. The very notion of planning is alien to adjudication. . . . The courts are mainly dependent for their impact information on a single feedback mechanism: the follow-up lawsuit. This mechanism tends to be slow, erratic, unsystematic. Courts have no inspectors who move out into the field to ascertain what has happened. They receive no regular reports on the implementation of their policies. . . . The judges have no grapevine extending into the organizations and groups whose behavior they affect. Judicial properties foster isolation rather than contact. Neither do the courts learn about the effects of their decisions by conducting investigations or planning exercises.

Study Questions

1. Horowitz begins his essay by noting that adjudication is focused and that judges frame problems in terms of rights and duties. As a result, they pay

relatively little attention to costs. Can you think of environmental problems where this approach might be desired? Are there other problems where it might not?

2. Not all cases get litigated to the bitter end; indeed, the vast majority are settled along the way. What role should the judge play in encouraging the parties to settle? How is he or she constrained institutionally?

3. In the classic essay "The Forms and Limits of Adjudication" (92 *Harv. L. Rev.* 353, 1978) the late Lon Fuller distinguished between cases that raise polycentric problems and those that do not. A polycentric problem is one that is so "many centered" that a pull at any one point distorts everything else. He argued, much like Horowitz, that courts are not well suited to resolving such problems. Are environmental problems like Storm King polycentric? What does the succession of lawsuits in the Storm King case reveal about the capacity of the judiciary to address the substance of such a controversy?

4. What resources does a judge have at his or her disposal to analyze any particular issue raised in a case? If the parties fail to develop the evidence adequately, how might a judge gather information independently?

5. Are judges well suited by training or experience to resolve the value conflicts that lie at the heart of most environmental disputes? The great majority of federal judges are white, middle age, upper middle-class males with relatively little experience with environmental problems. Would policy making be better served if there was more diversity on the bench?

THE SCOPE OF JUDICIAL REVIEW

Trial judges do not possess unlimited discretion to decide regulatory cases. Carefully crafted rules of administrative law restrict the scope of a judge's inquiry in any given case. These rules are intended to ensure that courts do not usurp executive decision-making authority. At the same time, however, these rules can also prevent courts from ever coming to grips with the substantive issues that typically lie at the heart of environmental disputes.

Procedurally, much important environmental litigation arises as a challenge to some action by an executive agency. For example, in the Storm King case, the Scenic Hudson Preservation Conference originally filed suit claiming that the Federal Power Commission had violated the terms of its enabling act by approving Con Ed's license application. (An *enabling act* is a statute that defines the scope of an agency's powers. Often it prescribes procedures that the agency must follow in conducting business as well as criteria for determining when the agency should act.) Courts are justifiably reluctant to second guess agencies in matters of substance, especially where the decision in question requires the agency to exercise special expertise. Accordingly, the law sharply limits the scope of judicial review in such cases. Typically, a court will only set aside an agency decision if the agency failed to follow proper procedures, if the agency incorrectly interpreted the law (deciding questions of law is an area where the court, not the agency, possesses special expertise), or if the agency abused its discretion or acted arbitrarily or capriciously. If the court merely disagrees with the agency on a question of policy, it lets the agency's decision stand.

In short, it is much easier for a plaintiff to prevail by showing that the agency in question committed an error of procedure than it is to show that the agency simply made a bad policy decision. As a result, much environmental litigation is ostensibly directed at narrow procedural and legal issues, instead of the underlying policy question. A group trying to stop construction of a nuclear power plant might bring a suit to set aside its license on the grounds that the regulatory agency acted without holding the requisite public hearings. Were they to prevail in the suit, however, they would not necessarily kill the project. Instead, all they would win would be the right to get a new hearing. Perhaps they would be able to introduce new evidence and arguments at that point, but the agency would not be required to reverse its earlier ruling. Similarly, many environmental cases deal with the narrow issue of whether an impact statement is required for the project or, if one has been prepared, whether it is adequate. Even if the project opponents succeed in court, the most that a judge can do for them is to require the agency to follow the appropriate procedures. In some environmental suits, one substantive issue is really a surrogate for another. Environmental groups may go to court ostensibly to protect some endangered species of plant or animal, when the paramount question is the general desirability of the project that presents the threat. Many environmentalists who fought the Tellicoe Dam would still have opposed it even if it did not endanger the snail darter.

In all such cases, the legal issue that is presented to the court merely provides the plaintiff with leverage to challenge a broader agency decision. It is ironic that the system operates in such a way that disputants often find themselves arguing shadow issues. The policy questions that are at the heart of most such controversies—should the nuclear power plant be allowed to operate, should the dam be completed, and so forth—are rarely addressed by the courts. As a consequence, environmental lawsuits seldom resolve the real differences between the contending parties.

Negotiation as an Alternative

A consensual approach to environmental dispute resolution offers a number of distinct advantages over the conventional adversary process. In contrast to litigation, negotiation relies upon the principals to create the terms of the final outcome. These principals bring to the bargaining table a much deeper understanding of the technical and institutional dimensions of environmental problems than is generally possessed by judges. Often, they are in a better position to explore different solutions and analyze their consequences. A judge who lacks any formal training in environmental science or policy may only see one or two similar cases in his or her entire career.

Because the negotiators usually will have to live with their settlement (for better or worse) they may also be more sensitive to implementation concerns than would be a judge whose involvement with the case typically ends with the issuance of the final decree. Moreover, because the outcome of a bargaining process usually represents a meeting of the minds, negotiation is more likely to produce results that accurately reflect the preferences of the parties. Opposing negotiators usually conclude their work on better terms than do opposing litigants. Because relationships between negotiators tend to be better and because they have a greater investment in a settlement than in a court-imposed order, the prospects for successful implementation should also be enhanced. Even when subsequent problems do arise, the earlier negotiation experience may serve as model for their expeditious resolution. Lawsuits, by contrast, often seem to breed more lawsuits.

Perhaps the strongest argument in favor of negotiation of environmental disputes, however, is that it makes it far more likely that substantive issues will be addressed. Although narrow standards of review often prevent judges from second guessing the policy decisions of administrative agencies, parties who meet face-to-face can bring these issues to the forefront.

As the cases in the following chapters document, consensual resolution of environmental disputes already takes place. The benefits outlined previously have attracted environmentalists, developers, and government officials to negotiation, mediation, and kindred processes. But although negotiation appears to have some advantages over litigation, it also may have some shortcomings as a method of policymaking. Some of these problems may be inherent in consensual processes, but others may be amenable to solution. To the extent that some of these obstacles can be removed or diminished, the road to negotiation will be made smoother. Much of the material in this book is intended to identify these obstacles and to analyze various means around them. Among the most serious issues are:

1. *Who should be included in the negotiations?* Unlike litigation, negotiation is usually an ad hoc process; therefore there are no firm rules governing who can participate. To return to the solid waste problem discussed earlier, do citizens or officials of neighboring towns have a right to take part in talks about constructing the new incinerator? What if the community tries to exclude such people?
2. *What incentives need be offered to induce the parties to bargain in good faith?* People usually will not bargain unless they think that it is in their

best interest to do so. Is some sort of mechanism needed to provide technical support to groups that feel that they lack the scientific information or the bargaining skills to negotiate effectively?

3. *How should complex technical issues be resolved?* Environmental disputes often raise complex scientific questions. Can these issues be negotiated, or should they be resolved by some kind of expert tribunal?

4. *At what point should the parties seek the services of a neutral third party to facilitate negotiation?* Mediators can play a useful role in suggesting alternatives for consideration, maintaining channels of communication, and reducing final agreements to writing. When should the parties look to a mediator for help? In what ways can a mediator simplify or complicate the bargaining process?

5. *How can an agreement be made binding?* It is one thing to negotiate a settlement, but it is quite another to implement it successfully. Uncertainty over the prospects for enforcement of a potential agreement may cripple negotiations. Why might parties to a negotiated agreement later breach it? What kind of steps might be taken to reduce the risk of breach?

As these questions suggest, negotiation has great promise, but it is not without its difficulties. The case studies that constitute the core of this book illustrate these and other problems as well as the methods that various disputants have used to overcome them.

2

Dispute Resolution Theory

Introduction

Negotiation is a fundamental method of dispute resolution. After all, even most lawsuits are not decided by judges or juries. Instead, they are settled out of court by the parties themselves. Negotiation is also central to other forms of dispute resolution. For example, mediation (a device sometimes used for settling environmental disputes) is basically negotiation that is carried out with the assistance of a third party.

On one level, all of us are familiar with negotiation. We may bargain over trivial things, like what to order at a Chinese restaurant, or we may haggle over important items, such as the price of a house. Sometimes we bargain for ourselves; in other cases, we may represent clients or organizations. This sort of firsthand knowledge of bargaining is supplemented by observing negotiations that are carried out in the public arena. The bargaining over the hostages in Iran, the battle over the nuclear power plant in Seabrook, New Hampshire, the air controllers' strike in 1981—all such exchanges regularly provide us with lessons in how (and how not) to negotiate.

Yet, as commonplace as negotiation is in our personal and professional lives, few people have a coherent understanding of the negotiation process. Bargaining often is seen as an art—not a science—and perhaps a "black" art at that. Until very recently, only a handful of law, business, and planning schools have offered courses in the theory and practice of negotiation. Serious interest in negotiation is on the increase, however, and there is now a substantial scholarly literature on the subject. Economists, psychologists, and policy analysts have long studied negotiation, and they have been joined, if belatedly, by lawyers and other professionals whose work brings them into the field. Although it is impossible to summarize negotiation theory in a single chapter, we believe it is essential to introduce some of the most fundamental concepts and analytic tools before we turn to the practice of environmental dispute resolution.

A theoretical framework allows us to identify and judge a negotiator's key actions. Environmental disputes typically involve many parties and issues. Without some sort of conceptual chart, one is unlikely to fathom complex cases. Consider, for example, the Brayton Point case described in full in Chapter 8. In this case, a power company in southeastern Massachusetts had the capacity to sell its excess electricity to a consortium of New England utilities. When the price of oil rose in the early 1970s, the company was very interested in coverting to coal, even though that meant possible violations of the federal clean air act. Air pollution standards, moreover, were not the direct responsibility of the federal Environmental Protection Agency; rather, they were the job of the Massachusetts Department of Environmental Quality Engineering. At the same time that the company was starting to negotiate with these agencies to get some relaxation of the standards, it confronted the attempts of the Federal Energy Administration to compel it to burn coal. It resisted these efforts, fearing that it could be required to use costly low sulfur coal and to install expensive scrubbing devices. Thus, while the company was seeking permission from two agencies to convert to coal burning—on its own terms—it was fighting the attempts of a third agency to mandate conversion. Also affected by the dispute were area residents who stood to suffer from increased air pollution but who were not, in fact, served by the power plant. Citizens in nearby Rhode Island also had much at stake, but they could hardly depend on Massachusetts officials to give their concerns the highest priority. Obviously, in complex cases of this sort, some sort of analytical framework is a prerequisite to identifying the interests, strategies, and tactical decisions of the parties.

A conceptual model also identifies the factors that encourage or inhibit negotiation. It thus can teach us how to revise legislation and invent new procedures so as to stimulate consensual dispute resolution. We are all familiar with disputes that linger months, often years, before they are resolved, even though the ultimate terms of the agreement were within the parties' hands from the start. It is important to ask why the dance of negotiation so frequently takes so long. In some instances, of course, one of the negotiators may want delay, but often the passage of time is expensive for all concerned. As you consider the following material, try to identify the conditions that lead to stalemate and delay. Can you devise solutions for these problems?

Negotiation Analysis

There are several perspectives from which negotiation can be studied. For example, much can be learned from careful descriptions of negotiation experiences. The extensive case studies in the following chapters are intended to illustrate the issues that arise and practices that are followed in environmental negotiation. Negotiation can also be studied experimentally. Over the years, behavioral psychologists have conducted revealing research into the way in which people act when they negotiate. Although this book does not emphasize a psychological approach, some of the problems that appear later in this chapter could be used as simple tests of bargaining behavior. Negotiation is also studied from an institutional perspective. Laws can be analyzed to see how they encourage—or discourage—consensual dispute resolution. In litigation, for example, the consent decree serves as a mechanism that enables parties to give greater force to their agreements. Similarly, the social and political contexts in which negotiations take place are significant. The perspective may be broad (national political agendas are relevant) or narrow (a young lawyer, eager to make his or her mark, may be intent on litigating rather than negotiating a case). Institutional analysis is emphasized in this book.

All of these methods have value, but at the outset we wish to introduce another approach that may not be as familiar: *decision analysis*, the application of which can greatly clarify complex negotiation situations. Decision analysis grew out of game theory, an abstract but informative examination of the strategy of competitive choices. In its purest form, decision theory can be highly mathematical and removed from common experience; yet its applications have been felt in economics, management, and foreign policy. The best introduction to this discipline remains Raiffa's *Decision Analysis* (Reading, Mass.: Addison-Wesley, 1968).

A negotiation presents an intricate sequence of choices. Initially, a prospective negotiator must decide whether bargaining is likely to be worth the effort, and if so, when it should begin. A negotiator also must select a basic strategy; for example, should one be competitive or cooperative? Once negotiation is under way, a participant must make countless tactical decisions: Should an offer be made? Is it necessary to gather more information? Is it time for a private caucus? Finally, the parties must decide if they should settle.

Because at least two parties are involved in any negotiation, the process is all the more complex. A negotiator's fate is never completely in his or her own hands. The results of whatever decisions are made depend also on the decisions of the other parties. To take the simplest of examples, two pedestrians "negotiating" their way down a crowded sidewalk will collide unless each moves in different direction. A prospective buyer of real estate may make a reasonable offer but does not have a deal unless the seller independently decides to accept it. A negotiator who is considering demanding the inclusion of a particular term in the settlement agreement must weigh whether this will provoke the other party into insisting on something else.

In negotiation, the decisions of all the parties interlock, and outcomes are interdependent. If any one party could unilaterally control his or her destiny in all respects, he or she would have no need to negotiate. Instead, however, negotiators have to practice what game theorists call *reflexive reasoning*; that is, when they are contemplating an action, they have to gauge the other parties' reactions. This is the heart of strategic thinking.

Decision analysis requires several steps. First, the parties must be identified. Next, the range of choices they confront must be defined; in all but the simplest situations, choices may be linked in a lengthy chain. Finally, the consequences of those choices must be estimated.

In environmental disputes, as in other negotiations, identifying the parties can be somewhat difficult. In the Brayton Point case, outlined earlier, the power company was clearly a party; likewise, the various government agencies were parties. Thereafter the matter becomes more difficult. Local residents certainly had a stake in the dispute, but they were not directly represented in the bargaining. Other power companies and their customers also were affected. Defining the boundaries of the negotiation is important to the participants as well as to observers trying to understand the process. Obviously, the number of parties to a negotiation and the nature of their relationship is crucial to the conduct and outcome of the bargaining.

As to the second consideration, *options*, the parties usually face different choices. In the Brayton Point case, for example, the power company choose to go ahead and purchase millions of dollars worth of coal—before it had obtained permission to burn it. Perhaps the company was confident of gaining approval, but its decision can also be seen as an attempt to force the government's hand. The EPA, in turn, could have chosen to be more aggressive with the power company, but in a time of oil embargoes, the agency was leery about looking unreasonable. Often, none of the parties can be absolutely sure of the consequences of their decisions. A political change in administrations can mean stricter or more lenient application of environmental laws. An antipollution device may fail to live up to expectations. Negotiators, then, must make decisions in an atmosphere of some uncertainty. Their attitudes toward risk, whether they can afford to take chances or need to be cautions, can shape their negotiation strategy.

In addition to identifying the parties, their interests, the choices they must make, and the outcomes they confront, it is important to understand the context of the negotiation. Two-party bargaining can differ markedly from multiparty negotiation. (In the chapters that follow we shall see examples of both.) Likewise, bargaining over one issue often puts the parties in an adversary stance, whereas the presence of a number of items on the agenda may open opportunities for joint problem solving. A negotiation may be independent of other problems or it may be linked to other disputes. The environmental controls the government requires in one instance may establish a precedent of sorts for other situations. Negotiation is quite different when it is between strangers than when it is between people who know one another. Similarly, the style and substance of bargaining usually is different if the negotiation is conducted privately instead of publicly.

The material in the rest of this chapter will present several different applications of decision analysis to negotiation. It is an approach, moreover, that is developed through the entire book. Readers should be aware, however, of the limits of decision analysis. It is primarily a presciptive tool; that is, it identifies how people should act, not how they really behave. Moreover, it is premised on rationality in this context—the notion that people act so as to promote their own interests. These interests, of course, need not be narrowly selfish. A negotiator may be more interested in being thought of as being open and fair than in maximizing his or her financial position; nevertheless, he or she may still be regarded as trying to promote self-interest, albeit in the currency of reputation. In truth, of course, many people are irrational and engage in conduct quite contrary to their stated interests. Sometimes, such people have not fathomed the consequences of their actions; strong emotions may have overwhelmed their intellectual capacities. One who is irrational, however, is not always at a disadvantage in negotiation. A rational person, after all, is vulnerable to threats. The madman or the dunce cannot be reasoned with. In a hostage dispute between cool professionals in the state department and religious fanatics, who wins?

Decision analysis is sometimes attacked for allegedly depicting negotiation as strictly an adversary process rather than one in which joint problem solving may be central. This criticism is misplaced. It is true that much of the early game theory literature, from which decision analysis evolved, set out problems in which the participants are labeled *party* and *opponent*, designations that certainly suggested competition instead of cooperation. In certain zero-sum games, moreover, the race goes to the individual who can commit himself or herself quickly, or who can use a forcing move to limit the other side's options. Even in certain nonzero-sum games, the game theorists seemed to be saying that rational strategies must be pursued, even though they would lead to mutually undesired outcomes. There are also, however, bargaining games that have been developed to demonstrate the dynamics of cooperative behavior. Where parties ultimately do come to a settlement that gives some advantage to all, their individual decisions to agree can be reasonably interpreted as advancing self-interest, whether that interest is mercenary or highly principled, manipulative or altruistic.

Decision theory is sometimes wrongly faulted for suggesting a static rather than a dynamic approach. If this is ever true, however, it is only for the simplest, most abstract of the classic two-party games. Indeed, one of the great virtues of applying decision analysis to negotiation is that it takes into account the important variables of uncertainty and time. The choices that a negotiator faces at the beginning of bargaining may change significantly before the process is over. Opportunities may develop or be foreclosed. A strategy that made sense at the start may later have to be abandoned lest it prove fatal. This is particularly true in the environmental arena. Coalitions of interests may come together and then drift apart. Technological developments, the passage of revised laws, and changing economic conditions may radically alter the possible outcomes.

In sum, negotiation is ultimately a consensual process: There can be no settlement if the parties do not all choose to agree. By identifying the choices that parties confront and the incentives and disincentives that constrain them, we can see negotiation from the parties' points of view and from a broader perspective. In addition to clarifying complex relationships, decision analysis teaches two important lessons. First, if one wishes to change the likely result of a negotiation, one must alter the incentives of one or more of the parties in order to encourage different bargaining decisions. Second, decision theory provides a basis for understanding reflexive reasoning and the strategic thinking on which it is based. Third, decision theory illustrates some important paradoxes of bargaining. As we shall shortly see, for example, even if all the parties in a negotiation behave rationally so as to promote their own particular interest, the collective result may be harmful to all. As Thomas Schelling has provocatively demonstrated in his book *Micromotives and Macrobehavior* (New York: W. W. Norton, 1978), the pursuit of self-interest, though utterly rational, may lead to an unwanted outcome. Finally, decision theory enables us to better understand the importance of the manner and context of negotiation. The parties' capacities to communicate with one another can be an important determinant of negotiation. As we shall see in the chapters on mediation, the key to resolution may sometimes be to get the parties to communicate less, not more.

Those interested in a far more extensive introduction to the rigorous study of negotiation are encouraged to read Howard Raiffa's *The Art and Science of Negotiation* (Cambridge: Harvard University Press, 1982).

INCENTIVES TO NEGOTIATE

Negotiation is a consensual process from beginning to end. Any party can elect not to participate, or, having once entered negotiations, any party can drop out. Moreover, a negotiator does not settle a case out of compulsion. One settles because settlement appears better on balance than nonagreement. That does not necessarily mean that the negotiator is pleased by the outcome; it may simply be the least of a variety of evils.

In their book *Getting to Yes: How to Negotiate without Giving In* (Boston: Houghton Mifflin, 1981), Roger Fisher and William Ury introduce the concept of a negotiator's BATNA—the best alternative to a negotiated agreement. To be acceptable, any proposed settlement must be at least a little bit better than the alternative of not settling. This should be self-evident. But what does *better* mean? As we shall see in the next three chapters—in which incentives to negotiate are explored at length—*better* can mean a variety of things. People who live near a factory and bring a nuisance suit seeking recovery for the damage done by its pollution may settle out of court if the company offers more money than they expect to get from a jury; for such plaintiffs *better* also could mean more significant actions to abate the nuisance. Other plaintiffs, however, may rationally agree to settle for *less* than they think they will win in court. Dockets in many states are crowded, and it can be years before a case will come to trial. It may be necessary financially to accept less now rather than waiting years for more. Likewise, people may be very optimistic about winning a lawsuit; yet, there is almost always some uncertainty. Judges and juries can err and unexpected evidence may appear. Some plaintiffs may be reluctant to take chances, even if the odds are very much in their favor.

Better thus can mean more, sooner, or with less risk. It can also mean *cheaper*. Lawsuits are expensive. Lawyers must be paid, investigations conducted, and expert witnesses obtained. Different phases of litigation bear different costs. It is not very expensive to file a lawsuit, but discovery—the taking of depositions and the production of documents—can be costly, particularly in suits where the facts are complex and disputed, as is often true in environmental cases. A person who can afford to start a suit may not necessarily have the means to keep it going.

Negotiation may be expensive as well. At the very least, the time of the participants should be regarded as an expense, a considerable one in protracted cases. In highly technical cases, it may be costly but essential to gather relevant scientific information. (If the case is being litigated at the same time, this may not necessarily be an added cost.) Environmental groups may find it expensive to inform and organize their constituencies. As we shall see in the Grayrocks Dam case in the next chapter, there may be other sorts of costs. A party may avoid negotiating with a long-time adversary, not wishing to give them or their claims any implicit legitimacy. On the other hand, the very process of negotiating sometimes may carry a positive value that is wholly apart from any agreements that may be reached. Good will may be important. A negotiation, though failing to produce agreement, may establish a useful precedent for the handling of future disputes.

It is important to remember that the factors that induce a person to come to the bargaining table may not be precisely the same as those that induce settlement. A person may rationally agree to negotiate, even though she or he sees no hope of reaching agreement. By the same token, a person conceivably may decline to negotiate even if there is an acknowledged possibility of settlement if the costs of negotiating seem too high. Focusing on incentives (and disincentives) is central to understanding negotiation. It helps explain the actions of individual negotiators. It also underscores the factors that must be manipulated if we wish to encourage people to seek consensual resolution of their differences. Incentives to negotiate are thus a central theme of this book.

We shall encounter two different sorts of negotiations. In one, if the parties do not settle, then a court, an arbitrator, or some other official will impose a resolution. This occurs in any lawsuit in which at least one of the litigants is intent on seeing it through. There is a second category of cases, however, in which the consequence of nonagreement simply is that there is no deal. If a conservation group is trying to buy a tract of beautiful land from a developer, the parties either will be able to come to terms or not. If they cannot agree, then the developer will look for or find other potential purchasers, and the conservation group will explore other ways of using its resources.

How are the incentives to negotiate—and to settle—different in these two different kinds of cases?

Obstacles to Consensus

People may decline to negotiate because they do not wish to recognize the legitimacy of other parties, because they seek delay, or because the costs of negotiating seem to outweigh any expected benefits. There are also instances in which people may be able to see the great need for consensus. Yet, a divergence between individual and collective incentives prevents them from reaching accord.

The "commons problem," described by Garrett Hardin in "The Tragedy of the Commons," is the classic example of this type of situation. *Tragedy*, as he uses the term, is not necessarily intended to connote sadness; rather, it connotes "the remorseless workings of things." In Hardin's view of the commons, of course, there is also a strong sense of doom:

> Picture a pasture open to all. It is to be expected that each herdsman will try to keep as many cattle as possible on the commons. Such an arrangement may work reasonably satisfactorily for centuries because tribal wars, poaching, and disease keep the numbers of both man and beast well below the carrying capacity of the land. Finally, however, comes the day of reckoning, that is, the day when the long-desired goal of social stability becomes a reality. At this point, the inherent logic of the commons remorselessly generates tragedy.
>
> As a rational being, each herdsman seeks to maximize his gain. Explicitly or implicitly, more or less consciously, he asks, "What is the utility *to me* of adding one more animal to my herd?" This utility has one negative and one positive component.
>
> 1. The positive component is a function of the increment of one

animal. Since the herdsman receives all of the proceeds from the sale of the additional animal, the positive utility is nearly +1.

2. The negative component is a function of the additional grazing created by one more animal. Since, however, the effects of over-grazing are shared by all the herdsmen, the negative utility for any particular decision-making herdsman is only a fraction of −1.

Adding together the component partial utilities, the rational herdsman concludes that the only sensible course from him to pursue is to add another animal to his herd. And another; and another. . . . But this is the conclusion reached by each and every rational herdsman sharing a commons. Therein lies the tragedy. Each man is locked into a system that compels him to increase his herd without limit—in a world that is limited. Ruin is the destination toward which all men rush, each pursuing his own best interest in a society that believes in the freedom of the commons. Freedom in the commons brings ruin to all. (Garrett Hardin, "The Tragedy of the Commons," *Science*, 162 [1960], p. 1244)

One need not ransack history to find examples of the commons problem at work. In the nineteenth century, frontiersmen slaughtered countless American bison, often taking only the highly prized tongues and leaving the rest of the carcass to rot on the prairie. Within several decades, the vast herd was reduced from many millions to just a few dozen. Perhaps some of the more perceptive hunters saw that they were both decimating the bison and eliminating their own occupation. Yet, there was nothing that any one individual could do to halt the trend. The exhaustion of the whale fishery is another example of the same phenomenen.

Hardin (1960, p. 1245) recognized that the commons problem applies not only to the consumption of resources like pasture land, bison, and whales but also to the pollution of air and water.

> Here it is not a question of taking something out of the commons, but of putting something in—sewage, or chemical, radioactive, and heat wastes into water; noxious and dangerous fumes into the air; and distracting and unpleasant advertising signs into the line of sight. The calculations of utility are much the same as before. The rational man finds that his share of the cost of the wastes he discharges into the commons is less than the cost of purifying his wastes before releasing them.

The multiperson prisoners' dilemma is a revealing variant of the commons problem. It succinctly illustrates the obstacles that can exist to negotiation, even, in cases in which all the parties can see the benefits of agreement. The best known of all game theory exercises, it draws its name from its two-person version in which two defendants involved in the same crime must independently choose between confessing and remaining silent. Their best collective outcome occurs if they both are silent; their worst, if they confess. Although it would seem obvious for the prisoners to conspire to be silent, under the terms of the game, the prosecutor can induce each of them to breach any bargain by the offer of a little leniency in sentencing. The prisoners' dilemma and some of its implications are discussed more fully in chapter 2 of Thomas Schelling's *The Strategy of Conflict* (Oxford: Oxford University Press, 1960). Lest you think that the prisoners' dilemma is wholly an abstract exercise, read George V. Higgins's description of the strategy of Watergate prosecutor Earl Silbert. "The Judge Who Tried Harder," *The Atlantic Monthly* [April 1974]: 83, 90–92.)

The multiperson prisoners' dilemma describes a range of situations in which each party wants to pursue one course of action but hopes that everybody else will do the opposite. The polluter who breathes dirty air most likely wishes everyone else would buy scrubbers and converters; then his foul contribution would not be noticeable. The apartment dweller who shares the building's heating bill wishes his neighbors would turn down their own thermostats but does not obtain much savings if he does so himself. The owner of a house in a blighted neighborhood may understandably be reluctant to invest in improvements if others on the street are going to let their property deteriorate. If, however, they fix up their houses, the parcel will increase in value even if nothing is done.

In all such cases the payoff to an individual depends largely on what all the other parties choose to do. The so-called dilemma arises because it is never in any one person's interest to take the step that will lead to improving joint welfare. It is not a true dilemma because rational choice always dictates one decision: confession, consuming, or polluting. Perhaps the phrase *prisoners' paradox* better captures the fact that rational individual action can produce an outcome that is preferred neither by the group nor the individual.

Whatever the game is called, it also illustrates the importance in negotiation of communication, promises, and the capacity to ensure future compliance. As we shall see in the chapters to come, the inability to guarantee future performance is often a major obstacle to consensus. The issue is not merely whether you can trust the other side to live up to the agreement, but how to get them to trust you.

Problem 1

You are the owner of a vacation house on a quiet lake in rural New England. Your property is presently worth $75 thousand. It would be worth $100 thousand were it not for the fact that the lake is so seriously polluted that it cannot be used for fishing or swimming. This pollution is caused solely by the antiquated septic systems of the hundred houses—yours included—that ring the lake. The problem could be eliminated totally if all the residents were to install new holding tanks. The cost of installing and operating a single tank is $10 thousand. Everyone who lives around the lake is distressed by its condition, yet even though everyone realizes that all would be much better off if the tanks were installed, nothing has happened. Why?

For the sake of simplicity, you may make the following assumptions: Everyone's house and lot is identical; the installation of any one tank reduces the original pollution in the entire lake by one percent; and a partial reduction of the pollution increases the value of all the houses accordingly. For example, if half the homeowners install tanks, the value of everybody's property increases from $75,000 to $87,500.

1. What solutions can you invent to break the impasse?

2. If 40 owners go ahead and install tanks, will that be enough to induce the others to join in?

3. What kind of private agreements could the parties fashion in order to ensure compliance with an agreement to install the tanks?

4. If none of the homeowners has the incentive to install tanks unilaterally, how would you expect them to vote on a referendum to require such installation? Would it make any difference whether the voting was at an open town meeting or was by secret ballot?

5. Is this a matter that is best addressed by private agreement or by government regulation? Is the level of the government relevant to your answer?

6. Finally, how do your answers change if we remove the simplifying assumptions, that is, that we acknowledge that some people contribute more to the pollution than do others, that some people feel the cost of the pollution more than do others, and that benefits of pollution control are unlikely to increase proportionately to expenditures?

In the problem example, people pollute the lake because the cost to them of doing so is less than the cost (to them) of not polluting. Because others also feel that cost, however, everybody is worse off collectively. Some economists have argued that pollution occurs because, until recently at least, the price for using the environment has been less (often nothing) than its true value. The best-known expression of this view is the Coase theorem, which is summarized in the following Kennedy School of Government Note.

> Drawing on an analogy between environmental problems and other overuses of common property, Ronald H. Coase attributes the undervaluing of environmental quality to the state's failure to define property rights clearly. He suggests that definition of property right (whether these rights were given to the sources of pollution or to the recipients) would permit bargaining be-

tween pollution sources and recipients that would lead to an optimal price for environmental damage. If the source were given the right to pollute, recipients of the pollution would be willing to compensate the source for reducing pollution at a rate equal to the value of the cost of the marginal damage from the pollution. If the amount of compensation exceeded the benefit of pollution, the source would accept the payment and reduce waste discharge. If the recipients held the property rights, the process would be similar: recipients would demand payment equal to their value of the cost of the marginal damage, and the source would be willing to pay for the right to discharge wastes until the fee exceeded the benefit from discharging wastes. Economists regard this result as optimal, since the marginal private benefit from the discharge equals the marginal cost of pollution to society. (William B. Marcus and Laurence E. Lynn, Jr., "Note on Environmental Enforcement Program," *Kennedy School of Government Note* [1977:1])

The application of the Coase theorem can be illustrated by the following example. Visualize a neighborhood divided into a 3 × 3 grid of 9 parcels, an acre each. As a house lot, each parcel is worth $25 thousand. The owner of the central lot, however, is planning to establish a piggery, and there is no zoning regulation in place to stop him. Used in this manner, his property will increase in value to $100 thousand, but the smell and noise will decrease the value of each of his neighbors' parcels to $10 thousand. Thus, from a collective viewpoint, the farmer's gain of $75 thousand is more than offset by the neighbors' loss of $120 thousand (8 × $15 thousand). In economic terms, the external costs imposed on the neighbors make the proposed use inefficient. If the owner of the central parcel does have the legal right to go ahead, then, according to Coase, the neighbors should pay him to stop. They should be willing to offer more than $75 thousand (but less than $120 thousand). Everyone would therefore be better off than if the piggery was established. Alternatively, if the law gives the neighbors the right to veto pollution (through a nuisance suit, for example), then the farmer can operate only if he can buy the neighbors out. Because the proposed use is inefficient in this example, he will not be able to offer enough money to induce the neighbors to waive their rights. If the piggery is to be much more profitable, however, the farmer would be able to offer more than the $120 thousand it would take to compensate the neighbors for their losses.

Although the Coase theorem yields interesting insights about property rights and efficiency, it does not address, let alone answer, other important issues, most of which are central to negotiation. First, although it posits negotiation among the neighbors, it does not determine how negotiation will proceed. As we shall see in the next section, the mere fact that there is a potential bargaining range does not necessarily mean that the parties will be able to agree on a settlement figure. Second, the illustration speaks of bargaining between the developer and the affected neighbors, but it does not consider the implications of the bargaining among the neighbors themselves. As in the multiperson prisoners' dilemma, each of the neighbors may look to the others to solve the pollution problem. For any individual, the best solution is to get the benefit of the bargain without having to pay for it. The free-ride factor may cause potential agreements to unravel. Third, the Coase theorem does not speak to questions of equity. What, for example, if the neighbors lack the liquid assets to buy out the developer? In any event, where does private bargaining end and extortion begin? Finally, there is the matter of transaction costs. It may be difficult and time-consuming to get all the neighbors together and work out an agreement.

In recent years antiregulatory advocates have invoked the Coase theorem in support of allowing the free market to set the price of pollution. Whatever the merits of deregulation, however, this argument ignores the considerable obstacles to negotiation. Transaction costs frequently are significant, and in cases in which each of us feels the effects of a particular polluter only slightly, it is unlikely that we shall band together to negotiate a more efficient use of environmental resources. Government regulation is, in part, a mechanism for working around the problem of transaction costs. Regulation does not, of course, eliminate the need for negotiation; rather, it reconfigures the context in which negotiation occurs. We will see a number of instances in this book where modification of the normal regulatory standards produced outcomes that were beneficial to all. By redefining entitlements, however, regulation does alter the balance or power in environmental disputes.

Zero-sum and Nonzero-sum Disputes

It is common to think of bargaining as a process of haggling back and forth in a situation in which one person's gain necessarily means an equivalent loss for the other side. Whatever goes into the rug merchant's till comes out of the customer's pocket. Such exchanges are called zero-sum games because the gains and losses of the bargainers exactly offset each other; that is, they add up to zero.

In practice, however, there are few conflicts that are purely zero-sum. A man who must pay alimony to his former wife at least can reduce the bite by claiming a federal tax deduction; if the wife is in a lower bracket, even after she pays taxes, she will effectively receive more than he has paid out of pocket. (The alimony game is zero-sum if the United States Treasury is considered a player.) In labor disputes, a union may value particular fringe benefits more highly than a straight raise, whereas management may be preoccupied with preserving its control over the workplace. Environmental disputes almost always involve a range of issues, the importance of which may vary among the parties. For example, if the battle is over the development of a tract of land, the environmentalists may feel that a certain portion of its is especially fragile and needs protection. Carrying high-interest costs, the developer may be under greater pressure to come to any reasonable accord. If the negotiators are perceptive enough to recognize their contrasting priorities, they may be able to trade concessions in such a way that the gains far exceed the losses. This is a nonzero-sum game.

There is nothing inherent in the *structure* of zero and nonzero-sum games that makes one more difficult to negotiate and settle than the other. Stalemate may occur in both instances, even when there are possible settlements that all the parties would prefer to impasse. It is possible, however, as Lester Thurow argues in his provocative book *The Zero-Sum Society* (New York: Basic Books, 1980) that our political system is poorly equipped to reach resolutions where gain to one segment of society must impose some loss on another. As Thurow demonstrates, the calculation of benefits and the allocation of costs of environmental protection is exceedingly complex, particularly when citizens attach markedly different preferences for clean air and water, jobs, energy costs, and transportation. He contends that environmentalism in general is closely linked to fundamental choices of income distribution, and thus it tends to be zero-sum in nature—at least in a stagnated economy. Even if this is true for general policy, however, most of the cases described in the following chapters show that specific environmental disputes can be decidedly nonzero-sum.

In any event, it is important to understand that although zero-sum and nonzero-sum disputes are amenable to settlement, they differ somewhat in their underlying dynamics. Consider first a zero-sum game.

Problem 2

Assume that a farmer is about to retire and sell his beautiful tract of land. A real estate developer who plans to build a subdivision has made a bid that the farmer is inclined to accept, but a local greenbelt group has organized a serious effort to buy the land in order to preserve it in its present state. The farmer knows he can sell the property to the developer for $300 thousand. The conservationists have raised $400 thousand to purchase the land.

1. What price do you expect the farmer and the conservationists to settle on? Why? If you were either party, would you be completely satisfied with this price? What might you do in order to make it even more attractive from your point of view?

2. Should the fact that there is a clear bargaining range (see Figure 4) facilitate a prompt resolution, or will it tend to prolong negotiations?

3. Would it facilitate settlement if each side knew the other side's "bottom line"? (One's bottom line—or resistance point or reservation level, as it is called

DISPUTE RESOLUTION THEORY

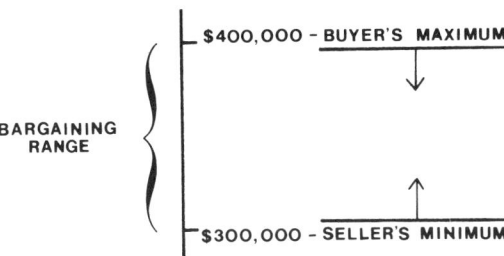

FIGURE 4. Single issue bargaining range.

in the economic literature—is usually calculated with reference to the BATNA, Fisher and Ury's acronym for the best alternative to a negotiated agreement. Here the farmer's bottom line is determined by the competing offer of the developer; if the deal falls through, the farmer can still realize $300 thousand by selling to him or her. Fisher and Ury caution against setting a rigid bottom line, noting that, in the course of negotiating, the parties may discover other terms or compensation that can be incorporated into a deal to make it more attractive even if the dollar amount proves less than the other offer.

4. Deadlocks are sometimes broken when negotiators agree to apply some "fair" principle, such as splitting the difference. Do you have enough information to decide whether that would be a fair resolution in this case? To the extent that splitting the difference is common practice, how does the principle affect the overall strategy of the negotiator?

5. What outcome would you expect if the conservationists knew the farmer's bottom line, but he was in the dark about theirs? What does your answer tell you about the way in which negotiations are then likely to proceed?

6. As posed, this is strictly a zero-sum dispute, but are there other issues that can be introduced to give the matter a nonzero-sum quality. What such issues might be latent here? What other parties might be interested in the outcome of the negotiations. What influence might they be able to exert?

PROBLEM 3

When there is just one issue in contention, a dispute can be illustrated in a simple two-dimensional diagram. In Problem 2 that issue was defined simply as how many dollars would it take to buy the farm. The resistance points of the buyer and seller constitute the end points of the bargaining range. In cases in which the seller demands more than the buyer can afford to pay, obviously there is no figure acceptable to both of them; hence there can be no deal.

If there are more issues involved, the model must become more intricate. Consider a case in which the environmental group wants a power plant to reduce its pollution of the air; specifically, its emissions of sulfur dioxide (SO_2) and particulates. From the environmental group's point of view, the ideal resolution would be total elimination of each pollutant. The worst of all worlds would be no reduction of either one. For the environmentalists to establish an agenda for settling a suit against the company, however, they must clarify their attitudes about the host of possible outcomes between these two extremes.

In Figure 5 the two axes represent the percentage reduction of the respective pollutants. The worst outcome is at 0,0 in the lower left; the best, at 100,100 in the upper right. The environmentalists will likely prefer outcomes closer to the latter over those near the former.

Were they asked, moreover, the environmentalists could probably identify a specific point as being marginally superior to taking their chances in the lawsuit. For example, they might draw the line at a promise by the company to reduce each pollutant by 40%; anything less than that would not be acceptable. On further reflection, they should be able to identify other potential solutions that they regard as no better but no worse. They might, for example, be willing to surrender some improvement in sulfur dioxide pollution for a still greater reduction in particulate pollution. As between their original 40,40 resistance point and a 35,50 outcome, they might be indifferent. Indeed, there should be a number of such combinations that are regarded as no better but no worse than one another. The line connecting all such points is called an indifference curve, and is indicated in Figure 6.

As defined here, the indifference curve also happens to be the environmentalists' reservation level. In the diagram, the environmentalists would ultimately

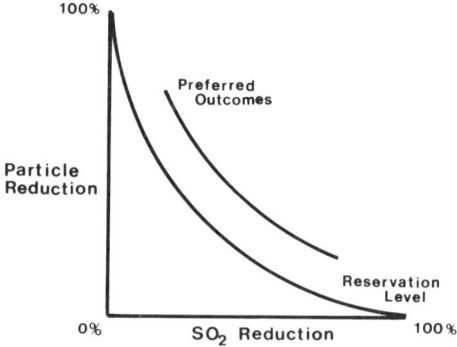

FIGURE 5. Two-dimensional bargaining.

DISPUTE RESOLUTION THEORY

FIGURE 6. Reservation level with two issues.

accept any proposed settlement that lies on that curve but would prefer settlements located above and to the right. We can imagine a second indifference curve connecting points that are equally attractive among themselves but that are all preferable to an outcome on the reservation level. Indeed, there may be an infinity of such curves. Proposals below and to the left of the reservation level would not be acceptable.

1. In such a case, what shape do you expect the environmentalists' indifference curves to take: straight, convex, concave, or irregular? (By definition, indifference curves can never intersect.)

2. The owners of the power plant, mindful of the expense of reducing air pollution, likely will want to be obligated to reduce particulates and sulfur dioxide as little as possible. Assume, that, if pushed to the wall, they would accept a 65,50 solution. Anything more stringent would be less desirable than fighting the environmentalists in court. How might the technology and economies of pollution control affect the shape of the company's indifference curves?

3. In the same way that the resistance points of the farmer and the conservation group defined the bargaining range in the zero-sum example, can you conceive of a way in which the reservation levels of the environmentalists and the power company here can be merged to define an area of possible settlement? You may find it helpful to sketch a resistance curve for the company, but keep in mind that it regards the status quo (0,0) as the best outcome and total elimination of pollution (100,100) as the worst.

4. Imagine that the parties reach tentative agreement on a solution that falls on the intersection of their resistance levels: How might *both* parties do better? In such a case must there always be a better outcome?

5. We have moved from one-dimensional bargaining to two. In many instances, of course, there will be far more than two issues. It is hard for most of us to think in more than three, or at the most, four dimensions, so we will excuse you from drawing a descriptive diagram of such cases. We have seen that two-dimensional bargaining often offers opportunities for joint gains not possible in one-dimensional situations. Do you suspect that this opportunity is present in multidimensional cases?

Bargaining Strength

Bargaining strength is a familiar but poorly understood commodity, largely because it sometimes works exactly the opposite from what we might suspect intuitively. A negotiator with limited authority usually has more power than one with complete discretion. A negotiator who does not have the capacity to receive messages from the other side may find that this is an advantage, not a shortcoming. The irrational negotiator may prevail over the rational one. This section will briefly explore these seeming paradoxes.

The source of bargaining strength usually lies outside the negotiation itself: one's power within the negotiation depends on the impact of possible failure of the negotiations. As Fisher and Ury (1981) stress, the consequences of nonagreement determine the relative attractiveness of settlement. To enhance your bargaining power, then, work to improve the consequences of nonagreement. For example, the farmer in Problem 2 will strengthen his bargaining hand with the conservationists if he can get the developer to up his bid to $350 thousand. The competing conservationists will now have to meet or exceed that offer. Note that the farmer has enhanced his bargaining strength by pursuing a deal he really does not want to make; moreover, he has done so outside the negotiation with the conservationists. Bargaining strength thus is not necessarily a constant. It can be manipulated by the parties in some cases. In others, it may be affected by events beyond their control. The farmer's position will be significantly changed if the developer independently decides to drop out of the bidding.

Thomas Schelling in *The Strategy of Conflict* (New York: Oxford University Press, 1960), explains bargaining strength as a function of commitment. Drawing on Schelling's example, picture two teenage drivers barreling down the road at each other, playing the deadly game of chicken. The first to swerve saves his neck but at the price of appearing cowardly. A driver who can wrench his steering wheel off and heave it out the window commits himself to going straight ahead. By committing himself, he forces the other driver to swerve, and he wins the game.

Commitment can take many forms. It may be aggressive, as in the game of chicken, or it may be decidedly conciliatory. The soldier who lowers his gun or the potential litigant who lets the dealine for filing suit pass has made commitments that may be small or large, depending on the circumstances. Commitment to one road is often made by deliberately burning bridges to others.

Making commitments can involve risk. The soldier who lowers his or her gun is exposed to the enemy who does not. People who forfeit their right to go to court may have their rights exploited. There are also significant risks to committing oneself to an aggressive strategy. The driver who tries to win the game of chicken by throwing the steering wheel out the window is doomed if the opponent does the same thing simultaneously. Likewise, the gambit is useless if the opponent does not see it or concludes that the driver has some other way of controlling the car.

The game of chicken may seem far removed from most of our lives (though there are chilling parallels to some international confrontations), but the lessons have broad application to everyday negotiation. Think back to the Brayton Point coal conversion case, mentioned in the beginning of this chapter. (This case will be discussed fully in chapter 8.) There the generating company spent millions of dollars on coal before it received EPA permission to burn it. Because of high transportation costs, the investment would have been substantially lost if the government had said no. By committing itself in this way, the company put more pressure on the EPA to deviate from the general air pollution standards. Whether or not it did so wittingly, the company took some risk beyond the immediate financial one. By acting unilaterally, it could have been accused by the EPA of acting in bad faith and jeopardizing future relationships.

Bargaining strength is thus related to the options each party faces and to the parties' abilities to commit themselves to act (or to refrain from acting) on them. Commitment, in turn, often is dependent on the parties' capacities to communicate. Making a commitment of any sort is unlikely to affect the negotiation strategy of the other parties if they are unaware of the step that has been taken. A negotiator who cannot receive messages may be immune from threats (but he or she is also deaf to promises).

Commitment also raises the question of the first move. There are negotiations in which the race goes to the swiftest. The first party that can commit itself to a course of action preempts all the others. Some unscrupulous developers have been known to dump fill in wetlands before seeking conservation commission approval. Such a tactic can breed ill will, but it often moots the question of protecting fragile ecosystems. Developers may have to pay a fine for illegal dumping, but they will get a permit. Had the developer asked first, he or she might have been denied permission. There are also conflicts, of course, in cases in which each side tries to wait for the other side to make the first move.

Commitment can also be affected by whether the negotiation is public or private. One common form of commitment is public declaration. The environmental group that proclaims that it will never accede to the development of a particular tract seeks to enhance its bargaining power by tying its own hands. If the group is later pressed to compromise, it must take into account the loss of credibility that will come with retreating from its prior stand.

Commitment is not always self-imposed. The lawyer or executive representing a company may profess sympathy with neighbors who are claiming damages for air pollution, but truthfully state that she or he has no authority to settle for more than a given amount. The client or employer has, in effect, been committed to a low settlement by refusing to give the authority to agree to a higher one.

In sum, commitment is one of the tools that a negotiator may use to increase bargaining strength. Even if one objects to the manipulative aspects of using commitment in this way, negotiators must understand how others may seek to use it against them.

Problem 4

Assume six identical apartment houses sit on six lots of equal size on a city block. The most profitable use for each lot is multifamily housing; each property is worth $500 thousand. If a developer could assemble all six parcels as a site for an office building, however, they would be worth $5 million in toto; that is, $2 million more than they are worth separately. Keep the concept of bargaining strength in mind as you consider the following questions.

1. How should a prospective developer approach the six owners of the apartment buildings: individually or collectively?

2. Given that the developer could sell the six lots as a single parcel for $5 million, how much should she or he be willing to pay for each one? If the developer has paid $600 thousand for the first five of the parcels, how much should she or he be willing to pay for the sixth, to complete the deal? Who is in the position of bargaining strength in this situation: the developer or the last owner? Can you imagine a situation in which the developer would rationally, if regretfully, agree to pay more than $5 million for the six?

3. Do your answers to the first two questions give you any guidance about whether as an apartment house owner you would want to be the first to deal with the developer or the last? What risks go with the strategy of waiting to be the last?

4. As the developer, what strategies can you devise to protect yourself from possible exploitation? (For a description of negotiation strategy and land assembly, see C. Trillin, "U.S. Journal: Atlantic City, New Jersey," *New Yorker*, January 8, 1979, p. 4; and P. Hellman, "How They Assembled the Most Expensive Block in New York History," *New York*. February 25, 1974, p. 31.)

5. In the Holston River case, which will be described in chapter 5, the EPA negotiated a pollution discharge permit with the Tennessee Eastman Company. One of the issues in that case was the capacity of the river to absorb the pollutants; there were other companies that discharged their waste into the river. The EPA had to allocate the carrying capacity of the river among the various dischargers. In this circumstance, if you represented one of these other companies, would you want to be the first to negotiate with the agency or the last?

Conclusion

This chapter has merely sketched a theoretical framework for understanding negotiation. The framework will be expanded and built upon in the chapters that follow. The case studies of actual negotiations will give you the opportunity to apply and test the theories that have been introduced. You should regularly ask yourself whether environmental disputes are sufficiently different from other kinds of conflicts that they require their own theories and practices.

3

INCENTIVES TO NEGOTIATE

INTRODUCTION

The process of negotiation is a chain of interlocking choices. The parties must first decide if they are going to negotiate at all, and if so, whether they will adopt competitive or cooperative strategies. They also must choose when and where to meet, whether to confer publicly or privately, and what should be on the agenda. Likewise, each negotiator must decide when to tender an offer and how to respond to a demand. Ultimately, the parties must decide whether to ratify a possible settlement, continue negotiating, or even leave the bargaining table. (In fact, as we shall see later in this book, after an agreement is reached, the parties may be faced with choices about whether to breach it in whole or in part.) All these choices are interlocking in that the decisions made by one party often affect the options available to the other.

If negotiation is a chain of choices, they a key to understanding the process is identifying the incentives—and disincentives—each party faces. The way in which an individual evaluates the possible consequences of going to the bargaining table will explain his or her decision to negotiate.

Incentives may be thought of as costs and benefits, but it is important to remember that many of the important consequences of negotiating and settling (or failing to do either) are not monetary. One person may feel great emotional relief at settling out of court. Another may pay considerable legal costs in order to win public vindication from a jury. That a benefit (or a cost) is subjective and hard to quantify, however, does not mean that it is any less substantial than hard cash. Furthermore, the way in which anyone balances emotional tranquility or reputation against dollars is a matter of personal values, and these may vary considerably from individual to individual. To an outsider, a steadfast refusal to negotiate may appear irrational, but it may be the product of a careful deliberation. (With negotiation, as with anything else, of course, it is possible for people to make miscalculations that they would grossly regret if they recognized them.)

We are broadly concerned here with incentives to negotiate, but we must recognize that the decision to enter negotiations is quite different from the decision to come to agreement. A person may be willing to explore the possibility of settlement without committing himself or herself to a consensual resolution. Likewise, the decision not to initiate negotiation may be different from the decision not to accept someone else's invitation to meet, just as the decision to withdraw from negotiations once they have begun is somewhat different from the decision not to talk in the first place. All such choices require a careful weighing of the consequences. As the consequences are likely to differ, so is the nature of the choices.

Sophisticated negotiators recognize that at least some of the incentives that they and their counterparts face are subject to control. A bargainer who initially is unable to get the other side to talk need not throw in the towel. Instead, he or she may succeed first by identifying the incentives to negotiate *as they are perceived by the other side* and then by trying to change those perceptions. That change may come about by persuasion—by demonstrating to the other side that they have underestimated the net benefit of talking—or by producing some sort of sweetener that induces them to come to the table. The other side of this coin is the tactic—often a dangerous one—of establishing a precondition before one negotiates.

We shall be considering incentives to negotiate throughout the book. This focus serves several functions. First, it will allow us to evaluate the decisions of individual negotiators, whether or not we encounter them in case studies that we read or watch them in the simulation exercises. Second, looking at incentives is a way in which we can understand the leverage that the parties may have on one another. As we shall see, this leverage has a dynamic quality, often shifting in the course of a negotiation. Finally, this approach allows us to take a broader perspective. If, as a matter of public policy, society wants to encourage more out-of-court settlements, how can it effectively manipulate carrots and sticks?

Incentives to Settle a Lawsuit

In considering incentives to negotiate, it is useful to distinguish two situations: those in which a failure of the parties to agree will result in a decision imposed by a court (or some other adjudicator), and those cases where nonagreement means that a deal has fallen through and that there will be no resolution among the parties.

Although lawyers often engage in commercial bargaining, there has been more attention in the legal literature to negotiation of lawsuits. Mnookin is the author of the felicitous phrase *bargaining in the shadow of the law*. Divorce lawyers trying to hammer out a separation agreement must always be mindful of the range of likely outcomes should negotiation break down and the case go to court. Recent judicial decisions help set a bargaining range for the parties.

Consider the following questions:

1. What if both parties in a pending lawsuit knew precisely what the judge would order: Would they have an incentive to negotiate?
2. In some jurisdictions, it can take years to get a court date for trial. This can be an important inducement to negotiate. Is the effect of court congestion felt equally by all litigants; whose bargaining power is enhanced by delay?
3. In practice, some measure of uncertainty is present in almost every case. (Even if the plaintiff has an overpowering case on liability, the amount of damages may be open to question.) To what degree can a party reduce that uncertainty; how valuable is it for him or her to do so?

Special Problem

The following quotation is from an article entitled "Some Suggestions Concerning the Judge's Role in Stimulating Settlement Negotiations" (Rubin and Will, 75 *Federal Rules Decisions* 89). The authors, themselves judges, describe different ways in which negotiation may commence, one in which they call the "Lloyd's of London" calculation. Do you agree with their analysis?

> The plaintiff says the most likely judgment is $100,000 and he has a 70% chance to win; he has appraised the case at $70,000. He would accept $70,000 "insurance" for his case. The defendant says the likely judgment value is $60,000 and the plaintiff has only a 50% chance to win. He has appraised the case at $30,000. He would pay $30,000 for an insurance policy that would indemnify him for this case. If both appraisals are reasonably informed and accurate, the parties ought to be willing to discuss a settlement midway between their own Lloyd's figures—here $50,000.

1. Why might parties to the same lawsuits have different assessments of the probabilities for success in the courtroom?

2. If the person or agency bringing the suit estimates its chances of success higher than does the defendant (as occurs in the previous hypothetical case), how likely is a negotiated settlement? Is one necessarily prohibited?

INCENTIVES TO NEGOTIATE IN OTHER CONTEXTS

Lawyers often negotiate in situations in which there is no judge or arbitrator to impose a solution if the parties cannot reach one on their own. A "greenbelt" association, trying to obtain a conservation easement from a farmer, must either come up with an acceptable proposition or forego the property because no third party is empowered to impose a price.

If negotiation of lawsuits takes place in the shadow of the law, what establishes the parameters for other kinds of transactions? In *Getting to Yes* (1981), Roger Fisher and William Ury identify the importance of a negotiator's BATNA—his or her "best alternative to a negotiated agreement." The farmer considering an offer from the green belt association weighs it against the possibility of a bid from someone else. The offer must be superior to the BATNA to be acceptable. In some cases, the farmer may have a firm bid in hand, whereas in others, the market may be weak or uncertain. The process of weighing a proposed settlement involves consideration of one's attitudes toward risk and delay. We are cautioned that a bird in the hand is worth two in the bush, but the axiom does not give us any help if the choice is between the bird in the hand and three or four in the bush.

The notion of bargaining in the shadow of the law and the concept of BATNA are both expressions of decision analysis—a technique that facilitates the evaluation of competing options. In each case, the consequences of settling are compared with the implications of not accepting the deal. Seldom are the consequences clear. Even in the case of settlement, there may be some uncertainty about whether it will be fully honored or what sort of precedent it will set. Uncertainty about the consequences of nonagreement is likely to be more pronounced. Declining one proposal, even a supposedly "final offer," does not necessarily mean that negotiations are at an end. A negotiator may turn down a proposal that is superior to his or her BATNA if he or she thinks that a still better proposal will follow.

Consider the following questions:

1. Fisher and Ury contend that a negotiator's bargaining power is enhanced by improving his or her BATNA. In which situation is the negotiator more likely to be able to do so—in the context of a lawsuit or operating in the market?
2. Why is it that the prices of some things are negotiable and some things not? Foreign cars, for example, are more likely to be sold at sticker price than are domestic cars. Is this simply a function of a strong market for foreign cars? Is not a dealer better off if he or she trims the profit margin a bit rather than lose a sale?
3. In his inaugural speech in 1961, John Kennedy said, "We will never fear to negotiate, but we will never negotiate out of fear." What was this statement probably meant to signify and who was its intended audience? Why is it better rhetoric than policy?

Case Study: Grayrocks Dam

The following material is based on a case study prepared by Julia Wondolleck.

The Grayrocks case illustrates how incentives to negotiate and settle can change over time. As you read it, focus your attention on how the various parties perceived the consequences of nonagreement. What made these perceptions change? What leverage did each party have over the others to get them to come to the bargaining table?

The Grayrocks Dam case has all the elements of a classic development/environment dispute. It pitted a large power company seeking to construct a dam to provide cooling water for a new power plant against a coalition of environmentalists, farmers, and state officials. To the power company the dam meant more electricity and jobs and higher revenues. To the farmers who lived downstream, it meant less water available for irrigation. And to environmentalists, it meant a threat to the habitat of the whooping crane, an endangered species. As the controversy unfolded, the battle over the dam was waged on many fronts, including the courts, Congress, and the state capitals of Nebraska and Wyoming. Ultimately, the parties got together and settled the dispute among themselves—the dam is being constructed, the farmers still have their water, and the whooping crane is still with us. This case describes how the parties fashioned a resolution that each of them prefered to continued litigation.

In 1970, six utilities formed the Missouri Basin Power Project (MBPP) for the purpose of constructing a $1.6 billion coal-fired power plant on the Laramie River near Wheatland, Wyoming. The plant was designed to provide power for expected industrial expansion in eastern Montana, Wyoming, Colorado, North Dakota, South Dakota, Nebraska, Iowa, and Minnesota. Two million customers will be served by the plant. To supply essential cooling water for the plant, the consortium proposed to build a dam and reservoir on the Laramie River, a tributary of the North Platte River. The dispute over the dam was first and foremost a dispute over water rights. Once operating, the project would divert 60,000 acre feet of water annually from the North Platte River. This diversion is in addition to the 70% reduction in streamflow that has occurred in the last 50 years due to construction of 43 dams and numerous irrigation projects on the same river. Conservationists worried that the additional reduction in streamflow would be "the straw that broke the camel's back" in its impact on North Platte River wildlife.

The focus of the conservationists' concern was the critical habitat of the whooping crane, an endangered species. The crane migrates annually between the Aransas Natural Wildlife Refuge in Texas and Wood Buffalo National Park in Canada. Two hundred and seventy miles downstream from the dam is a 60-

mile-long stretch of sandbars that serves as a major stopover for the crane on its yearly migration. Flood waters and ice from the annual snow melt scour the sandbars and keep them free from vegetation. The environmentalists worried that the additional reduction in streamflow occasioned by the dam would reduce the effectiveness of the scouring, thus permitting vegetation to overcome the sandbars and make them unsuitable for the whooping cranes.

Farmers downstream also worried about the impact of the dam on streamflow. Water is a very valuable commodity in the semiarid plains states; it is the lifeblood of their agricultural economy. As a result, it has been a source of conflict among the states for a number of years. Colorado, Wyoming, and Nebraska have feuded for years over the allocation of water from the Laramie and North Platte rivers. Entitlements to this water were defined by United States Supreme Court decisions in 1945 and 1956. Unfortunately, the three states all interpret these decrees differently. Nebraska has been able to liberally interpret its entitlement to North Platt River water because it is located farthest downstream. It has taken its share of the streamflow plus whatever has been left by upstream users. Nebraska officials opposed the dam because they worried that its construction would force them to reduce the state's water usage. Wyoming officials maintained that the Supreme Court allocation formula entitled them to the additional water that would be taken from the river by the dam. Because the Supreme Court rulings were ambiguous, none of the parties could be absolutely sure who was really entitled to the water at issue.

Informal negotiations began in 1973 when MBPP formed an environmental advisory committee to explore the potential impacts of the Grayrocks Dam. The committee solicited the views of concerned environmental groups and issued a report suggesting that future power needs could be met by a smaller plant with less environmental impact. MBPP apparently did not find the report persuasive. In the words of Robert Turner, the Wyoming representative of the National Audubon Society, the response of officials to the committee's advice and recommendations was "negative in every regard." The advisory committee was officially disbanded in 1976.

During this same period, Nebraska and MBPP officials were meeting over 30 times to discuss the water rights issue. The negotiations also yielded little in the way of agreement. The principals have different perceptions of what went on during these talks. William Wisdom, counsel for Basin Electric, a major interest holder in MBPP, asserts that the consortium made a number of offers of specific water levels to Nebraska that were all rejected. Paul Snyder, assistant attorney general of Nebraska recalled that MBPP refused to concede anything during these negotiating sessions. Snyder's view is that MBPP adopted a hard line in negotiating with environmentalists and Nebraska because it thought it had the political clout needed to head off any lawsuits, especially those brought by Nebraska, part of MBPP's service area. The utilities were "used to getting away with whatever they proposed; nobody had ever stood up to them before." Things would soon change.

Having been frustrated in its attempts to settle the dispute through negotiation, Nebraska fired the first salvo in a complicated legal battle: in 1976 it sued the Rural Electrification Administration (REA) alleging that the REA's loans to the project were illegal. Under the National Environmental Policy Act (NEPA), major federal actions that affect the quality of the environment, including loans and permits, must be preceded by an adequate environmental impact statement (EIS). Nebraska charged that the Grayrocks Dam EIS was inadequate because it said *nothing* about the impact of the dam on either Nebraska water supplies or the aquatic ecosystem of the part of the North Platte River that flows through the state. Nebraska pursued the same legal strategy in a suit that sought to enjoin the Army Corps of Engineers from issuing a 404 permit needed to dredge and fill a U.S. waterway.

Conservationists filed suit as well, likewise citing the allegedly inadequate EIS but also charging that the REA and the Corps failed to fulfill the requirements of the Endangered Species Act (ESA). The ESA requires that federal agencies consult with the U.S. Fish and Wildlife Service to ensure that their actions do not jeopardize an endangered species. Under this requirement, the requested permits may be refused, or mitigating procedures may be ordered. In effect, the conservationists sought to stop the dam until MBPP took steps to guarantee the habitat of the whooping crane.

The various lawsuits were combined into one consolidated suit. As the case progressed, the parties met a few times to discuss settlement. These efforts were futile in large part because each side felt confident of victory; hence, they had little incentive to negotiate out of court. This stalemate was broken when the court ruled against MBPP and enjoined REA from issuing the needed loan guarantees and the corps from issuing the 404 dredge-fill permit. It was at this point, in the words of Paul Snyder, Nebraska's assistant attorney general, that "the real negotiations started."

Although MBPP lost the first battle over the Grayrocks Dam, it was far from clear that they were going to lose the war. They appealed the decision and were confident that the injunction would be overturned. They also had friends in Congress. In an emotional speech, retiring Representative Teno Roncalio (D-Wyo.) pleaded,

> Do you want to send me back to Wyoming, after ten years as your friend and colleague, to face 2,000 unemployed people in Wheatland on account of a totally unjustified thing like this, the Endangered Species Act? (*Washington Post*, November 27, 1978, p. A1)

Roncalio's plea was warmly received. The House passed a bill exempting the Grayrocks Dam from virtually all federal regulatory requirements. The bill was amended in conference to limit the exemption to the Endangered Species Act provided that the newly created Endangered Species Committee gave its approval.

Thus, MBPP had hoped that it could achieve the victory in Congress that eluded it in the courts. But as MBPP had learned from the district court's decision, it could not be certain of a favorable decision. Because the Endangered Species Committee (known popularly as the "god committee" because of its power to make life or death decisions for both species and projects) had never decided an issue like the Grayrocks Dam, no one could be certain how it would vote. Moreover, the longer construction was delayed on the dam, the more expensive became the dispute for MBPP. MBPP officials estimated that they could lose close to $500 million if construction was delayed for a year; therefore MBPP had a clear incentive to find a quick way out of the morass.

Similarly, conservationsists were not enthusiastic about trusting the future of the whooping crane to the "god committee." Moreover, they did not relish the thought of an expensive court appeal. Because they never intended to stop the project entirely but merely intended to provide protection for the endangered crane, little was to be gained from fighting the battle to its bloody end. And Nebraska also was not inalterably opposed to the project; it merely wanted to protect its water. As a result, the dispute really was ripe for settlement.

MBPP initiated negotiations by proposing, through intermediaries, that all the parties get together to discuss a settlement. Nebraska and Wyoming quickly agreed as did the conservation groups. The initial meeting was held in Lincoln, Nebraska, in October 1978. Sixty people participated with the two governors serving as co-chairmen. In the view of one of the representatives of the conversation groups, the governors used the meeting primarily for "political posturing." Little of substance was accomplished, beyond agreeing to the date and format of the next meeting.

On November 2, 1978, a much smaller group reconvened in Cheyenne, Wyoming. At the Lincoln meeting the parties had agreed to continue discussions through six representatives: Nebraska's attorney general, Nebraska's director of water resources, Basin Electric's James Grahl, MBPP attorney Edward Weinberg, Patrick Parenteau of the National Wildlife Federation, and David Pomerly of the Nebraska Wildlife Federation. Each of these parties came to the Cheyenne meeting with "bottom-line proposals" developed since the first meeting in Lincoln. Each was also accompanied by legal counsel and technical advisors. When the size of the group proved unwieldly, the lawyers and advisors were shunted to a nearby room where they remained available for consultation. The remainder of the negotiations were conducted by the principals alone.

MBPP opened the negotiations by offering $15 million to the opposition groups to purchase water rights to maintain whatever streamflow they thought appropriate. It was MBPP's intention that some of this money be used, if necessary, to maintain the whooping crane's habitat artificially. MBPP officials arrived at the $15-million figure the same way most defendants calculate settlement offers; they estimated what they could afford to pay, how much they stood to loose if the case was not settled, and approximately what they thought it would take to satisfy the opposition.

What MBPP failed to assess, however, was how such an offer would be perceived by the plaintiffs. It was rejected for several reasons. First, Nebraska was extremely nervous about accepting any direct payment except for legal fees. Given the highly visible nature issues among Nebraska farmers, it was important that Nebraska not be perceived as selling out the interests of its water users for cash. Second, Nebraska was sincerely interested in maintaining the existing streamflow through the state and was not certain that the cash settlement would be adequate for this purpose. Third, like Nebraska, the conservationists were reluctant to accept cash, but they also were concerned about whether streamflow levels could be legally maintained through water rights purchases. Nebraska allocates water to users only if it will be put to "beneficial use." This includes agriculture, mining, municipal water needs, recreation, and the maintenance and propogation of fish and wildlife. Although it would seem that purchase of water rights to protect the whooping crane's habitat would fall within the fish and wildlife clause, there is a catch. A "beneficial use" must also entail "physical removal of the water from the stream." Because the water purchased to maintain the habitat would be left in the stream and not removed, it was not clear whether a Nebraska court would consider such a use *beneficial*. Thus, both Nebraska and the conservationists rejected the initial MBPP offer.

MBPP came back with the revised offer that cut the cash settlement in half and included varying guarantees for minimum streamflow for the North Platte River for different seasons. The $7.5 million was supposed to be used to purchase additional water rights when needed and to maintain the habitat artificially. Although the streamflow guarantees helped assuage some of the concerns of the opposition, Nebraska and the conservationists were still reluctant to accept any cash. As a result, the parties spent much time discussing how much a settlement could be consumated and no time discussing the size of the settlement. After much discussion, Patrick Parenteau of the National Wildlife Federation suggested using the money to create a trust fund for the preservation of the whooping crane and its habitat. Nebraska agreed to the settlement on the condition that the fund be governed by an "independent" board of trustees. Thirty days later the parties affixed their names to a formal, binding agreement that includes a monitoring provision to ensure implementation. The agreement establishes a perpetual trust fund with the interest used for protective measures for the whooping crane and its habitat. On January 23, 1979 the god committee met for the first time. It ratified the settlement and thus ended the Grayrocks Dam controversy.

Study Questions

1. At the outset, what was the relative bargaining power of the parties? In Fisher and Ury's terminology what were the alternatives to a negotiated agreement?

2. The different environmental groups in this case were well organized and united; they seemed to share similar views on strategy as well as objectives. In many environmental disputes, this is not true. Suppose the various conservationists in Grayrocks had not been able to form a united coalition. Would this have affected MBPP's willingness to seek a negotiated settlement? In what way?

3. The great majority of all lawsuits are settled outside of court. Court procedures, however, can influence the timing of settlement. Major procedural events include the process of discovery (through depositions and interrogatories). Discovery gives both sides the opportunity to gather information that may cause them to revise their estimates of success. There also can be pressure to settle when the matter is called for trial, right after the plaintiff has presented a case and just before the judge instructs the jury. Some cases are even settled while the jury is out deliberating. Identify the important settlement opportunities in the Grayrocks case? When did they occur? How did the willingness of each of the parties to settle change at each of these stages?

4. Had water rights to North Platte River water been clearly defined, Nebraska and Wyoming would have had much less to argue about. Although the Supreme Court had twice ruled on the water rights issue, these decisions failed to define clear entitlements because of seasonal variations in both streamflow and the relative needs of both states. The resulting ambiguity has permitted each state to interpret the rulings in a way favorable to the interpreting state. Nebraska could have tried to stop the dam by seeking a clarification from the Supreme Court on the water rights issue. (Because the Supreme Court has original jurisdiction in suits between states, Nebraska could have gone directly to the Supreme Court with its complaint.) Similarly, Wyoming could have sought a declaratory judgment supporting its interpretation of the earlier rulings. Why do you think neither state pursued this strategy?

5. Did the MBPP take a hard line in this case? Were there any actions it could have taken that might have convinced dam opponents that it would never accede to their demands? Could the conservationists have taken a stronger stand in any way? Why do you think the parties did not escalate the dispute?

6. During the course of this dispute, the Supreme Court affirmed a lower court ruling that enjoined construction of the Tellico Dam on the Little Tennessee River because it threatened the snail darter, an endangered species. The

Tellico decision stood for the proposition that the Endangered Species Act required that development projects yield to any endangered species threatened by the project. This decision was somewhat of a Pyrrhic victory for environmentalists because it prompted Congress to amend the Endangered Species Act by creating the "god committee" to review cases of irreconcilable conflict between the preservation of a species and construction of a new development. Do you think that the creation of this committee will, in the long run, affect the incentive of parties embroiled in endangered species disputes to negotiate?

7. Suppose that MBPP was legally forbidden to pay compensation for environmental degradation. (This assumption is not as unrealistic as it may seem. Some states do not allow utilities to include compensation payments in their rate base.) Also assume that it was infeasible for MBPP to guarantee streamflow. Would the conservationists have had any incentive to negotiate?

Inducements and Obstacles

A Catalog of Incentives

People settle disputes when they perceive that the cost of continuing the dispute exceeds the cost of settlement. These costs (or benefits) can be hard to calculate, and they can change radically in the course of dispute. As new information becomes available, the parties reassess their estimates of the likelihood that they will prevail if they do not settle. In Grayrocks, for example, the lower court decision enjoining construction of the dam shocked the MBPP into recognizing that its case was not as strong as it had originally assumed.

A second factor that accounts for changing willingness to settle is changing the costs of continuing the dispute. Early success can sometimes be paradoxical. Having won half a loaf, a party may become more nervous trying to win still more. The conservationists in Grayrocks were reluctant to risk the gains they had won in the lower court by going to the "god committee." Further, the out-of-pocket costs of continuing to fight can be substantial. Even if there is something to be gained by going to court, it may be eaten up in fees. There is particular pressure to settle on the courthouse steps because trials are expensive. It is relatively cheap to maintain to lawsuit in its early stages, but a trial eats up money quickly in lawyers' fees, expert witness fees, the cost of obtaining a record transcript, and so forth.

Finally, a person's inclination to settle may change during the course of a dispute because of changes in values. The very process of negotiation can change attitudes because of the way in which it can restructure relationships. Saving face, for example, may be irrelevant at the outset but critically important after a protracted negotiation. Likewise, someone who initially has just a mild interest in settlement may come to have a feeling of understanding and even obligation to his or her counterparts. (Of course, the reverse can be true as well. Face-to-face negotiation may engender such hostility that a party will refuse to settle no matter how attractive the substance of a deal is.) Attitudes and values often are affected by external developments. A party who suffers a substantial financial setback during the course of the dispute may be forced by circumstances to agree to terms that he or she would have rejected earlier.

It is instructive to identify the incentives that lead people to negotiate. It is equally revealing to catalog the factors that can cause them to refuse to bargain at all. For example, because negotiation is a process of mutual accommodation, it necessarily implies recognition of the legitimacy of the opposition's demands and of the opposition itself. This is rarely an issue in environmental controversies, but it is a problem in both international negotiations and labor relations. Israel, for example, has refused to negotiate with the Palestine Liberation Organization. The PLO similarly has thus far refused explicitly to acknowledge Israel's right to exist, apparently regarding the simple fact of recognition as an act that requires some reciprocal concession.

Sometimes parties designate certain issues as *nonnegotiable* as a matter of principle. For example, some labor spokesmen have taken the position that health and safety on the job is a right that they should not have to bargain to obtain through concessions on wages. In the early days of the labor movement, management refused to bargain with labor, in large part because it did not want to surrender its unilateral authority for making decisions over the terms and conditions of employment. Even now, there are disputes over locating the line between work conditions (which are negotiable) and management prerogatives (which are not).

In cases where appearances matter, people sometimes refuse to bargain when they think doing so will be seen as weakness. This was the rationale cited by the United States in refusing to negotiate with the Barbary Pirates in the late 1700s. Charles Cotesworth Pinckney, then the ambassador to France, was credited with declaring, "Millions for defense but not one cent for tribute," an explicit expression that precedent and principle may be worth far more than money. (Some authorities contend that Pinckney actually said, "Not a penny! Not a penny!" which, if less memorable, rests on the same policy.)

Finally, sometimes people bargain with no thought of reaching settlement but simply to win delay. Here the refusal to bargain is tacit but no less real. In international relations, this tactic has been used to buy time to build up armaments in preparation for war. In lawsuits, the party who is benefited by the status quo may stall. A defendent who privately knows that he will be ordered to pay the plaintiff a substantial sum may nonetheless profit by postponing the inevitable if he or she can presently earn a good return on the money that eventually will have to be paid out.

Consider the following questions:

1. What other reasons might someone have for refusing to negotiate? List particular examples.
2. What options are available to someone who wants to negotiate when the other side will not participate?
3. How is it possible to tell when a refusal to negotiate or a refusal to discuss a particular issue represents a sincerely held value and when it is a mere bargaining ploy?

The Form of Compensation

People have different perceptions of what constitutes proper compensation. To MBPP the initial offer of $15 million was a legitimate means of settling the dispute and of addressing the merits of their opponents' case. To the opposition, however, it was unacceptable because it looked like a bribe. In making compensation offers, how the offer is made is sometimes as important as its substance. One man's gift may be another man's bribe. Offerors have to try to package their offers in a way that does not cast the recipient in an awkward light.

Often, when people oppose a project not out of self-interest but from a concern for a third party (e.g., the conservationists' acting on behalf of the whooping crane), they may be offended by a direct offer of compensation. In such cases, in-kind compensation can be a more useful way of helping the intended beneficiary. For example, if a development is going to destroy a highly valued piece of forest, the developer as compensation might offer to purchase comparable land for the creation of a perpetual wildlife preserve. In-kind compensation was arranged in the Grayrocks case through a grant to purchase water rights to offset diminished streamflow caused by the dam.

PROBLEM

Assume that you are the head of the South Carolina Coalition for the Environment, a group that represents all of the state's environmental interests. You have been invited to participate in an ad hoc attempt to negotiate regulations governing hazardous waste management. The following background information may help you decide whether you should participate in this effort, and if so, under what conditions.

In recognition of the fact that mismanagement of hazardous wastes might have catastrophic effects on both human health and the environment, Congress enacted the Resource Conservation and Recovery Act (RCRA) in 1976. RCRA required EPA to develop regulations to control all hazardous wastes from the point of generation to their final disposal. The act allows the states to take responsibility for hazardous waste management, provided that state regulations are at least as stringent as those established by EPA.

Shortly after EPA published its proposed hazardous waste management regulations, South Carolina moved to adopt regulations of its own in order to assume responsibility in this area rather than defaulting to the federal government. Such a move was consistent with the strong states' rights sentiment that prevails in the South. The job of drafting the regulations fell to the state's Department of Health and Environmental Control (DHEC).

In drafting the regulations, DHEC labored under a cloud of suspicion that lingered from its prior involvement in the licensing of a hazardous waste facility. In November 1977, DHEC had issued an industrial waste permit to Bennett Mining Company in Pinewood, South Carolina. The permit was issued without a public hearing. Shortly thereafter, the Bennett plant was sold to a larger waste management firm, SCA Services, which hired DHEC's chief of Special Environmental Programs to manage the facility. The public did not learn of the Pinewood license decision until some time later, and when the story broke, local residents were angry. Moreover, the hiring of the DHEC bureau chief, though legal, had the appearance of impropriety. After Pinewood, DHEC's efforts to draft new hazardous waste regulations would be closely monitored.

DHEC drafted its first set of proposed regulations in late 1978. Public hearings were held, and the business community complained vigorously that the regulations were excessively costly and impractical. DHEC nonetheless submitted the regulations to the legislature for approval, as was required by state law. In the face of strong opposition from the South Carolina Chamber of Commerce, the legislature asked DHEC to withdraw the regulations and to host "further proceedings . . . to insure that the regulations are fair and equitable to all affected parties." Subsequently, DHEC decided to convene all of the interested parties in an attempt to negotiate a set of regulations that would be acceptable to all.

Assume that the new head of DHEC has invited your group to select someone to participate on the advisory panel that will draft the state's hazardous waste management regulations. If this panel is successful, a consensus draft will be submitted to the state legislature for its approval. Even if approval is granted, of course, it is possible that the EPA might still find that the state regulations do not meet federal standards. At this point, what do you have to gain by participating on the panel? What risks do you run? Should you seek any conditions on your participation? What additional information would you like to have before making a final decision on this matter?

4

JOINT PROBLEM SOLVING

INTRODUCTION

In their book *Getting to Yes: Negotiating Agreement without Giving In* (Boston: Houghton Mifflin, 1981) Roger Fisher and William Ury contend that effective negotiators are those who can convert competitive bargaining into joint problem solving. The search for mutually beneficial outcomes is, in their view, the key to success.

The complexity of many environmental disputes may make them particularly appropriate for this approach: the presence of many issues allows negotiators to make efficient trades. This potential is not always easy to realize, however. The road to joint problem solving may be blocked by legal obstacles and interpersonal differences.

As you read the following case, consider why this dispute—which began with a lawsuit in federal court—evolved from a competitive confrontation to become an exercise in joint problem solving. To what extent might the experience here be transferred to other cases?

CASE STUDY: BROWN PAPER

This case was initially drafted by David Gilmore. It was extensively edited by David Kuechle of Harvard University.

Introduction

The Brown Paper Company is a significant economic force in the state of New Hampshire. Its pulp paper mill, located in Berlin, New Hampshire, is by far the largest employer in the northern part of the state—and the second largest in all of New Hampshire. In 1972 Brown employed 2 thousand hourly and salaried workers in Berlin whose population was just under 10 thousand. In addition, the company purchases nearly all the pulp wood, lumber, and waste products from the forest-related businesses located in northern New Hampshire. Of the 30 thousand people who live in this part of the country—from the White Mountains northward—close to 28 thousand were economically dependent on Brown. During the 1970s, Brown was also the largest source of sulfur dioxide air pollution in New Hampshire and a significant source of particulate emissions in the state.

In July of 1979, Brown Paper Company concluded negotiations with the United States Environmental Protection Agency (EPA) after more than 18 months of meetings. The negotiations were not always tranquil. At one point, a participant almost walked out. Yet, in spite of frustration, mistrust, and hostility, the negotiators finally worked out an agreement wherein Brown would spend more than $16.5 million on measures to reduce pollution of the air and the EPA would relax certain requirements and standards that might otherwise have cost Brown untold millions more and that might possibly have forced the company out of business in New Hampshire.

After the conclusion of negotiations, the two principal negotiators, Donald Shields, assistant general counsel of Brown, and Laurence Goldman, chief of the Enforcement Branch for Region I of the Environmental Protection Agency, met to recap their experience. Both men expected that their relationship would continue and that they would negotiate in the future regarding other environmental issues. Both were anxious to profit from their experiences regarding Brown in order to be more effective negotiators in the future.

In the Brown case, both Brown Paper Company and the EPA ended up better off than either expected when the dispute began, even though neither got everything it wanted. The parties had engaged in joint problem solving but not before having engaged in a zero-sum game in which both were seeking victories at the other's expense. The recognition that these negotiations were part of a continuing relationship perhaps contributed to establishing a joint problem-solving mode, but this certainly was not enough. There were times, especially at the beginning, when relationships were extremely fragile and prospects for a workable settlement were nearly shattered. What were the ingredients of success here? What generalities could be derived from the Brown experience that might have transfer value to future negotiation experiences? Were there special characteristics of these negotiations involving a major private corporation and a regulatory agency of the federal government that set them apart from other types of negotiations?

Background

The Brown Company's Berlin pulp and paper mill is located on the east bank of the Androscoggin River Valley in the northeastern portion of New

Hampshire. Although the plant itself is 1070 ft above sea level, several mountain peaks to the northwest, west, south, and east of the plant rise 600 to 1,000 ft higher. Mount Washington, the highest mountain in New Hampshire at 6,288 ft is 12 mi to the south; the Maine border is 7 mi to the east.

The Berlin plant is known as a Kraft pulp mill. The Kraft process involves cooking wood chips after the bark has been removed in a pressurized pot containing a white cooking liquor. The liquor, an alkaline substance consisting of an aqueous solution of sodium sulfide and sodium hydroxide, dissolves lignin contained in the wood, leaving cellulose.

Upon completion of the cooking, the contents are forced into a blow tank where a major portion of the spent cooking liquor is drained off. The pulp is then washed, and nonreactive chunks of wood are removed. Then it is bleached, pressed, and dried to form the final product.

The spent liquor, now black in color, is reprocessed through a series of steps designed to concentrate the liquor to a level of solids that will support combustion; it is then burnt in a furnace where chemical recovery can take place.

Initially, the black liquor contains about 15% solids. It is placed into a multiple-effect evaporator where steam is passed through it in a series of evaporator tubes. This process increases the solid content to between 40% and 55%. Further concentration is then effected in a direct contact evaporator. At Brown, this is a cyclonic scrubbing device in which hot combustion gas from a recovery furnace mixes with the incoming liquor and raises its solid content to between 55% and 70%.

The liquor concentrate is then sprayed directly into the recovery furnace where the solid organic contents are burned. The inorganic contents fall to the bottom of the furnace and are discharged to a smelt-dissolving tank to form a solution called *green liquor*. This green liquor is then conveyed to a causticiser where slaked lime (calcium hydroxide) is added to convert the solution back to a white liquor that is reused in subsequent cooks. Residual lime sludge from the causticizer is recycled after removal of water and calcined (oxidized) in a hot lime kiln.

The pulp-making operations at Brown required four steam-generating power boilers (numbers 6, 7, 9, and 12). Two of these (numbers 6 and 7) dated from 1934. In addition, there were two recovery process boilers, two lime kilns, and two lime slakers. In 1972, Brown burned relatively cheap, high sulfur (2.2%) fuel oil in the boilers. The power boilers required about 50.4 million gallons a year and the recovery boilers about 5.8 million gallons. These, in turn, emitted close to 9,700 tons of sulfur dioxide (SO_2) into the air each year.

In 1972, the state of New Hampshire promulgated a state implementation plan (SIP) calling for the burning of 1% sulfur fuel by weight. This was approved by EPA in May of 1972. At the time fuel oil was cheap: the 1% oil cost 15 cents per gallon; 2.2% oil cost 11 cents per gallon. On July 12, 1973, however, cognizant of the first Arab oil embargo and increased prices of fuel oil, the state submitted to EPA a revision of the SIP that would allow Brown and others to continue using 2.2% sulfur oil. The EPA published a proposed approval of the revision in February of 1974, but the agency took no final action to approve or disapprove officially because it decided that further data were needed. In spite of this lack of final action, William Adams, regional administrator for EPA's Region I, being mindful of the oil crisis, orally promised Brown and other users of industrial oil in Region I that EPA would not enforce the 1% sulfur oil standard. This was a great relief to Brown and others because conversion to 1% oil represented an enormous financial burden. Aside from the difference in per gallon cost, it would be necessary to secure a reliable long-term supply, to arrange for transport to the Berlin plant, and to build new facilities and equipment required by the different viscosity of 1% and 2.2% fuel.

For the next 6 years, Brown continued to use 2.2% oil; by January 1978, it cost 58 cents per gallon—a 427% increase in less than 6½ years. In 1976, the EPA ordered Brown to correct certain problems with its recovery boilers that according to EPA investigators, were resulting in violations of the national ambient air quality standards (NAAQS) for particulates. After several months of negotiation between EPA and the Brown Company, a control strategy was worked out. As part of that strategy, the EPA required Brown to establish an expanded ambient air quality monitoring system because EPA representatives suspected that sulfur dioxide violations were resulting from short stacks on the recovery boilers. The system would cost Brown more than $250,000.

During the winter of 1977–1978 this expanded monitoring system detected violations of the NAAQS for sulfur dioxide (SO_2), something the parties had suspected but not confirmed till then. In February of 1978 the state of New Hampshire issued a notice of violation to Brown and an order to comply. This was followed on February 22 by a notice from EPA to Brown that it would not officially approve use of the 2.2% oil. Then, on February 28, in an apparent reversal of the oral promises made in 1974, EPA formally rescinded its proposed approval of the revised state implementation plan (SIP) that would have allowed use of 2.2% oil. This action applied to Brown Paper Company alone. On March 3, EPA took an additional step affecting Brown when it designated Berlin, New Hampshire, a nonattainment area for sulfur dioxide and total suspended particulates (TSP). In so doing, EPA cited section 107(d) of the amended version of the Clean Air Act.

The state of New Hampshire, having proposed the revision to the SIP that permitted Brown to burn high sulfur oil, was not eager to charge the company with violating the low sulfur requirement. Also, the state was not inclined to alienate a company of Brown's economic clout. The burden of enforcement thus fell primarily on the EPA.

On April 21, 1978, EPA started formal action against Brown by issuing a

Notice of Violation for failure to use 1% sulfur oil "pursuant to requirements under Section 113 of the Clean Air Act." Officials of Brown Paper Company were incensed. They had been lulled for more than 4 years into believing that burning 2.2% fuel oil was permissible. Now they felt betrayed and discriminated against. Because they alone were cited by EPA, other competitive mills in New Hampshire and across the state line in Maine would gain an advantage. Timing was especially bad, because Brown, along with most U.S. paper companies, had experienced a severe economic downturn. Avoiding increased costs was particularly important because Brown's return on shareholders' equity had dropped from a 10-year high of 32.4% in fiscal 1974 to 3.7% in 1977 and 4.5% in 1978. Furthermore, the inflation rate between 1972 and 1978 had been severe. The All Items Consumer Price Index in May of 1972 was 124.7. In May of 1978, it was 193.3. Construction costs during that 4-year period had increased by more than 55%.

At the time, EPA believed that burning of high sulfur fuel was causing the violations, and it appeared that Brown would be required to cease using the high sulfur fuel. Even if the company did cease using 2.2% fuel and converted to 1% fuel, there was no certainty that acceptable standards would be met. By designating Berlin a nonattainment area, EPA had automatically invoked a standard required by the Clean Air Act that called for reaching the "lowest achievable emissions rate" (LAER). This was much stiffer than that which applied to attainment areas. Attainment areas required the so-called "best available control technology" (BACT), a far easier and cheaper goal than LAER.

The Brown Paper Company quickly responded to EPA's action by filing a lawsuit in federal court against EPA challenging the nonattainment designation of the Berlin area. Later, in May, Brown submitted reports required by EPA's notice of violation. These were accompanied by two strongly worded legal memoranda, one challenging the EPA's jurisdiction over the state implementation plan (SIP) revision of 1973 that allowed use of 2.2% fuel. The other argued that the SIP revision had been effectively approved by EPA by virtue of its inaction and by virtue of its regional administrator's oral promises. Brown charged that EPA was violating the equal protection clause of the constitution and section 301 of the Clean Air Act by enforcing the 1% sulfur requirement only against Brown. Therefore, according to Brown's memoranda, the enforcement action must be dropped.

It seemed to Brown's officials as if battle lines has been drawn. Brown and EPA were engaged in a zero-sum game. One party could gain only at the expense of the other's loss. Part of this perception was based on the belief that EPA was inflexible. It was not known then that EPA's position was simply an opening gambit—that the agency had reasons to negotiate. Brown's economic influence and the size and importance of the facility at stake would prompt EPA to give greater attention to this situation than it would to many others. The

JOINT PROBLEM SOLVING   61

Brown case might possibly establish a significant precedent. With talk of oil shortages and another "energy crisis," EPA had negotiated with Brown before—in the 1976 enforcement action—and they were likely to face each other in the future over other environmental matters. On the other hand, there was a severe pollution problem in Berlin, New Hampshire, caused by Brown, and EPA sought its correction.

The First Formal Negotiation Session

Laurence Goldman, chief of EPA's Region I Enforcement Branch, represented the agency, and Donald Shields, assistant general counsel, represented Brown. The formal Section 113 Conference had been preceded by informal technical meetings the day before. At these, Brown presented modeling data that showed that violations of air quality standards would continue even with low sulfur oil. These were derived from a standard model used by EPA to predict pollution levels in areas like Berlin, New Hampshire, where the source of pollution was located near mountains. The EPA model, called the *valley model*, assumed that plumes from smokestacks would rise to the height at which their temperature matched the atmospheric temperature and then level off. At Berlin, the plumes rising from Brown's stacks were generally lower than the difference between the altitude of the plant and that of several nearby mountain peaks. The valley model predicted, therefore, that there would be significant ground-level pollution of the peaks, and boiler stacks could not be raised enough to alter this prediction. Thus, according to EPA's own model, pollution would not be controlled by use of a lower sulfur fuel.

This modeling information was furnished during the conference by representatives of the Environmental Research and Technology Company (ERT), an independent environmental consultant firm employed by Brown. The information served two purposes. One, it focused attention of the parties on pursuance of solutions other than switching to 1% sulfur fuel oil. And second, it provided an entrée for ERT to introduce its own predictive model later in the negotiations—a model that was eventually adopted by the parties as one of the ingredients of their final agreement.

At the first formal meeting of the Section 113 Conference, Donald Shields of Brown made the following statement for the record:

> [We] have had several meetings in the last two days to discuss various matters involving this Notice of Violation. As a result of those meetings, Brown finds itself with a slightly different understanding of the EPA's position. We prepared certain documentation based on other assumptions. We feel that some of this material may be appropriate and some may not in view of the discussions that we have had.

Laurence Goldman, in turn, stated that EPA was willing to consider alternative methods of controlling SO_2 emissions. Goldman said the EPA's ultimate concern was SO_2 emissions levels, not the burning of high sulfur fuel *per se*. He placed the burden on Brown to search for alternative control strategies that addressed the problem, as he saw it. According to Charles Williams, Brown's vice president for engineering and environmental affairs, Brown had to "come out in the open" or there would not be any progress. On the other hand, according to Williams, if EPA was to bend the rules, Brown had to give them justification for doing so.

During the Section 113 Conference, an EPA spokesman summarized matters:

> All I can say at this point is that I encourage the Brown Company—and you can rest assured that EPA will offer our technical expertise to you to work through these problems in an expeditious manner—but I have to caution you that the burden is on the Brown Paper Company and its consultant to provide us with these data. . . . Hopefully, we can reach some resolution short of conversion back to low sulfur fuel, but I have to caution that, if all else fails, then EPA would have no other recourse but to require the company to burn low sulfur fuel. We cannot allow ambient concentrations to exist and persist in the Berlin area in excess of primary health-related standards. I believe you understand our position in that regard.

By allowing Brown to propose its own solution regarding SO_2 emission standard violations, the EPA opened the door to fruitful negotiations. However, the opening was narrow at first, with the entire burden on Brown. In the beginning, the parties focused on the possibility of raising the stacks. Goldman of EPA stuck closely to bureaucratic rules and regulations during this discussion, stating:

> It is my understanding, that based on our discussions . . . we are going to pursue . . . the installation of a taller stack as long as the installation of such a facility is consistent with the requirements of Section 110 [regarding state implementations plans], EPA regulations regarding power stacks, and Section 123 of the [Clean Air] Act. In that regard we feel at this point in time that there is a potential for resolution of this problem. However we are lacking [a] considerable amount of technical data. In addition, we feel that further studies may be necessary, and we are prepared today to highlight for you what additional information and studies may be required. . . . Assuming we can operate within these constraints [sections 110 and 123 of the Clean Air Act], EPA could withdraw the proposed disapproval [of the SIP revision that it published on February 28, 1978] and issue a proposed conditional approval, the conditions being the control strategy that we can agree to.

Because two of Brown's boilers (numbers 6 and 7) were old, dating back to

1934, and would need to be replaced within 2 or 3 years, Brown was reluctant to build higher stacks for them. There was a real possibility, however, that they could convert one of the newer boilers (number 9) to burn waste tree bark rather than oil. This had been considered in late 1977, and in January of 1978 Brown received a permit from the state of New Hampshire to convert the number 9 boiler to burn bark. Potentially, such action had two important advantages. One, although some amount of oil would still be required in addition to the bark, total SO_2 emissions would likely be reduced to almost 50% of existing levels. And two, at existing prices for 2.2% sulfur fuel oil, the company would save more than $2 million in fuel expenses. On the other hand, it would cost at least $15 million to convert the number 9 boiler, and there was no assurance that conversion would result in satisfactory emission levels. This was especially true if EPA stuck to their "nonattainment" designation for Berlin, New Hampshire, thus requiring achievement of the "lowest" achievable emissions rate" (LAER). LAER would require Brown to burn low sulfur oil.

The morale of Brown's negotiators at that stage was low because they perceived EPA negotiators to be unnecessarily strict. Charles Williams, Brown's vice president for engineering and environmental affairs, felt that two of EPA's representatives who were responsible for modeling data requirements were especially intransigent. They were Norman Beloin and Marvin Rosenstein. These two men stuck steadfastly to the EPA's valley model, not seeming to care about the consequences. According to Williams, "Their intransigence could scuttle the proposed settlement." Williams felt that the EPA was taking unfair advantage of Brown's difficult economic circumstances and told them so. At one point, he threatened to walk out on a meeting. This threat, according to some, provided a shock that promoted further discussion.

In spite of uncertainties, anger, and low morale, Brown had the incentive to solicit bids for the conversion of the number 9 boiler. The burden of proof was on Brown, but the economic incentives to pursue bark burning were so great that the company went ahead and approached several equipment manufacturers in July of 1978. It was during this process that the Brown Paper Company received an unsolicited proposal for construction of an entirely new bark boiler (number 14).

Negotiation of the Number 14 Boiler Control Strategy

The number 14 boiler proposal appeared better than the number 9 conversion, both economically and environmentally. Based on 1978 prices it was $2.5 million cheaper than conversion of the number 9 boiler, and the new boiler would last longer and run more efficiently. Although any bark-burning boiler—whether new or converted—would reduce SO_2 emissions significantly, the new boiler would reduce them slightly more than the number 9 conversion because the number 14 boiler could operate without any infusion of oil as fuel.

The number 14 proposal was so attractive to Brown that on August 8, 1978, it submitted an application to the state for permission to construct the new boiler. A week later Brown presented the number 14 proposal to the EPA.

Laurence Goldman of EPA, who earlier had taken a relatively passive role, attended this meeting, and the interchange was lively. Charles Williams pointed out that the number 14 boiler promised considerable advantages for all parties. In addition to saving costs and presenting more of an environmental advantage than the number 9 conversion, it also would allow the scrapping of one of the oldest boilers (number 6) and the use of the other (number 7) for emergency standby only.

Several bureaucratic snarls presented themselves, however. In June 1978, EPA had issued new regulations that required Brown and other companies in New Hampshire to obtain PSD (prevention of significant deterioration) permits from EPA, rather than from the state. These regulations were promulgated as a result of the state's failure to incorporate new PSD regulations into its state implementation plans (SIP). On one hand, this bureaucratic snarl seemed to be a disadvantage to Brown because it was thought that the state would be more receptive to their needs than to those of EPA. On the other hand, it placed EPA in the role of a permitting agency rather than an enforcement agency, perhaps making EPA representatives feel more inclined to cooperate with Brown in guiding the company through the regulatory maze.

Goldman of EPA seemed to play a new role at this meeting—a factor noted by representatives on both sides. In effect, he assumed a neutral position between the Brown Paper Company and various sections of the EPA that had direct responsibility in particular substantive areas. His role was most clearly demonstrated in the case of modeling data. Beloin and Rosenstein had been unreceptive to Brown's efforts to convince them that the valley model might be inappropriate for the locale, but Goldman focused parties on the big picture and expressed a willingness to compromise on data requirements on the control strategy for the number 14 boiler. He did not take a rigid position regarding the valley model.

During negotiations over the next several months, Brown kept its state permit for conversion of the number 9 boiler. This permit would expire the following March, but it was valid until then. Because it had been issued by the state in January of 1978, it was exempt from the subsequent PSD regulations issued by EPA. As a consequence, Brown's negotiators felt they had some leverage by keeping the number 9 conversion project open.

The parties' first efforts after the possibility of a number 14 boiler was put forth were directed at determining the appropriate regulatory requirements. The EPA's position was stated in a memorandum to the file concerning the August 15, 1978, meeting:

JOINT PROBLEM SOLVING

EPA wants Brown to schedule construction of a bark boiler using LAER and power boiler stack improvements using good engineering practice, and to pay penalties for any schedule and ambient air quality violations. The goal of further meetings between Brown and EPA should be to embody the objectives of each party in a consent decree to be filed in a federal court.

In succeeding weeks, the parties had two primary concerns. First, they were concerned about the feasibility of meeting the LAER requirements. Second, they worried about the time it would take to get approval for the number 14 proposal if they adopted it rather than the number 9 option. Brown had set a corporate deadline of August 1979 for the issuance of purchase orders. This was necessary in order to get the boiler on line by their target date of 1981. Both parties were anxious to get the project underway as soon as possible. Brown especially was anxious in this regard because construction costs would almost certainly increase during the next several years at an annual rate of more than 16%.

To overcome both of these concerns, Brown tried to persuade the EPA negotiators to characterize the number 14 boiler proposal as a modification of the number 9 boiler project, so that it would be covered by the existing permit. EPA flatly rejected this idea.

In early September of 1978, Brown's negotiators concluded that it would not be economically feasible for them to build the number 14 boiler if it was subject to LAER standards. The parties appeared to be at an impasse.

In spite of the apparent stalemate, the parties kept talking. Goldman focused the negotiators' attention on ways to get out of the dilemma while remaining consistent with the new PSD regulations. As with most bureaucratic rules and regulations, the PSD regulations contained certain exceptions for special conditions. One of these allowed a permit to be issued requiring only the "best available control technology" (BACT) if the source could show on paper that when the project was completed the area would be in attainment. Brown's negotiators argued that ERT's (their consultants) analysis showed that the SO_2 problem resulted from downwash from the short boiler stacks. They claimed that high particulate measurements occurred when vapor emissions from the lime slakers captured and carried road dust along with particulates from the slakers themselves. Both of these problems would be cured by the number 14 boiler, according to Brown.

In addition, ERT introduced a new predictive model that they said was more accurate than the EPA's valley model. This was called the *rough terrain dispersal model* (RTD), and it assumed that airflow in the valley where the Brown Paper Company was located would be distorted around and over the mountain peaks.

ERT introduced data collected at two monitoring stations near Berlin on 41 days during which meteorological data were available for 24-hour periods. Each of these 41 days showed significant SO_2 levels at one of the two stations. Then ERT applied their RDT model and found that it was a better predictor than the EPA's valley model. RTD predicted actual SO_2 concentrations at one of the monitoring stations for the 5 days with highest concentration by factors ranging from 1.4 to 2.5. The highest valley model predictions overestimated monitored data at the same station by a factor of 8.1. At the other station, the valley model predicted no impact, whereas the RDT model overpredicted the two highest observed levels by factors of 1.9 and 4.5.

The EPA technical experts were impressed with ERT's findings. Although they were not willing to scrap the valley model, they were willing to let modeling demonstrations proceed based upon ERT's assumptions regarding downwash and high particulate emissions. These assumptions were valid unless and until the number 14 boiler was built and other measures had been taken. Thus, the parties were able to move ahead—to concentrate on modeling techniques.

As time passed, the ever-present problem of time itself became even more serious. EPA remained steadfast in its unwillingness to consider the number 14 project a modification of the already approved number 9 boiler conversion. The agency did agree to do everything it could to expedite processing of the new PSD permit, but this did not satisfy the Brown negotiators. They wanted assurance that their permit for the number 9 project would not lapse before Brown had obtained a PSD permit for Number 14, just in case the new permit was denied. By way of compromise the EPA, therefore, agreed to consider whether or not construction had "commenced" on the number 9 project. If it had, the existing permit would remain valid beyond March 19.

In January of 1979, Merrill Nash, president of Brown Paper Company, paid a personal call on William R. Adams, Jr., Region 1 administrator of EPA, in Boston. Neither man had been actively involved in the negotiations, but each had been fully briefed by their chief negotiators. Nash asked Adams for quick approval of the Number 14 project, pointing out that everyone would be a loser if matters were not expedited. In addition, he told Adams that contracts had been signed and parts had been ordered for the number 9 boiler. Adams, in turn, took an active role behind the scenes to try and cut through the bureaucratic maze. On March 15, 1975, he formally notified Donald Shields that the EPA had concluded that construction had indeed "commenced" on the number 9 project. The approval, according to Adams, was based on detailed information about progress on the construction process provided by Brown. Thus, the parties were able to continue talks on the number 14 project unencumbered by the March 19 expiration date on the number 9 construction permit.

Modeling Discussions

While discussions were taking place between Goldman, Williams, and Shields regarding the issuance of a PSD permit, Brown's consultants (ERT) continued to work with EPA technical experts. These technicians agreed to use ERT's new rough terrain dispersion model (RTD) instead of the EPA's valley model. ERT had shown repeatedly that the RTD predictors were closer to actual monitored data in the Berlin area then were the valley model's predictors. However, this did not end discussions on the technical data. It was then necessary for ERT to study whether the number 14 proposal would comply with national ambient air quality standards (NAAQS) using the RTD Model. By February of 1979, the technical staff of the two parties had agreed on how to do the compliance study. By mid-March, ERT had completed it.

The study was based on the following assumptions:

1. That stacks on power boilers 9 and 12 would be raised to the limit of "good enginnering practices" (GEP);
2. That boilers 6 and 7 would be shut down;
3. That filters would be installed in the line slakers; and
4. That construction would proceed on the new number 14 bark boiler.

Based on these assumptions, ERT predicted a 24-hour primary NAAQS violation for total suspended particulates in the Berlin area. This caused Goldman and Shields, with their respective technical and special assistants, to meet on March 19 to discuss further control measures. They considered three options:

1. Constructing a taller (GEP) stack or vent emissions from the lime kilns and slakers;
2. Increasing the height of the stacks of recovery boilers number 8 and number 11 to meet GEP; and
3. Decreasing plant production.

ERT believed, based on its studies, that option 1 alone would enable them to demonstrate attainment of the primary 24-hour NAAQS for particulates but that secondary and annual particulate standards would not be met under any of the options.

Option 1 was relatively palatable to Brown because ERT's tests clearly showed the lime kilns and slakers to be the source of particulate emissions detected at one of the monitoring stations. The parties were sharply divided on further control measures regarding the secondary standards. It became clear during the discussions that ERT's assumptions regarding background particulates made the difference between Brown's being able to demonstrate compliance with the secondary standards under option 1 and not being able to do so.

The background level of particulates is that level that exists in the air before considering the impact of the source (Brown) on the pollution level of an area. It is determined by taking actual monitored ambient air data. But different judgments are involved in selecting the monitoring stations and the days of monitored data to include as background. The days are important because the concentrations of pollutants measured by the monitors are significantly affected by meterological conditions. Brown, relying on ERT, disagreed with EPA over which stations to include. Brown wanted to include a station 6 miles away, but the EPA refused, saying the distance was too great. Brown also argued that the average background levels on days with meteorological conditions that create the highest pollution should be used. whereas EPA representatives argued for the maximum on those days. ERT, representing Brown, proposed a level of 30 micrograms per cubic meter ($\mu g/m^3$) based on the "average" levels, and EPA proposed 50–60 μg—the "maximum" levels. Several weeks of intense discussions were devoted to this issue, and some bad feelings between the parties resulted from these. Eventually, however, the parties agreed on a background level in the high 40s—a compromise from their extreme positions but not an agreement that was satisfactory to those technical experts on both sides who believed strongly that scientific data should not be bartered.

Nevertheless, the chief negotiators on both sides prevailed, and Brown was required, as a result of these negotiations, to raise the stacks on recovery boilers numbers 8 and 11 in addition to building a new stack for the lime kilns and slakers. This raised Brown's costs by $1 million. Because boilers 8 and 11 were obsolete already and Brown had planned to replace them with a new recovery boiler within 10 years, they considered the added cost to be a waste. They "bit the bullet," however, because of the economic benefits of the number 14 boiler proposal. The way had been paved for a final settlement.

Final Settlement

In spite of bad feelings among members of the technical staffs, they quickly agreed on the details of ERT's final modeling demonstrations. The legal staffs, primarily Rowena Conkling for the EPA and Donald Shields for Brown, began to discuss the form of the final agreement. The EPA wanted it to be embodied in a consent decree filed with the federal district court so as to maximize its enforceability. Under a consent decree, the court would have jurisdiction over implementation of the terms of settlement and could quickly deal with alleged noncompliance. Noncompliance would constitute contempt of court for violation of the court decree in addition to being a violation of environmental requirements.

Brown preferred that the agreement be represented by a contract or an administrative order by EPA. Donald Shields argued that Brown "clearly [did] not fit the category of recalcitrant sources whose agreements require judicial enforcement and supervision."

The EPA held fast on the issue, and Brown finally agreed to the consent decree of May of 1979. The parties then negotiated over the terms of the consent decree from May through July of 1979. They disagreed over four issues:

1. The extent to which EPA would bind itself to review the attainment/nonattainment designation of metropolitan Berlin;
2. Whether Brown would withdraw its suit agaist EPA challenging the EPA's nonattainment designation of Berlin;
3. The duration of any consent decree; and
4. The amount and nature of penalties (if any) for noncompliance with the decree.

On June 6, 1979, while negotiations over these issues proceeded, Harley Laing, assistant regional counsel for EPA, raised an objection to the planned order of construction of the number 14 boiler, the new stacks, and other required controls. Brown had planned to build the number 14 boiler first and begin its operation before building the stacks on the other boilers and kilns and before implementation of the dust control program. Laing argued that regulations would be violated if Brown operated the new boiler before other measures had been taken to bring the area into attainment status. As a result, Brown agreed to change the planned order of construction. The PSD permit was issued on August 3, 1979, and judgment was entered on the consent decree by the U.S. District Court of New Hampshire one week later.

Terms of settlement were:

1. The new number 14 bark-burning boiler would be constructed. Estimated completion date was January of 1981. Estimated cost was $12 million.
2. ERT's rough terrain dispersion model (RTD) would be used to demonstrate effectiveness of the parties' control strategy.
3. The PSD permit for the number 14 boiler would require BACT—the best available control technology, not the more stringent LAER, because modeling demonstrations by ERT in the Berlin area indicated that this would be an attainment area for SO_2 and total suspended particulates upon completion of the new boiler and other control measures.
4. The EPA would withdraw its proposed disapproval of the state of New Hampshire's SIP revision allowing use of 2.2% sulfur oil.
5. Brown would undertake a total environmental cleanup program to cure particulate emission problems and to eliminate SO_2 emission violations. Steps toward accomplishing cleanup would be the following:
    (a) a new combined stack for boilers numbers 9 and 12 to be built in accordance with good engineering practice (GEP), that is, it would be tall enough to prevent downwash;
    (b) a new combined stack also meeting GEP for recovery boilers numbers 8 and 11, the related smelt-dissolving tank, and lime slakers numbers 1 and 2;
    (c) a new combined stack meeting GEP for lime kilns numbers 1 and 2;

(d) scrubbers to be installed on lime slakers numbers 1 and 2;
   (e) shutdown of the number 6 boiler and use of number 7 for emergency standby only; and
   (f) various dirt roads and traffic areas in the plant to be paved.
6. Brown to expand its pollution monitoring network;
7. Brown to pay a penalty of $66,000 for past violations of the Clean Air Act; and
8. Brown would withdraw its suit against EPA challenging its nonattainment designation of Berlin, New Hampshire, without prejudice.

Total cost to Brown Paper Company of the settlement was estimated to be in excess of $16.5 million.

Study Questions

1. Identify the interests of each side as they appeared at the outset. What were the parties' respective priorities. What were their strengths and vulnerabilities?

2. Chapter 3 explored incentives to negotiate. What such inducements obtained here? Were the incentives likely to be peculiar to this case, or might they be present in other disputes?

3. Brown originally took a hard line with the EPA by quickly filing a lawsuit challenging the agency's attempt to enforce the stricter sulfur fuel requirement. Assuming that Brown could always act so swiftly, what risks are inherent in such a tactic? If the EPA believed that Brown was in clear violation, why did it not respond in kind?

4. Parties in negotiation often have trouble convincing the other side of the sincerity of their commitment to joint problem solving. Here, Brown officials were skeptical that the EPA would consider control strategies other than reversion to low sulfur fuel. Brown's skepticism could have prevented productive negotiation from ever commencing. How did the EPA persuade Brown of its willingness to consider alternative solutions?

5. Time often is a factor in negotiation that may not affect the parties equally. Why was Brown anxious to settle the case quickly? What accounts for the fact that it took a year and a half for Brown and the EPA to reach an accord?

6. The author of the case study observes that negotiation between the parties evolved from zero-sum to nonzero-sum. In what way was this so? Was it by design or by happenstance? The positions the parties took with one another changed. Did their underlying interests change as well?

7. The negotiations between the parties were not always tranquil. At one point a participant almost walked out; yet, in spite frustration and some hostility, the negotiations continued. Why?

8. The EPA insisted on an expensive monitoring system, one that cost a quarter of a million dollars just to install. Brown felt this was overkill. To the extent that this system added to the company's expenses, it might seem to have made settlement less attractive; yet, in another way, it may have enhanced prospects for resolution. How so?

9. At the close of negotiations, the EPA raised an additional issue, the order in which Brown planned to construct and operate the number 14 boiler, and new stacks, and controls. Making Brown wait to use the bark boiler obviously imposed costs on the company; still, the EPA insisted. Why?

10. The cost of the environmental program undertaken by Brown totaled $16.5 million. Why, then, did the EPA require the company to pay a penalty of $66 thousand? Also, what was the function of incorporating the agreement in a consent degree when the terms had been negotiated entirely out of court?

11. Some of the issues in this case might be characterized as technical, whereas some others are legal, and still others are more overtly policy or politically based. Do these different categories of issues argue for different dispute resolution procedures?

Problem 1

Fairly early in the negotiations, Brown was able to persuade the EPA to explore alternative control technology rather than switching to the much more expensive low sulfur oil. You have already been asked to consider why the EPA did not insist on immediate compliance. Consider hypothetically what would have happened if the EPA had taken a hard line and forced the conversion to one percent oil, but then discovered that the air pollution problems remained substantially the same. (This, in fact, is what the company warned when it presented the "downwash" theory.) How would such a development have subsequently affected the bargaining position of both the EPA and the company?

Problem 2

The prospect of the new number 14 bark boiler was an important breakthrough in the case and may have paved the way toward eventual settlement. At first glance, it appears to be that sort of development that benefits both sides: it helped Brown in that it was cheaper, more efficient, and more durable than any converted boiler; it helped the EPA in that it would reduce SO_2 emission significantly, even more than would a converted boiler.

Reread the section of the narrative dealing with this development and consider how the introduction of this factor changed the balance of bargaining power. Installation of the new boiler required a PSD permit from the EPA, and the agency could condition the permit on a total environmental cleanup—which it could not require for a mere conversion. Brown officials ultimately felt that this gave the EPA great leverage, and indeed, there was resentment that the EPA had exercised this advantage unfairly.

Consider the dynamics of power in this situation: Did the EPA really have the leverage to dictate a solution? What constraints was it operating under? Could Brown have made any credible threats to counter the EPA on this point?

Problem 3

Some of the major parties in this case had confronted one another before, and they recognized the likelihood they would meet again. In what specific ways does this perception affect the conduct of the parties? To ask the question another way, would the Grayrocks Dam dispute (described in chapter 3) have gone differently if the participants there had seen themselves as being involved in just one in a series of continuing conflicts? Is this a factor, one over which the parties can exercise some control?

Personnel shifts are common in business and government. If Brown Paper and the EPA have to bargain again, but if different people are involved, should such an encounter be regarded as a one-shot or repeat negotiation?

Postscript on Brown Paper

In December of 1980 the Brown Paper Company was acquired by the James River Corporation, a Virginia-based firm with 1980 sales in excess of $374 million. James River of course, assumed all legal responsibilities incurred by Brown and moved forward to comply with the EPA agreement.

The number 14 boiler was put into operation on January of 1981. Soon thereafter, the new stacks were installed on the numbers 9 and 12 boilers, the numbers 8 and 11 recovery boilers, the smelt-dissolving tank, and lime slakers numbers 1 and 2. Scrubbers had been installed on the lime slakers in early 1979.

According to Ray Danforth, director of technology for the James River Corporation's Berlin, New Hampshire, operation, the pollution monitoring network had detected no violations of SO_2 standards for the 8-month period from January through August of 1981 at any of its stations. One of these, at Lancaster

JOINT PROBLEM SOLVING

Street, had shown frequent violations before. In addition, the monitoring system showed no violations of particulate emissions standards except when the number 7 boiler was used.

Regarding background particulates in the Berlin area, Danforth reported a less favorable experience. The city of Berlin had been engaged in extensive road building during 1981, and this involved a considerable amount of blasting. This, combined with much more wood burning throughout the area and virtually no snow cover during the winter, apparently caused background particulates to increase well beyond expected levels. The predictions had suggested that particulate levels would increase in the spring, stabilize in the summer, decline in the fall, and then remain fairly low during the winter. During the year ending September, 1981, however, background levels were consistently high.

The Search for Solutions

The introduction of the possibility of a new, cheaper, less polluting boiler may seem like a lucky development, but often the key to successful negotiation is the ability to recognize opportunity in what others see as disaster, a lesson that is taught by the following bit of political history.

In 1912 former President Theodore Roosevelt was running for the White House on the Bull Moose ticket with California Governor Hiram Johnson. As a promotional device, his supporters printed three million copies of a stirring Roosevelt speech called "Confession of Faith." Just before they were to be distributed, a campaign worker noted that the cover photograph of Roosevelt and Johnson had been copyrighted by the Moffett Studio in Chicago, but no one had received permission to reproduce it.

The worker rushed to George Perkins, the brilliant financier and Roosevelt manager, to warn him of the dangers. If the pamphlets were released, a fine of one dollar per copy could be levied. The campaign would owe three million dollars and Roosevelt would be accused of being a lawbreaker. Yet, if the pamphlets were destroyed, the printing bill would still have to be honored, and valuable time would be lost. Clearly, Moffett's permission to use the photograph had to be obtained, but at what cost?

Before reading the concluding paragraph of this note, stop to consider what you would do if you were Perkins, keeping in mind his apparent bargaining strength relative to Moffett's.

Perkins barely hesitated before dictating the following telegram to the photographer:

> WE ARE PLANNING TO ISSUE AN EDITION OF THREE MILLION COPIES OF
> ROOSEVELT'S SPEECH, WITH PICTURES OF ROOSEVELT AND JOHNSON ON

THE FRONT PAGE. THIS WILL BE A GREAT ADVERTISEMENT FOR THE PHOTOGRAPHER. WHAT WILL YOU GIVE US TO USE YOUR PICTURES? RUSH ANSWER.

Perkins' gambit was rewarded with a quick response: "WE HAVE NEVER DONE THIS BEFORE, BUT UNDER THE CIRCUMSTANCES WE WILL GIVE YOU $250."

Comparison Case: The White Flint Mall

In the Brown case, the problem that the parties solved was primarily a technical one: the development of a boiler system that would be clean enough to satisfy the government, yet cheap enough to be affordable to the company. Not all environmental problems are technical, however. In the White Flint Mall case, described in Malcolm Rivkin's *An Issue Report: Negotiated Development: A Breakthrough in Environmental Controversies* (Washington, D.C.: The Conservation Foundation, 1977), residents near the site of a proposed shopping center were concerned that noise and traffic would cause their property values to fall. The developer put up a bond to assure abutters that a 14-ft-high earth berm would be built and landscaped, but still the neighbors were anxious. Ultimately, the developer contracted to indemnify the neighbors for any loss in property value over the succeeding 5 years. The developer was more confident than his new neighbors that his project would not have negative impacts, so he was willing to become the insurer for whatever risks the shopping center meant for others. By engaging in joint problem solving, the parties were able to identify their fundamental interests and to take steps to protect them.

The Collective Bargaining Model

There are several important parallels between environmental negotiation and collective bargaining involving unions and management. Both situations are often nonzero sum, that is, there are a number of issues, at least some of which are weighed differently by the contending parties. Efficient trades can generate benefits for both sides. There are also environmental disputes in which the particular parties have negotiated with one another before, as is often true in collective bargaining, and in which they work with the shared expectation that they will likely meet again. This was true in the Brown case.

There are, however, major differences. Although collective bargaining is mandated by federal law, there is no comparable legal obligation to negotiate over environmental problems. The lessons of labor law are nonetheless important for both theorists and practitioners. In fact, many of the private companies that find themselves in environmental disputes have extensive collective bargaining experiences that may well affect their attitudes and expectations. The concept of "good faith bargaining," for example, has been extensively defined and applied in the labor field; though not legally obligated to do so, those managers who work with this norm may bring it to the environmental table (For a more extensive discussion of the parallels and contrasts between collective bargaining and environmental negotiation, see Lawrence Susskind and Alan Weinstein's "Toward a Theory of Environmental Dispute Resolution," 9–2 *Boston College Envir. L. Rev.* 311–351, 1980–1981.)

5

Data Negotiation

Introduction

At the heart of most environmental controversies lies a dispute over the likely future consequences of a proposed action. In the Grayrocks Dam case, the parties argued over the probable effect of the dam on wildlife and farming downstream. In the Brown Paper case, the parties argued over the necessity of installing expensive air pollution control technology. In the Holston River case that follows, the Environmental Protection Agency and the Tennessee Eastman Company had very different opinions about both the cost and the beneficial impact of the agency's proposed water pollution control efforts.

Making predictions about impacts that will occur well into the future necessarily involves a fair degree of scientific and technical analysis. To forecast the effect of construction of the Grayrocks Dam required knowledge of seasonal variations in the North Platte River waterflow, the hydrological characteristics of the dam, and the migratory habits of the whooping crane. Similarly, in order to isolate the effect of construction of a new bark boiler on the Berlin, New Hampshire, air quality region required a sophisticated understanding of the distribution of smoke plumes in rough terrain. Although we can predict the operation of a few natural systems quite accurately—the rise and fall of the tides is a good example—our understanding of how most ecosystems operate is fairly limited. As a result, our predictions are at best approximations of reality.

Regulatory decisions more frequently than not turn on mathematical models that are based upon simplifying assumptions. This produces a situation ripe for conflict. Because modeling is expensive, there is a trade-off between accuracy and cost. Government models are constantly subject to challenge by outside experts who claim that their industry-funded models are more accurate. Moreover, because different people are inclined to make different assumptions, environmental disputes often become battles between experts hired by the opposing parties to defend a particular set of premises.

DATA NEGOTIATION 77

The Holston River case that follows is a good example of such a dispute. At the root of the conflict between EPA and the Tennessee Eastman Company was a dispute over the validity of the models used by EPA to estimate the capacity of the river to absorb pollutants. In reading the case, think carefully about how access to information and expertise influenced the bargaining strength of the parties. Would the outcome have been any different if Tennessee Eastman's experts worked for EPA and vice versa?

The Holston River Case

This case was originally prepared by Alexander Jaegerman. It has been substantially edited and revised.

Introduction

In October, 1972, Congress passed the Federal Water Pollution Control Act Amendments over the veto of then President Richard Nixon. The law declared that all pollution discharges into U.S. waters were illegal unless specifically authorized in kind and quantity by a permit issued by the EPA. Under the law, the agency is required to set standards and issue permits by reference to the technology available to control pollution. By July 1, 1977, the National Pollution Discharge Elimination System (NPDES) was supposed to reduce discharges to the level achievable through application of the "best practicable technology (BPT) and operating practice, taking into account costs of implementation and benefits derived." By July 1, 1983, discharges are to be reduced further to the level attainable through use of the "best available technology" (BAT) economically achievable.

The NPDES standards vest a tremendous amount of discretion in the hands of EPA officials. Determining BPT and BAT requires not only an assessment of the state of the art of pollution control technology but also a balancing of the costs and benefits of alternative control strategies. In practice, these are difficult decisions that are scrutinized closely by industry and environmental groups alike. The stakes are high for all parties; controversy is not uncommon.

In late 1972, shortly after adoption of the Clean Water Act amendments, the Tennessee Eastman Company submitted an NPDES application to EPA. Tennessee Eastman is a major chemical processor. Its plant in Kingsport, Tennessee, occupies over 400 acres on the Holston River and employs close to 12 thousand people (see Figure 7). The plant produces an array of chemical products, including Kodel polyester fibers and Kodak films and chemicals. Chemicals are processed in vast quantities; on an average day, over 700 million pounds of materials are handled in the plant. Not surprisingly, Tennessee Eastman is

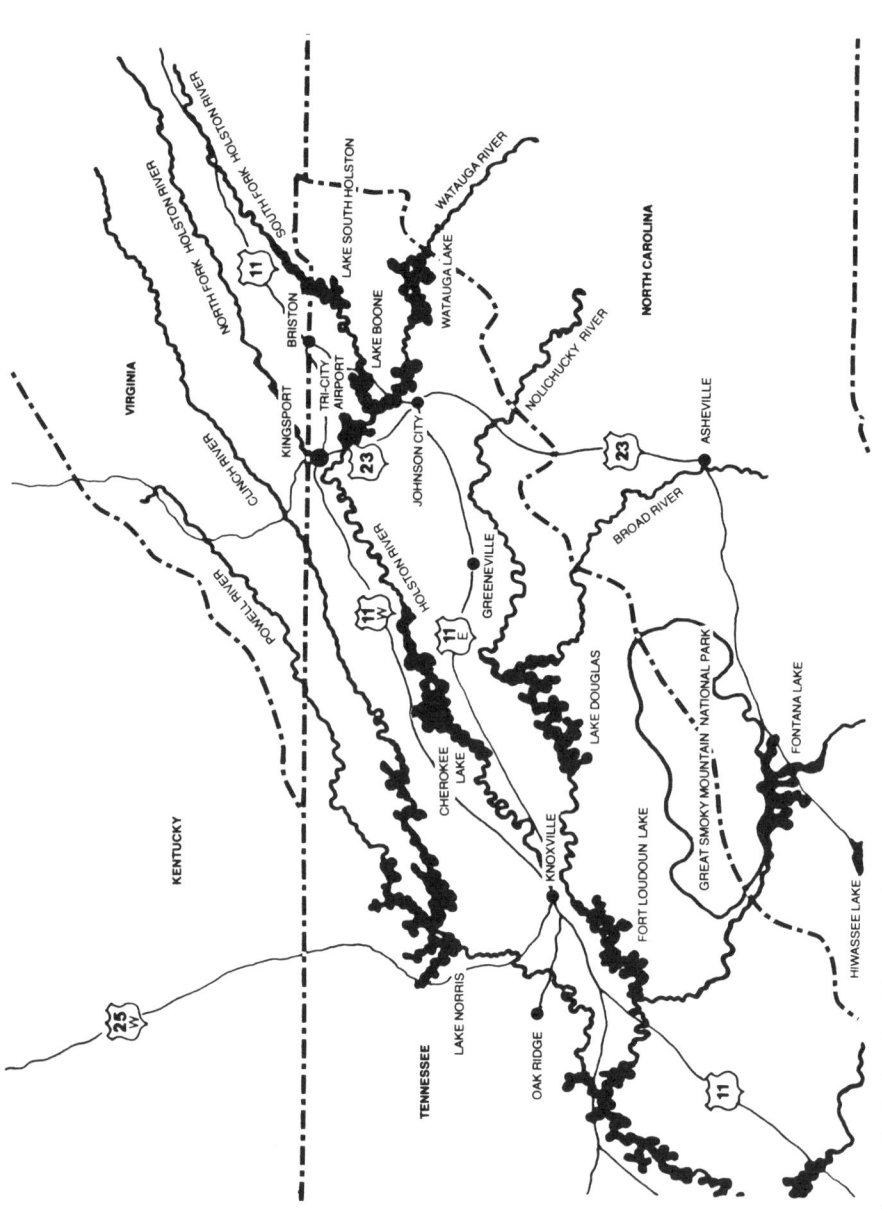

FIGURE 7. Northeastern Tennessee. (From "Water-Borne Effluent Limits," by Tennessee Eastman Company, September, 1978, III.2.)

also a major producer of chemical waste. At the time of its application for an NPDES permit, the company was discharging 400–500 million gallons of treated wastewater into the Holston River daily—about the same amount of wastewater produced by a city of 5 million people, and at times equal to the entire flow of the river.

The Tennessee Eastman NPDES application posed a number of difficult problems for EPA. First, the company was one of five major dischargers along this stretch of the Holston River, albeit by far the largest one (see Figure 8). Effluent limits had to be established in an equitable manner.

Second, setting effluent limits involved making complex determinations about the capacity of the Holston River to assimilate pollutants. EPA's models were certain to be challenged by Tennessee Eastman's experts. Third, Tennessee Eastman had substantially greater technical resources at its disposal than did EPA. It employed over 1,800 scientists, engineers, and support personnel in five laboratories located on site. Moreover, it had the capacity to hire the best consultants in the field. Fourth, the criteria specified in the statute for granting NPDES permits were inherently ambiguous. Although this ambiguity gave EPA a fair amount of latitude to fashion a permit, it also gave Tennessee Eastman room subsequently to challenge any permit it regarded as excessively strict on the grounds that EPA had misinterpreted its statutory mandate. Finally, as a major employer essential to the economic health of the region, Tennessee Eastman was capable of mustering substantial political support in favor of its position if the NPDES permit discussion blossomed into a full-scale public dispute.

Setting Effluent Limits Under the NPDES

The purpose of effluent limits permits under the NPDES is to ensure maintenance of minimum water quality in the nation's waterways. Water quality is generally measured along a number of dimensions, including color, odor, turbidity, and the presence of toxins, pathogens (e.g., viruses), surface scum, oil, or foam. The presence of dissolved oxygen is also an important indicator of water quality. When organic matter is discharged into a stream, a decomposition process occurs in which microorganisms digest the waste, breaking it down into its essential elements—generally nitrogen, phosphorous, and carbon. During this process, which is called *waste assimilation*, the oxygen that is dissolved in the streamwater is consumed; as more waste is assimilated, more oxygen is drawn from the stream. If waste enters a stream in large quantities, the oxygen supply will be depleted in the decomposition process. As the level of dissolved oxygen falls below three to five parts per million (ppm), fish are adversely affected. If the oxygen level drops to zero, anaerobic digestion occurs, killing all fishlife and causing odorous gases to be emitted.

The oxygen available for waste assimilation at any point in a stream varies

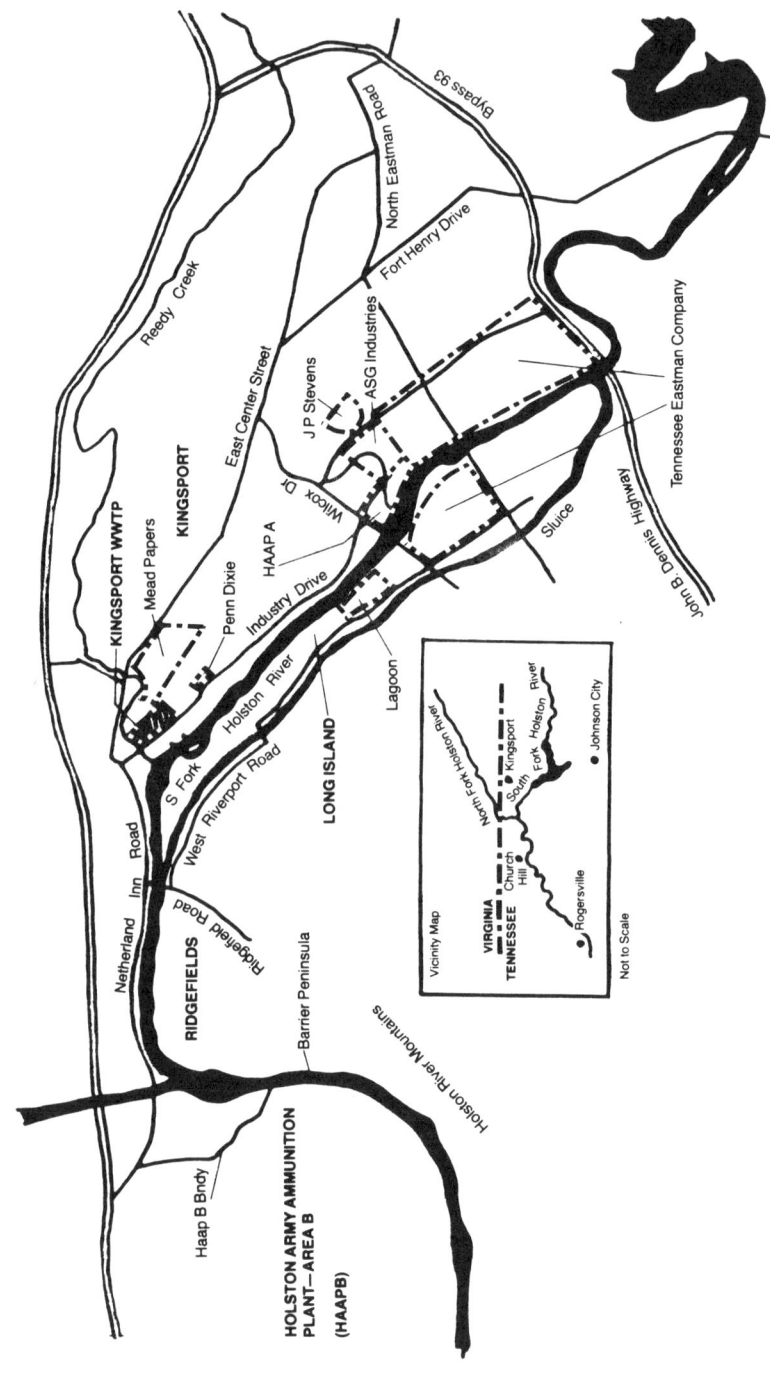

FIGURE 8. Location of industries on Holston River. (From "Water-Borne Effluent Limits," by Tennessee Eastman Company, September 1978, III.13.)

with a number of factors, including temperature (colder water supports more oxygen), turbulence, depth, upstream conditions, bottom deposits, aquatic weeds, and the like. The precise way in which these factors interact to influence the level of disolved oxygen is not perfectly understood. As a result, competing models exist to describe the impact of waste discharges on the availability of disolved oxygen downstream.

The responsibility for modeling the water quality of the Holston River in the Tennessee Eastman case fell to Douglas Lankford, a sanitation engineer employed by EPA with a recent masters degree in engineering from Vanderbilt University. Lankford used a waste load allocation model to specify discharge limits for the various firms along the Holston River. The basic assumption of such a model is that the stream can assimilate a certain waste load and still have an adequate supply of dissolved oxygen to maintain water quality. Given the biological oxygen demand (BOD) represented by a particular discharge, the model will describe the downstream effect on the oxygen deficit in the river, including the minimum oxygen concentration known as the *oxygen sag*. Later, during the months of negotiation, Lankford found himself defending his model from vigorous attack by several of his former professors.

EPA began to assemble the information needed to run the model with three trips to the Tennessee Eastman plant in late 1972. Of particular interest were the magnitude and composition of Tennessee Eastman's waste discharges. In giving consent to EPA to do sampling at the plant, James Mitchell, executive vice president for manufacturing, indicated that Tennessee Eastman would collect and analyze duplicate samples to those taken by EPA. In so doing, Tennessee Eastman was sending a clear signal to EPA that it was prepared to challenge the technical underpinnings of any effluent limitations that it believed to be unacceptable.

EPA returned to the Holston River in January and February to gather more data on heavy metals, monitor stations, BOD generation, and previous pollution control measures. This information was used by the Region IV EPA staff to formulate the permit limitations. The results of these information-gathering and modeling efforts were published in a report entitled *Waste Source Investigations—Kingsport, Tennessee."* The report was distributed to Tennessee Eastman as well as state water quality officials. At the same time, EPA sent to the state (but not to Tennessee Eastman) a preliminary draft of the NPDES permits for the Kingsport area with a request for review and comment.

Negotiations Begin

When a copy of the previously mentioned report was received by Jim Edwards, manager of the Clean Environment Program at Tennessee Eastman, he responded quickly with a letter to EPA. The text of the letter set the tone for many of Tennessee Eastman's subsequent communications and emphasized the complex nature of determining effluent limitations; it urged that Eastman be given opportunities to discuss the permitting process, particularly before EPA went to public notice with the permit conditions. Tennessee Eastman was insistent in the months that followed that they and EPA should resolve their differences through discussions before bringing the matter before the public.

In July, 1973, EPA sent to Tennessee Eastman a NPDES Fact Sheet and Public Notice that delineated most of the permit details. Edwards responded by phone and by mail to Howard Zeller, chief of EPA's permit branch, suggesting again that a meeting should be arranged to discuss the proposed effluent limitations that "present a serious and urgent situation for us and the communities of northeast Tennessee." He went on to say that

> the proposed limitations . . . would require a major reduction of Tennessee Eastman Company's employment and production. We know of no wastewater treatment system that is technically and economically feasible which will reduce the water-borne wastes to levels comparable with the proposed effluent limitations . . . Therefore, we request a meeting with you and your staff to determine procedures and schedules for developing and presenting factual data and meaningful information concerning the proposed effluent limitations. . . . It is essential to complete this discussion before public notice.

The insistence upon staff-level discussions, out of the public eye, reflected a desire by the company to resolve difference on scientific or technical grounds. Although the effluent limitations certainly raised issues of social choice—for example, whether it was reasonable to provide such a high degree of protection for waters that were inaccessible to the public—most of the subsequent discussions centered on technical questions such as the validity of the model, the effects of nutrient discharges on weed growth and dissolved oxygen variation, and the reliability of a particular treatment technology for removing nitrogen from wastes.

There are a number of reasons why Tennessee Eastman may have decided to restrict its challenge to technical issues. Budding trial lawyers are taught that the first line of defense in any case is the facts. If the facts are not with you, then argue the law. And if the law fails, argue policy. By trying to shape the discussion in technical terms, Tennessee Eastman was adhering to this old adage. Although the company was prepared to argue that it was noneconomic to protect a river that was inaccessible to the public (one Tennessee Eastman report contained a map showing how far downstream the riverbanks are either on industry-owned property or are restricted for military security reasons), the law clearly provided for the protection of inaccessible rivers. Moreover, conducting the debate on technical grounds favored Tennessee Eastman because of its vast technical resources and its access to experts in the field.

As requested by Edwards, a staff meeting was held in Atlanta between Tennessee Eastman and EPA at which the company argued that the proposed effluent limitations were intolerably and unjustifiably strict. During this meeting, Tennessee Eastman agreed to submit a counterproposal to EPA that would address the subject of long-term effluent limitations from the company's perspective. Tennessee Eastman prepared this report in house and submitted it to EPA in the fall of 1973. The report, entitled *Water-Borne Effluent Limits*, quite predictably proposed effluent limitations that were substantially higher than those contemplated either by EPA or the Tennessee Water Quality Control Board (TWQCB). (Tennessee Eastman was required to obtain permits from both EPA and the TWQCB. Although the EPA kept the state informed of its activities, the two permitting processes proceeded separately, notwithstanding Tennessee Eastman's efforts to telescope them into one single regulatory process.)

Table 1 compares the initial bargaining positions of the three principals. The table is revealing in at least three respects. First, for effluents in which both proposed limitations, the state and EPA were in close agreement (with the exception of nitrates and nitrites.) Second, the differences in the substances to be regulated and the way in which they were to be measured indicate the lack of coordination between the state and EPA. Finally, the major disagreements between EPA and the company centered on limits for BOD, the nutrients nitrogen and phosphorous, and on pounds per day of total suspended solids. There was substantial agreement on the other limitations.

Tennessee Eastman also hired two consultants, Peter Krenkel and Vladimir Novotny, who prepared a report entitled *The Assimilative Capacity of the South Fork Holston River and Holston River below Kingsport, Tennessee*. The report critically reviewed EPA's modeling efforts and pointed out a number of potential weaknesses in the EPA analysis: an alleged exaggeration of the depth of the river; a failure to verify the results of the model against observed values of the river water quality parameters; and a possible miscalculation of the reaeration coefficient—the rate at which oxygen is reintroduced into the stream from the atmosphere. The report also pointed out that the EPA model did not account for the effect of aquatic weed growth on dissolved oxygen levels. In respiration, aquatic plants at times consume oxygen and at other times release oxygen into the water, causing diurnal variations in dissolved oxygen (DO) levels. Lankford did not try to model this variation, in part because the EPA was imposing very strict standards for nutrient discharges—primarily nitrogen and phosphorous—on the assumption that reduced nutrient discharges from Tennessee Eastman would significantly reduce the magnitude of the nuisance weed problem and simultaneously eliminate the diurnal variation in DO. This assumption became a

TABLE 1. INITIAL POSITION OF THE PARTIES

Units	Effluent constituents	TWQCD proposed limitations (lb/day net)		EPA proposed effluent limitations		TEC proposed effluent limitations	
		Daily average	Instantaneous maximum	Daily average	Daily maximum	Max. monthly average	Max. day
pounds/day	Biochemical oxygen demand (BOD)	3350		2230	3350	10,000	17,000
	Total Kjeldahl nitrogen (TKN) as N	400		270	400	9,000	18,000
	Total nitrogen (TN) as N	3310		—[a]	—	—	—
	Ammonia nitrogen as nitrogen	400		—	—	—	—
	Nitrates plus nitrites as N	2910		100	700	7,500	11,000
	Phosphates as P	213		140	210	800	1,500
	Total dissolved solids (TDS)	1,200,000		—	—	650,000	1,200,000
	Total suspended solids (TSS) (gross)	—		2230	3350	6,000	15,000
	Zinc as Zn	250		125	250	125	250
	Phenols	100		10	20	100	175
	Iron as Fe	1000		—	—	300	1,000
	Manganese as Mn	—		100	200	100	200
		Daily average	Instantaneous maximum			Monthly average	Max. day
mg/l[b]	Total suspended solids (TSS)	40.0	50.0	—	—	40	100
	Chromium	0.05	0.075	—	—	0.05	0.1
	Copper	0.02	0.03	—	—	0.05	0.1
	Mercury	0.005	0.0075	—	—	<0.005	<0.0
	Lead	0.05	0.075	—	—	0.08	0.5
	Dissolved oxygen (DO)	≥5.0		≥5.0	—	5	—
ml/l	Settleable solids	0.5		—	—	<0.5	0.5

[a]For entries that contain a dash (—), no value has been specified.
[b]Measured by concentration in each process discharge.

DATA NEGOTIATION

source of heated debate in subsequent negotiations. Not surprisingly, if the consultants' assumptions were substituted for Lankford's, the model suggested that the river had much higher assimilative capacity than originally calculated by EPA.

At this point in the process, it appeared that the battle lines were drawn. In a letter to EPA's Howard Zeller, Tennessee Eastman's vice president for manufacturing, James Mitchell, indicated that the company stood behind its consultants' analysis of effluent limitations:

> It is not our intention to propose limits with the anticipation of negotiating them. This is not to say that we have closed minds on the subject; however, we have determined the best that we can do and what is necessary to protect the river and we are prepared to support our proposals. . . . The Tennessee Eastman river protection program will require a great effort on our part and will involve very substantial costs. It will also accomplish our mutual objective of providing proper protection for the river.

In response, EPA gathered more information, sent the Tennessee Eastman consultant's report around for critical review, and reassessed its own model in light of Krenkel and Novotny's criticism.

Five months later, on February 1, 1974, EPA made its second attempt to procede with Public Notice for the Tennessee Eastman permit. Upon receipt of a draft of the proposed permit and a tentative Public Notice date of February 25, Tennessee Eastman Vice President Edwards again appealed to EPA, stressing that Tennessee Eastman still had serious problems with the terms of the proposed permit and again asking for more technical discussions prior to Public Notice. Edwards succeeded. EPA postponed the Public Notice until after a technical meeting could be held between experts for both sides. The meeting was scheduled for early March in Atlanta, the site of EPA's regional office.

The Atlanta Meeting: The Experts Confer

The March technical meeting was a major event in the negotiating process. Tennessee Eastman assembled an army of consultants including Peter Krenkel (Vanderbilt University), W. A. Drewry (University of Tennessee), Wesley Eckenfelder, Jr. (Vanderbilt University), Carl Adams, Jr. (president, Associated Water and Air Resources Engineers), and Ruth Patrick (chairman, Academy of Natural Sciences). Tennessee Eastman also brought along an equal number of company scientists and engineers as well as a stenographer who kept a record of the proceedings. EPA was represented by its own bevy of experts, including four scientists from its National Field Investigation Center (NFIC) in Denver, three staff members from the regional office (including Doug Lankford, author of the EPA model), and several other EPA and state of Tennessee officials. Representatives of the Tennessee Valley Authority also participated at the meeting. Paul Traina of EPA presided.

The substantive part of the meeting began with George Harlow of EPA describing the permit and indicating areas where there might be room for compromise. Lankford followed with an explanation of the model and justification for the proposed permits. The model indicated that the river could assimilate about 8,000 lb of BOD and still have 5 mg/l of dissolved oxygen at low flow (800 ft^3/sec or cps). EPA had allocated the 8,000-lb capacity to the five major users on the river in a way that required each discharger to treat its waste to the same percentage of the difference between best practical technology and best available technology. According to this formula, Tennessee Eastman would be limited to 3,156 lb of BOD. Krenkel, the author of the report that was critical of Lankford's model, pointed out that the assimilative capacity of a stream is generally greater in the winter than in the summer because of increased flows and lower temperatures. Using a single year-round limit based on worst-case (summer) conditions would preclude utilization of the river's increased winter waste assimilative capacity. EPA's Traina responded that EPA would be open to consideration of variable winter/summer limits that were eventually written into the permit.

Generally, the BOD restrictions did not engender much heated discussion. The nutrient limits were another story. EPA was committed to relatively strict nitrogen and phosphorous limits in the belief that (1) nuisance weed growth could be brought under control, and that (2) the alleged eutrophication problem in the Cherokee Reservoir located about 50 miles downstream from Tennessee Eastman could be alleviated. Tennessee Eastman contested every point relevant to the nitrogen limits—the need for nutrient restrictions, the technology for treatment, and the legality of the proposed limits.

EPA's argument for restricting nutrient limits rested on a laboratory test of river water that suggested that nitrogen and phosphorous discharges were contributing to downstream weed growth. Patrick, one of Tennessee Eastman's consultants, disagreed, claiming that other limiting factors existed that had not been considered. For example, a plentiful supply of nutrients existed independently of Tennessee Eastman's discharge, both from bottom deposits and from the North Fork Holston River. These sources were not likely to be regulated in the future. Thus, Tennessee Eastman argued, it made little sense to impose expensive controls on the company if unregulated sources of nutrients would continue to produce weed and eutrophication problems.

The parties also disagreed on the feasibility of alternative strategies to control nutrient discharges. Because the actual permit limitations would be determined by EPA's interpretation of what constituted BAT and BPT, these disagreements were critical. EPA argued that treatment techniques that had been developed and demonstrated in municipal plants could be successfully transferred to chemical waste streams. In practice, several different systems exist for removing nitrogen from chemical wastes. If the nitrogen is in the form of ammonia, a chemical process called ammonia stripping is used. The ammonia is literally blown out of the waste stream. EPA had proposed a biological treatment, nitrification/denitrification, for the Tennessee Eastman Company. This treatment involves an activated sludge system in which any ammonia must first be combined with oxygen to form nitrates. Then, during denitrification, the treatment removes the nitrogen from the waste stream by converting it to nitrogen gas. These biological processes work better in warm temperatures—in Lankford's words, "in the winter the bugs get sluggish"—thus necessitating different winter and summer standards as with BOD.

Tennessee Eastman's consultants doubted whether these processes would be effective in TEC's waste stream because the organisms that achieve the necessary reactions are sensitive not only to temperature but also to the presence of inhibiting chemicals. The company was reluctant to invest in a treatment system that eventually might prove ineffective. EPA's solution was to set the nutrient limitations contingent on completion of a pilot plant that would demonstrate the feasibility of the treatment technology.

The Atlanta technical meeting ended with a glimmer of a possible voluntary resolution. EPA had indicated a willingness to yield on two major points: the importance of different winter–summer limits and the decision to make the nitrogen restrictions contingent on the effectiveness of a pilot plant. Although neither party made any other concessions during the meeting, the limits that were proposed on certain effluent categories had changed somewhat from those proposed earlier. For example, both sides had backed away from earlier positions on BOD to the point where they now stood only 2,884 lb per day apart. (Earlier they had been 7,770 lb apart.) With the possiblity of different seasonal limits, a resolution to the BOD dispute was within reach. The dispute over nitrogen and phosphorous limitations persisted, but the propsect of contingent limits provided some potential for compromise. And although Tennessee Eastman did not move from its initial position on total suspended solids (6,000 lb) EPA had narrowed the gap by increasing its proposed limit from 2,230 to 4,500 lb. These gains, however, did not come easily. The struggle had been uphill and had left the participants strained and tired. Arguements had been heated at times, often over minute details. In his closing remarks, Paul Traina of EPA acknowledged that some items remained in dispute, particularly the issue of nutrient limitations, and that at the very least the parties could "agree to disagree."

The regional EPA office was anxious to wrap up the Kingsport Holston River permits. The negotiations had gone two rounds. The parties had made some headway, but in small increments. Although Tennessee Eastman's array of consultants grew larger with each round and included some of the big names in the field, EPA had no additional technical resources upon which to draw. The state wanted a strict permit but was not actively assisting EPA. There were other permits to attend to in the Southeast, and staff time was limited. For whatever reason, the EPA chose to play a card that had previously been held back. On April 11, 1974, George Harlow sent Mitchell a copy of the latest draft of the NPDES permit, and with it, notification that EPA had gone to public notice on the permit.

This letter marked the third cycle of proposed permit discussion. Tennessee Eastman responded in much the same manner as it had previously (by indicating serious concern with certain aspects of the permit) and by registering dismay that EPA had gone to public notice while differences still remained. On May 7, 1974, Tennessee Eastman sent a letter accompanied by a 74-page document entitled *Comments by Tennessee Eastman Company*. The tone of the letter was severe and threatening. It covered the major areas of disagreement and closed with the following statement:

> The Tennessee Eastman Company position is environmentally, technically, and legally correct. Any more stringent limitations are not in accordance with the law, are not necessary to protect the environment, will waste valuable natural resources, and will cause adverse economic and social consequences in the region. . . . The Company has been advised by the foremost authorities in the field of water quality management and wastewater treatment technology. . . . They have stated that the proposals by Tennessee Eastman Company represent application of the best available technology economically achievable and are appropriate for protecting the South Fork Holston River and Holston River for fish and aquatic life as well as for industrial water supply. . . . Eastman is prepared to defend, to the extent necessary, the limits which the Company and recognized authorities have determined to be appropriate.

The letter concluded with a suggestion that EPA and the company resolve their differences before the public hearing. Traina sent copies of the letter to the EPA staff present at the Atlanta technical meeting and directed them to prepare the EPA's case for the May 29 hearing. Traina's memo to the staff stated that "this is a major discharger which we should be fully prepared to respond to and carry our case forward."

The Public Hearing and the Final Permit

An NPDES permit is a complex document that specifies all the limitations required of a plant and establishes a schedule for reaching them; it also includes information on requirements for monitoring and sampling. The real bite of the Tennessee Eastman permit, however, was embodied in the limitations that the company had to achieve by July 1, 1977. Tennessee Eastman included in its comments a copy of the draft NPDES permit that was marked up to correspond to the permit that the company considered to be acceptable. The more than 70 pages of comment also detailed every aspect of the permit that the company found unacceptable. Table 2 presents the main features of the debate on the effluent limitation in the permit. The table shows EPA's position, the company's response, and the limitations actually included in the permit after the public hearing.

The hearing was held on May 29, 1974. Apart from the usual newspaper articles and some specific notices to certain parties, EPA did not try to generate additional attendance. Tennessee Eastman, on the other hand, brought two of its consultants, Krenkel and Eckenfelder, to give statements. Local business and political personalities also spoke on behalf of the company. Aside from the EPA staff, only two persons testified in support of EPA's strict limitations. One of them, Phyllis Pierce of the League of Women Voters, explained why others did not attend:

> Many citizens—even well-educated ones—are intimidated by the mass of technical data, by the formalized procedure, and by the town and industry leaders in their suits and ties; particularly they are intimidated by the "experts" the industrialists bring along to study their case.

One might ask whether EPA also was intimidated by those experts. As Table 2 reveals, the company prevailed on every limit that was in dispute prior to the hearing. Certainly, there was much give-and-take throughout the negotiations, but EPA "gave" on the last interaction. There are a number of reasons why.

First, the technical questions and issues favored the company. The nitrogen debate had centered on the viability of a treatment process that depended on either isolation of the nitrogen waste stream or on elimination of chemicals that disrupt the biological neutralization process. By the end of the negotiations, it was clear that these processes were not economically achievable. Consequently, EPA's bargaining position was not legally defensible, given the NPDES effluent criteria.

Second, the differences over the other effluent limits were attributable to the debate over the validity of the stream models. If EPA had pushed much harder, it would have been forced to defend its model in an evidentiary hearing, and perhaps ultimately in court. Although neither side wanted to litigate the terms of the permit, this outcome was particularly onerous to EPA. Litigation would have tied up valuable staff time; it would have delayed the processing of permits for other, smaller dischargers; and perhaps most important, it would have further delayed the Tennessee Eastman NPDES permit. In the end, senior EPA officials decided to settle because the costs of continuing the dispute were just too great in light of the potential benefits.

STUDY QUESTIONS

1. This case is extraordinarily technical. (Indeed, in editing it, we have simplifyed it greatly.) Both sides worked hard to limit bargaining to technical

TABLE 2. FINAL DEMANDS AND TERMS OF AGREEMENT

Units	Effluent characteristics	EPA draft permit[72]		TEC draft permit[72]		Final permit[73]	
		Daily avg	Daily max	Daily avg	Daily max	Daily avg	Daily max
lbs/day	BOD_5 (summer)[a]	3,200	6,300	4,000	8,500	4,000	8,500
	BOD_5 (winter)[b]	4,500	8,350	6,000	13,000	6,000	13,000
	Total suspended solids	4,500	8,350	6,000	15,000	6,000	15,000
	Total dissolved solids	650,000	1,200,000	650,000	1,200,000	650,000	1,200,000
	Total nitrogen[b] (summer)	680	1,350	3,000	6,000	3,000	6,000
	Total nitrogen[b] (winter)	970	1,800				
	Phosphorous, total	150	300	300	600	300	600
	Zinc, total	125	250	125	250	125	250
	Phenols	50	65	50	65	50	65
	Iron, total	300	1,000	300	1,000	300	1,000
	Manganese, total	100	200	100	200	100	200
ml/l	Settleable solids	NA	0.5	NA	0.5	NA	0.5
mg/l	Chromium, total	0.05	0.1	0.05	0.1	0.05	0.1
	Copper, total	0.05	0.1	0.05	0.1	0.5	0.1
	Lead, total	0.08	0.5	0.08	0.5	0.08	0.5
	Mercury, total	0.005	0.0075	0.005	0.0075	0.005	0.0075

[a] Summer and winter as designated in the EPA draft permit were May 1 to October 31 and November 1 to April 30, respectively. The TEC draft and final permit differed by a month, with summer May 1 to September 30 and winter October 1 to April 30.
[b] There was a caveat on these limits to allow TEC to test the technology in a pilot plant. The final permit contained the following notation: "If TEC can demonstrate to EPA by July 1, 1975 that these total nitrogen and total phosphorous limits are unattainable by TEC's currently planned wastewater treatment plant and currently planned in-plant controls, these limits will be revised accordingly by EPA. The revised limits will make proper allowance for seasonal and operational variabilities." (Doc. 625) This wording closely resembles that suggested by TEC in the comments.

issues even though they recognized that these issues were often proxies for larger social choice questions. Why did the parties seek to define the dispute in purely technical terms? Why were they reluctant to discuss the real underlying question: How much should Tennessee Eastman be required to spend to clean up the Holston River? What were the consequences of restricting the agenda in this manner? In retrospect, would EPA have been better off with a broader debate?

2. Both sides claimed that what kept them at the bargaining table was a desire to resolve their differences before going public with the permit. What was so bad about conducting this debate in the public eye? Why was the threat of judicial intervention so onerous?

3. Analyze the strengths and weaknesses of the parties as they appeared at the outset of the negotiations. Was there a clear-cut winner in this dispute? Would the outcome have been different if EPA had been represented by Tennessee Eastman's experts and vice versa? Was the cost of nonagreement symmetrical? Were the differences between the parties ultimately resolved purely on the basis of technical considerations or did other factors influence the outcome?

4. Parties to negotiations often try to influence the willingness of their opposition to compromise by flaunting their own intransigence. Tennessee Eastman tacitly did this when it collected duplicate water samples to those taken by EPA, thus giving notice that it was prepared to challenge any analysis based on those samples. Contrast this rather clear signal with the company's later statement to EPA "that it is not our intention to propose limits with the anticipation of negotiating them." If you were the EPA official who received this letter, how would you have interpreted this statement?

5. Recall Thomas Schelling's discussion of commitment as a means of building bargaining power from chapter 2. We have a good example of this tactic in this case. When EPA notified Tennessee Eastman on April 11, 1974, that it had given public notice on the NPDES permit, it was, in effect, committing itself publicly to its last bargaining position. What risks were involved in this move? Did Traina misread the situation? What signals did this action provide to Tennessee Eastman? Did it provide any signals to anyone else? Would EPA have been better or worse if it had threatened Tennessee Eastman with public notice before actually going public?

6. Tennessee Eastman was not the only discharger located on the Holston River; it was just the largest. Would EPA have been better off trying to negotiate all of the NPDES permits for the river simultaneously? Should it have tried to negotiate with the smaller dischargers first?

Judicial Review of Technical Decisions

In the Holston River case, the parties were clearly reluctant to have a court decide the complex issues involved in modeling the assimilative capacity of the river. To get a feel for what both sides feared, read the following excerpt from *South Terminal Corporation v. EPA* in which the court succeeded in goring the ox of both sides. This case was decided while Tennessee Eastman and EPA were negotiating the Holston River NPDES permit. Although the case deals with compliance with air quality standards, the central issue—the accuracy of EPA's modeling efforts—is common to both cases. As you read the case, think about the following questions. Was the reluctance of the parties in the Holston River case to let the issues go to trial justified by the actions of the court in South Terminal? Was the court comfortable deciding the technical issues? Is it likely that the parties in South Terminal could have improved on the outcome dictated by the court if they had negotiated among themselves? Why do you suppose that they failed to do so?

South Terminal Corporation v. EPA
504 F.2d 646 (1st. Cir. 1974)

Under the Clean Air Act, EPA is charged with promulgating ambient air quality standards for each pollutant having an adverse effect upon the public health or welfare. Each state, however, has responsibility for designing a program to see that the ambient standards are met. Typically these state implementation plans limit allowable pollution from stationary sources like factories, power plants, and incinerators as well as from mobile sources like cars, trucks, and planes. The Clean Air Act provides that if a state fails to implement necessary measures to comply with the ambient air standards, EPA may impose an implementation plan on the state. When Massachusetts failed to adopt a transportation control plan to limit emissions from mobile sources, EPA stepped in to fill the gap. Through extensive modeling, EPA concluded that if Boston were to comply with national ambient standards, emissions of hydrocarbons would have to be reduced in metropolitan Boston by 58 percent, and carbon monoxide emissions by 40 percent. To achieve these reductions, EPA proposed that off-street and on-street parking spaces be frozen or cut back, and the construction of new parking facilities regulated. It also proposed special bus and carpool lanes and a computer car pooling system. A vehicle maintenance and inspection program was also mandated. South Terminal Corporation was one of a number of plantiffs that filed suit to overturn the EPA transportation control plan on the ground that the underlying technical analysis was deficient. The court's discussion of the adequacy of the modelling as well as the legality of the transportation control plan follows [All footnotes and citations have been omitted.]

I. The Scope of Judicial Review
   . . . The questions about the plan on review are of two types: the rationality of EPA's technical decisions [such as its determinations of local photochemical oxidant and carbon monoxide levels and the amount of reduc-

tions required to meet national standards] and the rationality of EPA's "control strategy," that is, the measures adopted to reduce emissions. The former present peculiar difficulties for nonexperts to evaluate. Yet "our inquiry into the facts is to be searching and careful," . . . and we must assure ourselves as best we can that the Agency's technical conclusions, no less than, others are founded on supportable data and methodology and meet minimal standards of rationality. . . .

Assuming EPA's technical determinations are reasonably based, we must decide whether the selected controls are arbitrary or capricious. In so doing, we must bear in mind that Congress lodged with EPA, not the courts, the discretion to choose among alternative strategies. Unless demonstrably capricious such as much less costly but equally effective alternatives were rejected or the requisite technology is unavailable, the Administrator's choices may not be overturned.

III. Whether EPA Committed a Clear Error of Judgment in Computing the Need for Emission Reductions.

. . . [The plaintiffs'] arguments can be divided into attacks on EPA's data and methodology as to (1) photochemical oxidants in the Metropolitan Boston Interstate Region; (2) carbon monoxide in the Boston core; (3) carbon monoxide at Logan Airport (East Boston).

1. EPA is said to have overestimated the photochemical oxidant problem in the Boston region. Most pertinent are petitioners' arguments that the key ambient air quality reading taken on one day at a monitoring device located at Wellington Circle must have come from a defective instrument. This single reading, inserted by EPA in its so called rollback formula (or "model"), was the basis for a region-wide estimate of the amount of hydrocarbon reduction required. If it was incorrect, so were the conclusions about how much reduction was necessary to achieve the primary standard. Petitioners point to a computer printout taken at that monitoring station: it contains a high number of "9999" readings which may indicate instrument malfunction. EPA's response is that the designations may also result from "instrument calibration, instrument zeroing, transmissions loss and depletion of span gas, all of which causes are unrelated to any malfunction." But petitioners contend that the irregular readings occurred too often to be attributable solely to innocent causes. On the present record, we cannot say with confidence that the use of a single reading from a machine as to which objective readings suggest a substantial possibility of malfunction is sufficient to support EPA's photochemical oxidant determination.

We find less persuasive petitioners' attack on the accuracy of the rollback model itself because of its purported failure to take account of local topography and meteorology. EPA's technical support document appears to consider these influences, and the only expert to stress Boston's unique features did not include gasoline in his analysis. Petitioners further claim that EPA incorrectly related oxidant concentrations directly to emission of hydrocarbons, relying in part on an extra record document never brought to the Agency's attention. Photochemical oxidants are a secondary pollutant derived from the reaction of two primary pollutants, hydrocarbons and nitrogen. To reduce oxidant concentration, it is therefore necessary to control hydrocarbon emission and EPA has advanced plausible reasons for choosing the ratio that it did. . . . Finally, petitioners object to the determination that regionwide controls, rather than controls in only a few heavily polluted sections, were necessary to bring oxidants down to a reasonable level. But background reports indicate that automobile use is heavy, particularly in the outlying manufacturing areas. The technical support document presents the view that the necessity for regionwide controls stems from the nature of the pollutant; petitioners' contention that contrary conclusions can be drawn from the data does not lead us to suspect that EPA committed clear error. To the extent different conclusions could be drawn, the Agency was entitled to draw its own.

2. Carbon monoxide data is attacked as unreliable. EPA determined that its national primary standard requiring the average amount of carbon monoxide in the air over an eight hour period not to exceed 9 ppm is not being met in the Boston core and will not be met by mid-1975. It did this by a series of calculations which have as their essential element an ambient air quality reading obtained on one day in 1970 from a monitor at Kenmore Square. Although petitioners attack use of the rollback model itself as unsophisticated, we are mainly impressed by the contention that the crucial figure for determining required emission reduction may be unrepresentative. At the time the plan was designed the next highest reading at Kenmore Square was nearly 50 percent lower than that utilized. EPA points to readings elsewhere even higher than that used in the rollback model, recorded after the plan was announced, as evidence that it may have "underestimated the extent of the CO problem." But petitioners claim these high readings are also freak events. . . . Here again, on the present record, we have no basis to say with judicial conviction that such a slender base, without further justification, is sufficient to support EPA's conclusion as to carbon monoxide in the Boston core.

3. In the best documented of the challenges to EPA technical data, South Terminal and Massport attack the carbon monoxide determinations at Logan Airport (East Boston). [EPA determined that it was necessary to reduce carbon monoxide emissions at the airport which is located across the harbor from downtown Boston without actually sampling at the airport. The same Kenmore Square air quality figure, inserted in the rollback model, was used to project the required reductions at Logan. Massport, which runs the airport, objected and conducted its own test which suggested that federal primary standards were being met. Moreover, the Massport report concluded that the concentrations of carbon monoxide at Logan were substantially lower than at other Boston sites. EPA responded by citing a different study which indicated that carbon monoxide levels at the airport were roughly equivalent to those measured elsewhere in the region, and exceeded

federal standards. After reviewing the conflicting evidence the court reached the following conclusion.]

The method of sampling at Logan, Massport's own testing, and the lack of monitoring in East Boston, collectively, on the present record, prevent us from holding that the data are sufficient to support EPA's conclusion as to carbon monoxide in East Boston.

4. While we are unable at this time to uphold EPA's conclusions as to photochemical oxidant and carbon monoxide levels and reductions, we do not say that they are necessarily incorrect. Petitioners forcefully contend that the Agency's measurements are without reliable foundation, and hence, in effect, arbitrary and capricious. . . . But as laymen we are in no position to know how much ultimate weight to give to these arguments, based as they are on technical assumptions. We can only say that the objections as to data and methodology seem too serious to us simply to pass by; they demand investigation and answer. While reviewing courts are not to substitute their judgment for an agency's, they are to establish parameters of rationality within which the agency must operate. A court would abdicate its function were it, when confronted with important and seemingly plausible objections going to the heart of a key technical determination, to presume that the agency could never behave irrationally. It has a duty to see that the objections are faced in a proper procedural setting and satisfactory answers provided demonstrating careful agency consideration. [The court consequently remanded the case to EPA for an explanation of the agency's measurement procedures.] . . .

V. Whether Transportation Controls are Arbitrary and Capricious

1. The "freeze" boils down to the requirement that no new parking spaces be created after October 15, 1973, in the more congested protions of Boston, Cambridge, and some other outlying areas. There are important exceptions: residential parking spaces adjacent to homes, apartments, condominiums, etc.), employee parking outside the Boston core (so long as it complied with the separate employee parking restrictions), and free customer parking. Our role, of course, is not to decide whether the freeze device is an ideal solution; Congress delegated to EPA the authority . . . to select the preferred means. We cannot say that such a freeze is arbitrary and capricious assuming EPA is able to support by credible data its position as to the magnitude of the need for carbon monoxide emission reductions in relevant segments of the region. Indeed, the enlargement of parking facilities in areas where the public health requires curtailing the flow of traffic would itself seem irrational. The exemption for residents, customers and, in part of the area, employees, would seem a reasonable attempt to ameliorate the hardship upon individuals and businesses.

[The court went on to uphold other aspects of EPA's plan that included a ban on on-street parking between the hours of 7 to 10 A.M. weekdays, a reduction in the availability of off-street parking, a regionwide 25% reduction in parking provided by employers, and a requirement that if parking was to be expanded at Logan Airport by more than 10%, such increases must be offset by retiring spaces elsewhere in the freeze zone. Finally, the plaintiffs contended that the entire transportation control plan was arbitrary because the EPA had paid too little attention to its economic and social impact. The court rejected this argument as well.]

The material portions of the Clean Air Act itself do not mention economic or social impact, and it seems plain that Congress intended the Administrator to enforce compliance with air quality standards even if the costs were great. Particularly in the case of primary standards—those set as "requisite to public health"—Congress' position is not extreme or unprecedented. Minimum public health requirements are often, perhaps usually, set without consideration of other economic impact. Thus, insofar as petitioners claim that either EPA or ourselves would be empowered to reject measures necessary to ensure compliance with primary air quality standards simply because after weighing the advantages of safe air against the economic detriment, we thought the latter consideration took priority, petitioners would be incorrect. Congress has already made a judgment the other way, and EPA and the courts are bound.

Courts and Technical Issues

As so often happens in lawsuits, the decision in the South Terminal case did not please either side. The court reprimanded EPA for what it viewed as shoddy technical analysis. EPA's numbers were called into question, and consequently, so was EPA's authority to regulate emissions of hydrocarbons. (If the reading taken at Wellington Circle proved inaccurate and Boston was in compliance with ambient standards for photochemical oxidants, EPA would lack authority to act.) On the other hand, the court rejected the petitioners' arguments that the severe measures ordered by EPA were either unnecessary, excessively costly, or otherwise illegal. To the contrary, the court ruled that EPA had broad discretion to fashion the appropriate response. In many ways, the court's decision was predictable. As we have noted before, the judges are reluctant to second-guess decisions of federal agencies.

As the South Terminal case suggests, courts are often uncomfortable rendering decisions in cases that turn on highly technical or scientific issues. Judges are first and foremost generalists. They hear an extraordinary range of cases dealing with issues as diverse as Indian land claims, antitrust matters, products liability actions, and civil rights complaints. On succeeding days, a federal judge may be forced to serve as an amateur historian, economist, sociologist, psychologist, or scientist. Except for those judges who sit in Washington, D.C., where a large number of regulatory cases are filed, most judges will hear only a handful of complex environmental cases in their careers on the bench. Thus, the challenge for an attorney arguing such a case is to teach the judge enough science so that he can understand the merits of the attorney's argument. Because the attorney is himself or herself usually a layman, this is a very difficult task.

Unfortunately, judges have relatively little opportunity to consult with experts in the field. Our adversary system leaves it to the litigants to call expert witnesses; they, in turn, inevitably offer testimony favorable to the side that has called them. Although the federal rules of evidence do permit a trial judge's own expert witness to be summoned, this procedure frequently does nothing more than generate a third expert opinion for the judge to consider (although it is the opinion of a disinterested party). Moreover, some cases like South Terminal are appealed directly from an agency to the Court of Appeals. Because appeals courts must base their decisions entirely on the written record developed during the course of the regulatory process (and the oral argument of counsel), appellate judges do not hear any expert testimony firsthand. Conscientious judges who would like to consult privately with experts often finds themselves thwarted by the canons of judicial ethics that greatly limit such discussions.

The one resource to which judges have ready access is their clerks. Federal District, Circuit, and Supreme Court judges each employ from two to five clerks to assist in legal research and drafting of opinions and orders; typically, these clerks are high-ranking recent graduates of prestigious law schools. Often judges take a liberal view of what constitutes legal research. (For example, when the Supreme Court was deliberating the Brown v. Board of Education desegregation case, a group of clerks was charged with the task of mapping out every home in Spartanburg, South Carolina, to see how readily the existing white and black schools could be integrated. Similarly, in environmental cases, the task of mastering the vast technical record often falls to the clerks. If a judge knows in advance that he will be hearing a lengthy and complex case, he may seek a clerk with special expertise, but this is a rare luxury. Clerks, like judges, tend to be generalists.

Largely because the courts lack the capacity to make substantive policy decisions, the law of judicial review limits the circumstances under which a reviewing court may overturn a decision of the executive branch. Administrative law attempts to draw a distinction between questions of substance and questions of procedure or law. Agencies have a comparative advantage in deciding the former, whereas courts are better equipped to decide the latter. Accordingly, the law admonishes courts to defer to the judgment of agencies on substantive matters and only permits judicial reversal of an agency decision if the court finds that the agency: (1) exceeded its jurisdictional mandate; (2) did not comply with a procedural requirement (e.g., the agency failed to hold a statutorily required hearing prior to rendering a decision); (3) violated a statutory duty (e.g., the agency ignored its obligation to consider alternatives that might be less harmful to the environment); (4) acted in an unconstitutional manner; or (5) abused its discretion or otherwise acted arbitrarily or capriciously. The latter requirement empowers courts to reverse only for gross errors of judgment and is rarely envoked.

Distinctions between substance and procedure and questions of law and questions of fact are more easily stated than they are made in practice. For example, in the South Terminal case, the question of whether EPA acted properly in imposing a transportation control plan on Boston was nominally a question of law; the agency had legal power to do so only if Boston was not in compliance with ambient air standards. But to make such a determination, the court was forced to review EPA's testing procedures, a highly technical inquiry that the court was clearly uncomfortable in performing. Similarly, many regulatory statutes are written in such a way that they thrust reviewing courts into the position of second-guessing the substantive decisions of agencies.

This is a situation that, like the weather, everyone complains about, but no one seems capable of rectifying. Two general types of reforms are commonly suggested: better precision in drafting of statutes by Congress and the creation of courts with special substantive expertise.

The first reform clearly stands little chance of success. In theory, if Congress was capable of being more precise in giving guidance to regulatory agencies, the courts would have less of a substantive nature to review. For example, had Congress been more precise in specifying the procedures to be followed in determining whether a municipality was in compliance with the Clean Air Act, the court in South Terminal would not have had to wade through a mass of technical material to decide the case. But, in practice, Congress appears incapable of greater precision for at least two reasons. First, the legislature frequently vests discretion in the hands of executive agencies like EPA precisely because they possess the expertise that Congress lacks. Just as judges throw up their hands in frustration in trying to determine the proper procedures for assessing air quality, so do senators and congressmen.

Second, Congress, for political reasons, is often not interested in being more precise. Acts of Congress represent the result of a political bargaining process that relies upon logrolling to achieve consensus. In this process, ambiguity and obfuscation often are helpful in building a coalition. For example, it may be much easier to gain support for a bill that charges an agency like EPA with setting air and water quality standards than it is to get legislators to support a bill in which the standards are specified. The second type of bill is unpopular because regulatees who are likely to be adversely affected by the specified standards will come out of the woodwork to oppose the bill. (Indeed, this is precisely what happened to the EPA in the 301(h) case discussed in chapter 7; EPA had to set standards for secondary treatment of municipal wastewater and every municipality that was affected by the proposed standards registered its objections.) Usually, it is easier for congressmen to delegate many of these difficult policy judgments to agencies. By so doing, they avoid direct responsibility for the decision, and they may still criticize the agency if the decision adversely affects their constituency. Unfortunately, this process also thrusts the courts into the position of reviewing agency judgments to insure that agency decisions conform to the vague guidelines set down by Congress.

The second reform—endowing courts with special expertise—has been adopted for other types of problems. For example, we have special tribunals for handling bankruptcy matters, tax cases, and claims brought against the federal government. Arthur Kantrowitz has advocated the creation of a national science court for resolving policy questions that turn on highly technical issues. The court, which would consist of scientists, would issue opinions on questions submitted to it by Congress and the executive branch. Similarly, from time to time, proposals surface for the creation of special environmental courts consisting of judges who would hear only environmental cases.

Do you think a science court would be a good way to resolve the kinds of technical issues that arose in the Holston River case? Do you think we would have fewer disputes of a technical nature if such a court existed? (For a thorough discussion of the advantages and disadvantages of specialized courts see "The Environmental Court Proposal: Requiem, Analysis, and Counterproposal," 123 U. Penn. L. Rev. 676, 1975.)

THE ELUSIVE NATURE OF FACTS IN ENVIRONMENTAL CASES

We began this chapter by noting that most environmental disputes involve disagreements over how ecosystems are likely to respond to various types of human activities. If policymakers posessed the proverbial crystal ball, the range of disagreement in environmental controversies would be narrowed substantially. Instead of arguing over the impact of Tennessee Eastman's discharge on the Holston River, we would simply debate whether the costs of achieving a given reduction in discharge were justified by the resulting benefits. Although this would still not be a trivial dispute to resolve, at least the parties would be arguing from the same basic set of facts.

In an article entitled "The Technical and Judgmental Dimensions of Impact Assessment," 1 *Env. Impact Assess. Rev.* 109, 115–120, 1980, Lawrence Bacow has suggested that policymakers typically overestimate the degree to which science can supply unambiguous answers to complicated environmental questions. Although we would like to believe that science is dispassionate and value free, Bacow has argued that, in fact, the process of modeling is often very subjective. Although the article is concerned with the role of subjective analysis in impact assessment, it also sheds light on how technical analyses often mask

important judgments in other types of environmental decisions. As you read the following excerpt, consider these questions:
1. If Bacow's thesis is correct, what are the implications for environmental dispute resolution?
2. What is the appropriate role for experts in the dispute resolution process?
3. Should everything be negotiated, including science?

   Conceptually, making predictions about the future consequence of a proposed action involves three distinct activities. First, the analyst must decide where to focus his attention. Since analytic resources are always in short supply, choices have to be made about which impacts will be documented in depth, which will be analyzed only briefly, and which will be ignored entirely. Second, a prediction must be made of how the ecosystem or social system under study will evolve over time in the absence of the proposed project. Finally, an estimate must be made of how the proposed action will cause these systems to depart from their normal evolutionary patterns. The difficulties encountered in specifying the impacts to be studied can best be illustrated by telling a story. For many years, the Massachusetts Department of Public Works has considered widening Route 2, a major artery linking Boston with its affluent northwest suburbs. Widening the highway from two to four lanes would affect the natural environment in a lot of different ways. Land would be consumed. Some flora and fauna would be lost. Noise levels would increase during the construction period. Increased traffic would generate more noise and more air pollution along the route. The highway would also have a number of less obvious effects. Increasing access to the suburbs would probably increase development on the current suburban fringe, thus affecting employment patterns among suburban construction workers. If the new development would have occurred elsewhere but for the widening of Route 2, then widening of the highway will have affected employment patterns in other parts of the Boston metropolitan region as well. Similarly, since Route 2 is an integral link in an interdependent transportation system, increasing its traffic capacity will also affect traffic density (and air pollution and noise pollution) in other parts of the transportation network. It is possible to keep working back through this maze of probable impacts almost indefinitely. It is like pulling on a loose thread of a knitted fabric; it just keeps unraveling. Although it is easier to illustrate interdependencies for impacts that affect social systems like transportation networks, ecologists are quick to point out that ecological systems are perhaps even more interdependent.

   Given the multitude of possibilities, which impacts should the author of an environmental impact statement address? It is tempting to say all of them. But the resources available to assess impacts are not limitless. Moreover, even if it were possible to produce a truly comprehensive EIS, its sheer size would ensure that it would never be read. Thus we must somehow define the boundaries of analysis for assessing impacts. If we are only going to assess a limited number of impacts, then the rational strategy would be to concentrate our efforts on "the most important impacts".

[Bacow argues that each constituency affected by a project is likely to have a different opinion of which impacts are the most important. If the modeling effort relies upon the modelers to scope the impacts, important value judgments will be masked, and the model is likely to be criticized for being biased or uninformed.]

Even if we are unanimous in our view of what is important, it still may not be obvious how to evaluate these impacts. Suppose in the Route 2 example, people are concerned about air quality and noise. Although we may be able to say that these conditions have changed as a consequence of the highway widening, it is often very difficult to state unambiguously whether they have gotten better or worse. For many environmental conditions, there is no single accepted index for evaluating the state of the condition. Consider the problem of assessing air quality. We care about air quality because air pollution affects human health, aesthetics, plant and animal life, and the durability of materials exposed to air. A given change in air quality will affect each of these conditions to a different degree. It is not possible to construct a single index for air quality unless we are first willing to weigh each of its components—a process that necessarily depends upon the preferences of the person constructing the index. Even if we cared about only one aspect of air quality—its effect on human health—it still would be difficult to construct a single objective index because of the complex way individual pollutants interact to produce air pollution. For example, the relationship between the airborne concentration of a pollutant and human health may be nonlinear. Similarly, two pollutants may interact synergistically. In some cases, controlling one source of pollution, such as carbon monoxide from internal combustion engines, may actually increase the level of another pollutant, specifically, nitrogen oxide. Assessing the environmental impacts of noise is even a more difficult task than evaluating air quality. Technically, noise is measured as the ratio of energy transmitted across a unit surface to the minimum energy that can be perceived in the air. What is bothersome about noise, however, is not just the amount of energy transmitted. The annoyance value of noise is determined not only by amplitude but by pitch, frequency of occurrence, the information content of the noise, background sounds, and the dispersion capacity of the physical environment. For example, it may be far more difficult to sleep if a truck rumbles by every 20 minutes than if there is a steady, uninterrupted stream of trucks. Similarly, although almost inaudible, a small scratch on an otherwise perfect recording of a Beethoven concerto is likely to be extremely annoying even to someone who is not an aficionado of classical music. The point to be made is that even the simple task of measuring change in the environment forces the analyst to make judgments about the relative importance of the different components of the change. Moreover, these are not trivial decisions: different indices can lead to different conclusions.

Reaching agreement on the impacts to be studied and the proper form of their measurement does not get us out of the forest. Before we can predict the impact of a development on the environment, we must first be able to describe how the environment is likely to evolve without the development. In practice, our ability to describe accurately the evolution of physical and social systems is limited by our understanding of how such systems operate as well as our ability to predict changes in technology, regulatory policy, market forces, and human preferences.

If nature were static, impact assessment would be a much easier task. But the natural environment changes considerably without human intervention. Species come and go as evolution runs its course. The elements both erode and create land. While some of these events occur gradually, and consequently are predictable, others occur with little warning and may change the character of the environment suddenly and radically. Forest fires, hurricanes, earthquakes, droughts, and volcanoes are all naturally occurring events whose incidence and effect can only be predicted imperfectly. Thus, although we can safely say that a hurricane of the magnitude of Dora may strike the East Coast once every hundred years, we cannot predict its specific environmental consequences without knowing its precise location, the distribution of development in the affected area at the time of the hurricane as well as the relative stability of the affected ecosystem. Consequently, our long-term predictions about the natural state of the environment are necessarily couched in terms that reflect our relative ignorance about future states of the world. We would expect that during the next 25 years a hurricane will strike the Gulf states with sufficient force to reduce the population of the Mississippi sandhill crane by at least 80%.

Our ability to predict the marginal impact of a particular development on the natural environment is also affected by our capacity to predict changes in the natural environment occasioned by the normal development of social systems. To go back to a previous example, if we are interested in predicting the increase in air pollution that would result in 1985 if Route 2 were widened, we have to be able to predict traffic density on Route 2 in 1985 given a highway of current dimensions. But such a prediction requires knowledge of the likely growth in suburban housing demand as well as suburban job opportunities—two large determinants of traffic density. At present, we only imperfectly understand what makes cities grow or not grow, so predictions about likely growth in traffic density will again be imprecise. Further complicating the analysis is our ability to predict changes in other conditions that influence traffic density. For example, traffic density varies as a function of the cost of driving relative to other modes of transportation— as the price of gasoline has increased, at least some people have left their cars at home and taken public transportation. So if we are to predict traffic density, we need to know not only the future price of fuel, but also the behavioral relationship that constitutes the demand curve for gasoline.

Moreover, in many cases it is difficult to predict the natural evolution of the environment without making some assumptions about the future impact of regulation and technological change. For example, future air quality in urban areas will be determined, in large part, by the success (or failure) of federal efforts to produce a nonpolluting car. Thus, our ability to predict the evolution of the atmospheric environment is directly related to our ability to predict the success of regulation or the rate of change in technological innovation.

Because our predictions of what the world would look like *without* any additional government intervention are so uncertain, it is difficult to isolate changes that are attributable solely to new projects. Furthermore, our capacity to make confident predictions about impacts varies in a rather perverse way with the controversialism of the issue. While we can state quite conclusively that the U.S. Air Force's new long-range radar station on Cape Cod will destroy 10 acres of flora, we have little knowledge of the long-term effects of prolonged exposure to low-level ionizing radiation—and that is what everyone on the cape is upset about.

In practice, it is unreasonable to expect impact statements to be anything more than synthetic documents. We rarely have the time available to do new research necessary to answer the questions that lie at the root of controversy over development proposals. Instead, we are forced to cull the available evidence to draw conclusions. More frequently than not, however, the available evidence is ambiguous; it can support a host of different conclusions. In some cases, we simply do not understand causal relationships well enough to draw inferences about stimulus and response. In other cases, the consequences of intervention are subtle and difficult to document. And, in still other cases, synergistic interactions make it hard to determine why something has changed. The kinds of inferences people are willing to draw from such ambiguous evidence varies with both their professional training and their personal stake in the outcome. Scientists tend to be a very conservative lot—they are reluctant to conclude, for example, that an observed increase in the cancer rate is attributable to exposure to a particular chemical unless they are at least 95% certain that the increase is not attributable to chance. In contrast, people at risk are far more willing to conclude that a hazardous condition exists on the basis of information that the scientist would deem inconclusive. Thus, it should not be surprising that the process of collecting information about impacts is divisive: the information collected is grist for the mill of both sides. Instead of looking for opportunities to resolve differences between competing interests, we have created a system that amplifies existing differences. Moreover, we have done so because we have underestimated the degree to which impact assessment is a subjective, judgmental, nontechnical activity.

6

Two-Party versus Multiparty Negotiation

Introduction

The environmental dispute in the Brown Paper Company Case (described in chapter 4) essentially involved two parties—the EPA and the paper company. By contrast, there were many negotiators (both groups and individuals) in the Grayrocks and South Carolina cases (chapter 3). Environmental problems are often of the multiparty variety.

The number of participants in a negotiation can markedly affect its character. One obvious problem is coordination. The more people there are around the bargaining table, the harder it likely will be to coordinate the negotiation. If each party is to have his or her say, the proceedings will be protracted. The coordination problem is tied to the question of representation. Who participates in the negotiation? Who is authorized to speak for affected constituencies? There may be factions within an organization that have different goals.

The fact that many participants are involved in a dispute necessarily expands the choices open to each negotiator. In a simple two-party case, a party must ultimately decide whether to settle or accept the consequences of nonagreement. By contrast, in multiparty cases, one party may have to weigh the attractiveness of agreement with all the others against possible deals with just a few. As a result, strategies are much more intricate. In some cases, coalitions may form, disband, or realign.

Though multiparty bargaining is more complex, it may also offer richer possibilities for settlement. Having a number of negotiators, each with a particular set of priorities, may enrich opportunities for efficient trades. When groups or individuals have to share costs or benefits, there can be bitter fights over the distribution.

This chapter explores the theoretical and practical implications of multiparty environmental disputes. Two short case studies introduce issues that are further explored by reflecting on cases presented in earlier chapters.

Case Study: The West Side Highway

This case was obtained from the article "Mediation: An Instrument of Citizen Involvement," by Willis B. Goldbeck, president of Public Policy Communications, Washington, D.C. It appeared in 30 *The Arbitration Journal* 241–252, 1975. Information was also obtained from "Mediating Environmental Disputes," by Laura M. Lake, 262 *Ekistics* 164–170, Sept. 1977.

New York City's West Side Highway runs along the Hudson River from 72nd Street down to the tip of Manhattan. When the elevated roadway was built in the late 1920s it represented the most advanced notions of design. By the 1960s, however, it was clear to transportation planners and automobile drivers alike that it had become obsolete. Lanes that were set out for smaller, slower cars could not accommodate the press of modern traffic. The structure itself was disintegrating.

In 1971, the Urban Development Corporation, a state agency with extensive independent authority, released a study of waterfront development in which it concluded that improvement and alteration of the West Side Highway was central to the solution of other problems. In response, Mayor John V. Lindsay formed the West Side Highway Project to develop highway alternatives. The effort was funded by city, state, and federal appropriations, and it won the cooperation of then-Governor Nelson Rockefeller. A steering committee representing 16 city agencies and all the planning boards in affected communities was created to monitor the project's work and to reach a consensus on the best alternative.

The West Side issue came to a head in late 1973 when a truck fell through the highway. Major sections of the road had to be closed, and traffic was routed to adjoining streets and avenues. Traffic along 10th Avenue increased 360%. With the traffic came noise, congestion, increased local air pollution, and cries of protest from area residents.

By the next spring, the project published its draft environmental impact statement describing five possible solutions to the highway problem: (1) reconstruct the road along its present design; (2) maintain the road basically as is, but with some safety modifications; (3) build an "arterial" road along the riverfront; (4) build an "inboard" limited access interstate using 90% federal funds; or (5) build an "outboard" interstate involving massive landfill along the river, again using 90% federal funds.

Of the proposals, only the fifth met the project's own previously developed criteria. Yet, because of the plan's magnitude and its relation to other controversial projects, it sparked significant opposition. Public hearings failed to develop clear support for any of the alternatives.

With much of the highway shut down and other projects hanging in the balance, the Regional Plan Association initiated mediation in an attempt to break the impasse. The American Arbitration Association, an organization with a long history in settling commercial and other private disputes, provided a mediator, its past president Donald Straus. The Regional Plan Association took responsibility for selecting the participants. Groups that had already been actively involved in the West Side Highway controversy were the first to be included. The RPA then classified these groups according to their constituencies—business, environmental, ethnic, labor, civic, and professional. When a category was underrepresented, the RPA tried to enlist organizations that could, in spite of their previous noninvolvement, advocate the interests of important affected groups.

According to Willis B. Goldbeck, this selection process, though well intended, had gaps:

> There was no labor participation even though the Building and Construction Trades Council of Greater New York was among the first to be invited. The Puerto Rican Community Development Corporation was another invitee which did not participate. No other specific minority organizations were invited. A third gap, identified by those who did participate, was the local special issue community groups. (Goldbeck, 30 *Arb. J.* 241, 243–249)

Five full-day mediation sessions were held during the fall of 1974. The RPA prepared a tentative agenda, and all participants agreed to the objective of the process, though with the caveat that participation did not bind any group to accept the conclusions. According to Laura Lake, thirty-eight representatives of twenty-three organizations sat around the boardroom table of the Rockefeller Foundation at the first session. All participants were allowed to state their positions. The West Side Highway Project staff attended all meetings to provide technical information. Transportation and planning consultants, supported by city and federal grants, assisted participating community planning boards.

Laura Lake observes that the participants initially shared a common interest. "Both the opponents and proponents of the highway realized that continued delay was against their interest, for local detour traffic would continue to be a serious nuisance, and construction costs would continue to rise with inflation"(Lake, 262 *Ekistics*, 164, 168, 1977). (Later, however, some environmentalists appeared to be stalling, waiting for the election of Governor Hugh Carey, who they thought—mistakeningly—would oppose any new highway.)

Mediation also exposed sharp differences in values and opinion among the various groups. In some instances, the differences were over priorities: which should be preferred—enhancement of environmental quality or stimulation of economic growth? There were also markedly different opinions over the impact of the proposed alternatives. According to Goldbeck:

> Debate was very heavy on the degree to which the various alternatives would increase or lessen traffic on existing streets. This issue was a perfect study of the conflict between technical information and community emotionalism. On one hand, statistics were apparently made to prove that the highway could both increase and decrease local traffic! On the other hand, communities which opposed highway construction evidenced no willingness to change their position no matter what the numbers showed. (Goldbeck, 30 Arb. J. 241, 245, 1975)

Positions on other policy issues often depended on the technical assumptions of the parties. New York City, for example, was under pressure to meet federal air quality standards, but plans to comply rested on expectations about the impact of highway alternatives on traffic patterns, and these expectations were subject to debate. The goal of compliance was itself controversial. One participant stated that "clean air doesn't get us anything."

The polarizing issue was whether to do anything more than to repair the West Side Highway. The participants split into two antagonistic factions: one in favor of new building, the other opposed. Goldbeck has stated that the "intransigency" of the groups "forced the mediator to 'lead' the coalition to agree to discuss alternatives of what to build rather than continue what fast became a repetitious and futile debate" (Id., p. 244).

Complex political and economic issues made the mediation all the more difficult. As Goldbeck observes:

> The city had no money to do anything with the highway. The state feared the highway would consume its entire transportation budget, which was both true and politically unacceptable. Federal funds were available on a 90–10 basis if the road became an interstate and 70–30 basis if designated as a primary or secondary urban road. (Ibid)

Moreover, under a recent federal law, half a billion dollars in highway funds could be designated for mass transit; the city wanted to use any new mass transit funds for the Second Avenue subway rather than for anything on the West Side.

The issues proved to be too formidable to be solved, at least in this setting. In spite of good intentions and significant technical and financial support, the mediation effort failed to produce agreement. Only years later was the deadlock temporarily broken, and then through conventional political decision making, not broad-based negotiation. In late 1981, President Ronald Reagan presented New York City Mayor Edward Koch with a "check" for half a billion dollars, representing the first installment of federal funds for the massive Westway that is to replace the old highway. Soon after work began, however, opponents revived their lawsuits in hopes of killing the project.

West Side Highway Questions

1. The history of the West Side case, even when summarized, raises a host of negotiating questions. The particular focus of this chapter, however, is on multiparty bargaining. The dispute affected countless parties—Manhattan residents, commuters from outside the city, businesses, unions, government agencies, and so on. Even if there is some sense that a negotiated settlement is desirable, how is it possible to get everyone to the bargaining table? The Regional Plan Association tried to solve this problem by sponsoring mediation among the various organizations that already had been involved in the dispute; it then tried to invite other groups that were underrepresented. What were the weaknesses of this procedure? Can you suggest alternatives?

2. What determines the bargaining power of the various groups and individuals at such a session? Does power have a bearing on who should be invited? For that matter, does it explain why some invitees might decline, as indeed some did?

3. Goldbeck notes that "mediation is an expensive process, and no element of the process represents a greater investment than the time spent by the participants" (*Id.*, p. 248). Not all the parties feel the same constraints. For example, it was no hardship for the transportation director of the Chamber of Commerce to take part; doing so was simply an aspect of his job. But what about people whose employment has nothing to do with the issue or who work for public interest organizations with limited assets? Is it possible for negotiators to operate on equal footing at the bargaining table if there are such disparities beyond it?

4. The Regional Plan Association prepared an agenda for the first session. With 38 participants, the need for some sort of structure seems clear, but an agenda can be a powerful tool for guiding discussion to a particular outcome. Can you imagine any efficient way in which the group could have contributed to drafting the agenda?

5. Access to technical information can raise similar issues: parties who are unable to hire scientific consultants may believe they are in a weaker position than those who can; hence, they may decline to negotiate. Can you think of specific ways in which data gathering and analysis in the West Side case could have been conducted impartially? Given the complexity of traffic impact studies and air quality science, should the technical consultants be mere advisers, or should they be regarded as full participants as well?

6. At the outset, the parties agreed that participation in the mediation did not commit anyone to a consensual resolution. Is this always a wise policy? Why do you suspect it was adopted here?

7. When the participants split into two camps, one in favor of new building, one opposed, the mediator nonetheless *led* them, to use Goldbeck's term, to discuss alternative projects that might be built. Why might a mediator foreclose discussion on an option (not building) preferred by some of the negotiators? What are the risks of this move?

Concluding Note

Additional questions on the West Side case are posed on p. 113 of this chapter. The description of the case, though brief, raises important issues that are examined in other parts of the book. We shall see examples of successful mediation, for example, in chapters 8 and 9.

As already noted, the case also raises the matter of data disputes, the subject of the preceeding chapter. Goldbeck expresses apparent impatience with what he calls "community emotionalism," specific opposition to highway construction in the face of statistical projections that showed that neighborhood traffic actually would be reduced. Is such opposition really irrational, however, when Goldbeck himself acknowledges that "statistics were apparently made to prove that the highway could both increase and decrease local traffic"?

Laura Lake makes the following assessment of the West Side mediation effort:

> While this experiment did not resolve the West Side Highway dispute, it did reveal the potential for compromise within the group, and illustrated how important the ground rules for organizing intervention can be to a positive or negative outcome. Several reasons for the negative outcome of this effort can be identified: participants commented in private that they could not enforce a settlement; they were sure that they would wind up in court and did not want to prejudice judicial proceedings. They also felt that the local community groups were not adequately represented in the mediation group. With hindsight, it is possible to speculate that under different procedural, organizational, and stylistic conditions (which were not available to mediator Straus), a consensus might have been reached on the knowledge generated during the sessions. This would have required a great deal of mutual trust."
> (Lukes, 262 *Ekistics* 164, 170, 1977)

NEGOTIATION PARTICIPANTS: REPRESENTATION

The following material on multiparty bargaining is from a doctoral thesis by Timothy John Sullivan, *Negotiation-Based Review Processes for Facility Siting*, (Kennedy School of Government, Harvard University, 1979). Although his focus is on the siting of controversial facilities, such as hazardous waste treatment plants and nuclear power stations, many of Sullivan's observations apply with equal force to environmental disputes generally.

A development conflict often requires multilateral negotiations. Each participating group has its own set of interests which it seeks to promote. Thus, reducing the number of negotiators becomes much more difficult than in a bilateral negotiation. However, development conflicts will generally see project proponents negotiating with project opponents and regulatory officials setting bounds on developer actions. Although the opponents may include many groups with different interests, often there is only one and many, instead of among many. This situation has parallels in the labor field where one management team negotiates with many different unions. Nevertheless, it presents a complex bargaining problem.

The involvement of governmental regulatory groups as parties in any environmental/developmental conflict usually adds another dimension to the negotiation. Negotiated settlements may require governmental approval, zoning variances, or other special considerations which neither project opponents or proponents can deliver. When regulatory groups have discretionary power, their active participation and support of negotiations can assist the bargainers to reach a settlement. [pp. 129–130]

Those instituting a negotiation-based review process must decide who will participate in the negotiations. . . . The first class of participants includes those who have a formal position to affect the development controversy. Those formal participants will include representatives of licensing and regulatory bureaucracies, representatives of state and local governments, and the developer.

The second class of participants in the negotiation will include individuals and groups affected by the development project but with no official status. This class may include community groups in the neighborhood of the site, regional groups concerned with impacts on the regional environment, and special interest groups whose interests are affected by the project.

Finally, a mediator will participate in the negotiations. [Editors' note: mediation may be the exception, not the rule.] The mediator, unlike the other negotiating parties, does not represent a specific constituency or viewpoint. His goal is to facilitate the bargaining process, help each side to reach an agreement and see that standards of due process are met. . . .

A viable negotiation-based review process must include those individuals who have power over the final development decision. These individuals include the developer who wishes to build the project, representatives of governmental agencies which must review the project, and local officials who may take action to expedite or retard a facility. Finally, at times the negotiators may wish to consult with an expert concerning either environmental, sociological, or economic aspects of the proposed project. [pp. 296–297]

Negotiation may [also] provide a major opportunity for public participation in the review of the project. Projects may generate particular interest among local community groups who share their neighborhood with a project, regional groups who may receive the benefits of a plant's services and bear the impacts of its operations, and interest groups who have a special concern over a particular technology, facility, or site. [p. 300]

Some groups, not geographically concentrated, may have a special interest in a proposed facility. These interest groups may oppose the project for a variety of reasons. The proposed project's location may affect a particular interest of a group. For example, the planned construction of an interstate highway through Franconia Notch in New Hampshire directly affected the Appalachian Mountain Club's interests in the preservation of the White Mountains and the preservation of the surroundings of its chief hiking center, which was located in the Notch. Other groups may oppose a project for more ideological reasons. Antinuclear groups may oppose nuclear power plants wherever they are planned because they oppose the deployment of this technology. . . . [p. 301]

For negotiations to take place, there must exist a system for recognizing groups as legitimate parties to bargaining and for determining who shall represent the bargaining groups. Choosing formal groups and accepting their representatives is a simple task. . . .

A problem arises over how to recognize non-formal groups and individuals as participants in the negotiations, and how to determine who legitimately represents these groups. Whenever the formal review limits the number of participants in negotiations, some process must determine which groups may negotiate and who shall officially represent them. Since unlimited participation may create cumbersome and unproductive negotiating sessions, our objective of efficiency suggests that we limit participation in some way. The process objective of fairness requires that the mechanism for limiting participation avoid arbitrary actions. The process objective of encouraging public participation requires a screening mechanism which does not impose heavy burdens on those who wish to participate. In the author's view, a qualifying petition offers a natural way of limiting participation and a simple way for groups to designate an individual to represent their interests. Although other methods may provide a practical solution to this problem, we examine only the petition process.

A. Limiting participation. Several considerations support attempts to limit the number of negotiation participants. If only a small number of individuals bargain, negotiation sessions may prove productive. Large numbers of bargainers may make the negotiation process unwieldy and difficult to manage. Negotiation sessions are unlikely to accomplish much when the number of bargainers is large. Additionally, in negotiations over environmental/developmental conflicts, many people will participate voluntarily. When the number of negotiators is large, the bargainers may feel that the groups will not miss their contributions, and that they have only a small effect on the final outcome.

In bargaining over development issues, many of the benefits of negotiation may arise only from an atmosphere of trust and understanding which develops through personal contacts between the disputants. Trust will not likely develop in a large group. Further, if participants can easily join and withdraw from negotiations, a climate of trust is unlikely to develop. People cannot constantly adjust to new faces.

If individuals are free to participate, the negotiations may attract "meeting gadflies." Personal and social rewards are one of the reasons people volunteer time and effort to community causes. Unfortunately, these rewards which make voluntary actions less onerous, may attract some people who make a career of attending meetings and speaking in public. . . .

B. Recognizing groups by petition. The legislation authorizing a negotiation/referendum review process could require the circulation of a qualifying petition. Those groups who meet the required number of signatures should be automatically recognized as legitimate participants in the negotiation process. The number of signatures may be set to limit the number of participants.

Legislators will face a tradeoff between negotiation advantages gained through the consolidation of interests and the barrier to participation which a high qualifying minimum represents. A low qualifying standard will facilitate participation, but in the extreme, it may produce an unwieldy number of participants. A low standard will enable many groups to generate the needed number of qualifying signatures internally, thus reducing the need of groups to reach out to others. The number of signatures needed to qualify should thus increase with the population of the state or town.

A petition process possesses several major advantages which support its use to qualify negotiation participants. Circulating a petition is a political activity, and this accentuates the fact that the review of development projects is not simply a technical matter. . . .

Petitions need not cost much money to circulate. Petitions generate only printing, paper, and certification costs. The major burden a petition imposes is the burden of circulation. Gaining the required signatures requires that those advocating a position spend time and effort to persuade others to endorse their views, but this requires no direct financial outlay. This may open the project review process to concerned groups that lack financial resources. . . .

C. Choosing representatives of nonformal groups by petition. The determination of legitimate representatives of competing interests may pose severe problems for anyone attempting to mediate a developmental dispute. Determining representatives of groups without organization structures can create great difficulties. If a mediator chooses representatives from informal groups, then his choice may affect the balance of power within the group. This choice may create leaders where none existed, and create conflicts within the group. These decisions are best left to the individual groups for resolution.

The circulation of the qualifying petition in the name of a representative individual and perhaps one alternate may offer a simple way for designating representatives. People, in signing a petition, could designate an individual to represent their group interest. The petition is an established way of consolidating support behind a candidate or issues, and current practice uses petitions to qualify candidates and issues for ballot consideration. [pp. 301–307]

Questions on Sullivan's Proposal

1. Sullivan's proposed use of petitions to certify informal groups contemplates a formal negotiation process that is under special legislation to foster facility siting. Can the proposal be extended to environmental disputes generally? In the West Side Highway case, would it have made any difference if neighborhood and environmental groups had been designated in this way, instead of by invitation of the Regional Plan Association?

2. When petitions are used in other contexts, signatures may be rather casual acts: A person who signs a candidate's nomination papers is not bound to vote for her or him in the election. Is it not necessary, however, that signers to Sullivan's petitions agree to be bound by their representatives' actions? (If not, then one disgruntled person could seek to overturn a negotiated agreement by means of a lawsuit.) Yet, in the earlier stages of conflict, when information is contradictory and incomplete and the issues are not fully formed, is it fair or realistic to ask people to bind themselves to the actions and decisions of a representative who himself may be little known?

3. Even if it is possible to designate representatives through petition, what relative status should they have at the bargaining table? Specifically, should the representative of a small community group have a vote that counts as much as the delegate from an environmental group with tens of thousands of members? Is it relevant that most of those members live nowhere near the proposed project? Do we need to be concerned about votes at all?

Case Study: The Snoqualmie Dam Dispute

This case study is adapted from a portion of "Mediating Environmental Disputes" by Laura M. Lake, (262 *Ekistics* 164, September 1977) and from a doctoral dissertation by Timothy John Sullivan, Negotiation-Based Review Processes for Facility Siting, (Harvard Univeristy, 1979).

The Snoqualmie River Valley is located in the western part of Washington State, just 30 miles from Seattle. Before 1959 the river had overflowed periodically, but without causing extreme damage. That year, however, a severe spring flood swept away crops and topsoil from lower valley farms and destroyed many homes and businesses in the town of North Bend. The country, backed by riverside residents, asked the United States Army Corps of Engineers to study the problem. The corps proposed building a dam. Environmentalists were opposed, fearing not just the loss of a free-flowing river, but possible suburban sprawl on the floodplain.

Before building a dam, the corps must by law obtain approval from the governor of the state in which it will be built. Washington's Governor Daniel Evans twice vetoed the proposed dam, in 1970 and 1973, but he acknowledged that there was a legitimate need for flood control. Gerald McCormick and Jane McCarthy of the University of Washington's Environmental Mediation Project had already had preliminary meetings with dam proponents and opponents. At McCormick and McCarthy's behest, Evans formally appointed them to mediate the dispute.

Working under a 6-week deadline imposed by the governor, the mediators identified 10 people who they felt had credibility with the conflicting groups and who represented a range of views on the project. These people did not represent their organizations; rather, they represented general constituencies. The mediation sessions helped the participants to overcome long-held stereotypes about one another. According to Laura Lake:

> These sessions began to educate and socialize the participants: the environmentalists learned that the farmers had no desire to sell their land to developers; the townspeople realized that continued development would ruin the quality of rural life they valued; and the environmentalists learned that even while they were resisting the dam, real estate development was occurring, despite flood hazards. (262 *Ekistics* 162, 167, 1977)

Timothy Sullivan notes:

> Dam proponents established that flooding caused them economic hardship by destroying their crops, and that continued flooding would not provide an acceptable solution. They made dam opponents believe that they would be held politically responsible for any damages from a future flood. They stressed that a flood would destroy the regional credibility of environmentalists and lead to the construction of a dam without any amenities or land-use restrictions.

The mediators satisfied the governor's 6-week deadline by reporting substantial progress: the participants had endorsed a general statement that acknowledged the need for some kind of flood protection and some form of land-use control. The governor gave the mediators two more months in which to come to final agreement. At one point, talks had to be suspended while the environmental groups caucussed to develop a unified position, but mediation was resumed in time to reach a tentative agreement by the deadline. Final details were approved several months later.

The agreement provided for: (1) a dam on the north fork of the Snoqualmie instead of the middle fork; (2) a series of levies and set-backs along the middle fork; (3) land-use and zoning restrictions on the downstream farmland; and (4) other measures, including the creation of river basin planning council and the purchase of development rights and floodway easements.

QUESTIONS ON SNOQUALMIE

1. As in the West Side case, the participants in the mediation were invited. Given the smaller scale of the dispute, might it have been more appropriate simply to open the mediation to all who wished to participate. If that proved too unwieldy, then could Sullivan's petition method be used? What bearing does the fact that the dispute had been stewing for more than a decade have on this selection issue?

2. Note that this was not an issue that divided everyone into two distinct groups, those favoring or those opposing the dam. Ten participants were needed, after all, to represent the range of views. How might coalitions have developed and changed in these circumstances? To consider this question, assume that representatives of the following constituencies were involved: farmers who were interested simply in protecting their operations, farmers who also were interested in enhancing the value of their land for possible future sale, residents who welcomed the prospect of future growth, area residents who wished to preserve the area's semirural qualities, environmentalists opposed to suburban sprawl, and canoeists and kayakers who wished to preserve open water. What alliances would you expect to be formed among such groups? What factors encourage or inhibit the formation of coalitions? Feel free to add to the suggested list of subgroups as you consider the problem.

Snoqualmie Epiloque

Timothy Sullivan (1979, 93–94) has written the following analysis of the Snoqualmie case.

> In this negotiation, several circumstances aided the mediators' efforts. The existing community infrastructure enabled the mediators to select people with sufficient influence and power to represent the conflicting groups. The commitment of Governor Evans to negotiation and his powers of office gave the bargaining efforts a special legitimacy. Governor Evans created interim deadlines to enforce progress.

Those fearing future flood damage were particularly successful in convincing the environmentalists that the citizens of Washington would hold them responsible for any flood damages. They argued that this would undermine the credibility of environmentalists throughout the state. This gave strong incentive to the environmentalists to negotiate.

Although the original conflict arose over the single issue of dam construction, the communication required in bargaining helped change the shape of the conflict. The negotiation changed from a yes/no dam issue into a search for environmentally acceptable flood control measures. Both dam proponents and opponents moved beyond their original misconceptions of the other side and dealt with each other's real needs and concerns.

The geography of the Snoqualmie River allowed the creation of an imaginative alternative which proved critical in reaching a settlement. The three-branch nature of the river proved critical in permitting dam opponents to maintain their early public stand against a Middle Fork dam yet still meet the farmers' needs. In the final compromise position, the North Fork dam will provide flood control to all farmers below the point where the three branches merge. Set-back levees along the Middle fork will provide a measure of flood protection to Middle valley residents yet still permit the Middle Fork to remain a free flowing river. These levees allowed dam opponents to retain their public stand against a Middle Fork dam while agreeing to a flood control project.

Although this solution will probably cost more than the original proposal, the Army Corps accepted it. This willingness to pay for a more expensive proposal permitted a widened set of alternatives, and changed the conflict from a binary decision to a question of design. This transformation provided an issue over which each disputant could make concessions and realize gains.

Multiparty Negotiation and Coalitions

Multiparty negotiations are fundamentally different from two-party negotiations in that they present participants with an overlapping network of possible agreements. A farmer negotiating with a greenbelt organization over the possible purchase of his land may either come to agreement or not. His bottom line or resistance point is often defined by the consequences of not agreeing. Farmers will compare the final offer they receive from the greenbelt group with what they expect they can get from someone else. If the bid is better overall, the farmer will take it; if not, the farmer will not settle.

In the case of multiparty (that is, more than two) bargaining, however, lack of consensus among all the parties does not preclude agreement between some of them. Certain situations may require unanimity, but many do not. For example, a factory that is being sued by its neighbors for nuisance may settle with those

TWO-PARTY VERSUS MULTIPARTY NEGOTIATION

whose demands are low or whose cases are strong but go to trial with the rest of the plaintiffs. In settling one case, of course, it must consider the impact on other claims.

Because of possible competition and cooperation among subgroups, multiparty negotiation is much more complex than two-party bargaining. The complexity is manifested in bargaining strength and strategy.

Problem 1

In problem 2 in chapter 2, we considered a simple two-party negotiation between a farmer and a conservation group. The farmer was entertaining a developer's offer of $300 thousand for his land. Because of the special beauty of the property, the conservation group was ready to pay as much as $400 thousand for it. We saw that any figure within this bargaining range would leave both the farmer and the group better off than if they made no deal. Left open, though, is how the buyer and seller will split the $100,000 "surplus." Social convention sometimes suggests splitting the difference, but there is no point of equilibrium.

Consider this problem again, but with an important variant. A second developer has appeared, but this one is ready to pay $400,000 for the land. For the sake of simplicity, assume that the sole objective of the farmer is to get as much as possible; the farmer does not care in the least what happens to the farm. The prospective purchasers are each interested in getting the land as cheaply as possible.

1. Who has the bargaining power in this situation?

2. As competing bidders, the developer and the greenbelt organization may seem like strange bedfellows, but is there anything that they can do to prevent the farmer from exploiting the situation? If you represented the conservationists, can you imagine an attractive proposition you could make to the developer that would leave you with the land in your name and totally under your control and that would cost you no more than $400 thousand in all? Can you imagine a deal in which you fail to get the land, but do not come up empty-handed?

3. Now, once again, who has the bargaining power?

4. If you represented the conservationists here, and you knew the priorities and resistance points of the other two parties, how would you prefer for the negotiations to proceed? Would you, for example, want a three-way meeting, or would you like to meet with one of the parties first; if the latter, which one?

5. If negotiations with the farmer have pushed you close to your resistance

point, how does that affect the attractiveness of a side deal with the developer? How do you evaluate one against the other?

6. Assuming still that you represent the conservationists, how is your bargaining strength affected if instead of one competing bidder there are 2, 5, or 10?

PROBLEM 2

The preceeding problem, though multiparty, is still relatively simple in that it is zero-sum; that is, there is a possible $100 thousand surplus to be divided among three parties. Frequently, of course, the fact of coalition may introduce nonzero-sum elements. Economies of scale may mean that three companies who discharge waste into a river may control waste more cheaply if they work jointly than if they act independently and duplicate one another's investment.

The following problem is a variation on an abstract exercise devised by Howard Raiffa of the Harvard Business School and described in his book *Negotiation Analysis* (Cambridge: Harvard University Press, 1982, pp. 262–269). His book includes a lengthy consideration of coalition strategy in general and this exercise in particular. For our purposes, assume that A, B, and C are three companies under legal compulsion to control their waste discharge into a river they all abut. They are not under any obligation to work together, but a rigorous study has shown that it is clearly advantageous for them to do so. The most savings will be realized if they act as a trio, but even if any one of the three stayed out, the other two could cut costs by banding together. Table 3 shows the savings (in thousands of dollars) that are possible through various coalitions. Each company's goal is simple to maximize its own savings.

For example, A might explore working just with C. Together they could save $84 thousand, but they still have to decide how to divide it. If C insisted on a 50–50 split, A could threaten to make a deal with B.

Professor Raiffa has his students do this exercise in class. He creates trios in

TABLE 3. COALITION PROBLEM

Coalition	Savings
A (alone)	0
B (alone)	0
C (alone)	0
A & B	118
A & C	84
B & C	50
A & B & C	121

which each student represents a company. People have 30 minutes to come to an agreement. Under his rules, there can be no prior communication until all three meet. Two players may arrange for a private meeting, and the third must not interrupt for at least 2 minutes. Please consider the following questions.

1. What strategies does this situation invite? How, if at all, would your strategy be different if you represented A, B, or C?

2. If you were an arbitrator in this matter, what result would you order? One resolution, for example, is to require all three to participate and to divide the $121,000 savings equally, that is, give each $40,333. Is that fair? Is fairness defined by the bargaining structure—that is, the savings that are obtained by various coalitions—or is it necessary to look at other factors such as the size of the companies, the degree to which they currently pollute, and the cost they will incur if they go it alone?

3. How is your analysis of this situation affected if we build in more realistic considerations? Let us imagine, for example, that the projected cost savings, albeit carefully calculated, are not infallible. What if, as is likely, that each company has a fairly good idea of the savings that it is likely to realize through the various coalitions but is less clear about the precise advantages perceived by the others? Does the introduction of other nonquantifiable factors, such as public relations, help or hinder consensus?

Problems of Cost Sharing

Introduction

Many environmental conflicts involve problems not of sharing benefits—as in the preceding problem—but of allocating costs. In some respects, this one is merely the obverse of the other: parties will jockey to form coalitions to minimize their costs instead of maximizing their benefits. In certain cost allocation situations, however, it may be possible to design a process in which the party who happens to draw the short straw receives some sort of compensation from his or her more fortunate cohorts.

The following excerpt is from Howard Raiffa's *Negotiation Analysis*, (Cambridge: Harvard University Press, 1982, pp. 311–313). Though highly simplified, it is inspired by a new Massachusetts law on hazardous waste treatment facility siting. The law is examined in detail in chapter 12, but Raiffa's abstracted example helps illuminate the complex bargaining relationships among various communities that may have to host a treatment plant. All agree that such a plant is essential, but none want to see it built in their backyards. The risks of illegal dumping of toxic wastes are currently shared by all. The negative impact of hosting a treatment facility would be felt principally by one community, whereas the benefits would be realized by all the others.

To keep things understandable, Raiffa makes some assumptions that actually are contrary to the Massachusetts procedure. For simplicity's sake, he also imagines a state with just five towns.

Negotiating Cost Allocation

Suppose a facility could be located in one of five towns: Aspen, Baileyville, Camille, Donneybrook, and Eaglestown. Contrary to reality, let's assume that each town is monolithic in its views and each is represented by a negotiator (A, B, C, D, and E, respectively) who has full power to commit his or her town. While each town wants the facility to be built (somewhere else) let's assume at first that the state has agreed to build and maintain the facility in any one of the five towns, but that they have to decide jointly just where it is to be built. If they can't decide, it will not be built.

The five representatives bicker among themselves and can't reach agreement. Someone proposes using a randomized procedure to determine the location of the facility, [and] all towns have an equally likely chance to be chosen. They all agree to this randomization procedure, and the unlucky "winner" is representative C. He can't, after the fact, suggest that he's having second thoughts about the procedure; but because he represents a rich town he is able to bargain with B, the penurious town of Baileyville, to accept the facility—for a price. B bargains hard and agrees to C's request, with a compensation sweetener of $100,000. D is furious. Why should the people of Camille get out of their obligation just because they're rich? Why should poor Baileyville always get stuck with dredge work of the society? "Hold on," says B, "Who are you helping? My town is not only poor, but you won't allow us to improve our position. That's double jeopardy. That one-hundred thousand will finance a long-needed library and shelter for abused unfortunates."

Society is schizophrenic about the morality of certain financial transactions. The rich are not allowed to buy themselves exemptions from the military draft; in a college dormitory people would think poorly of an affluent student if he were to financially entice a scholarship student to swap dormitory rooms that were assigned by random numbers. But it's permissible for workers to receive premium wages for hazardous jobs.

Assume now that the five representatives have agreed to use a random drawing, but the drawing has not yet been conducted. A knows that B would assume the obligation for $100,000, but since Aspen can only afford to pay $50,000 in order to shift the obligation to some other town, A forms a deal with E who thinks similarly. If the randomization designates A or E they each agree to pay $50,000 to B to assume this obligation. D has second thoughts. "I don't like giving or taking compensation for this obligation, but if this is going to be the accepted norm, then I would be willing to do it for $80,000," announces D.

"That's wonderful," responds C. "Let's each get up $20,000 to give to D."

But B intervenes: "Baileyville can't afford $20,000; but we'd be willing to lower our price for accepting the facility to $75,000."

Finally E comes up with a suggestion. She presents two numbers that describe her feelings as a representative of Eaglestown: (1) the amount of compensation that Eaglestown would be willing to *give* to another town that accepted the facility (rather than not have the facility built at all); and (2) the amount of compensation Eaglestown would *need* in order to accept the facility (rather than not have the facility built at all). She declares that Eaglestown would be willing to give $50,000 but would need $150,000 for acceptance.

"Let's see if I understand those two numbers," interjects C. "You see the benefits of the facility without any of the inconveniences as worth $50,000. But the inconveniences are sufficiently high that you need $150,000 to accept the facility, if the other alternative were no facility in any of our five towns. Is that it?"

"Yes, that's it."

The parties agree to call the first number CWG ("compensation willing to give") and the second number CNA ("compensation needed for acceptance"). Each agrees to write down their CWG and CNA and to let a reputable adjudicator, Mr. X., resolve their conflict based on the ten numbers. [Table 4 displays CWG and CNA values.] The adjudicator, Mr. X., observes that the facility cannot be built in Aspen, since Aspen needs $200,000 and the other towns are only willing to give $150,000 collectively. Baileyville needs only $50,000, and the others are willing to give Baileyville $190,000. The facility cannot be built in Camille; it can in Donnybrook and (just barely) in Eaglestown.

TABLE 4. COMPENSATION PROBLEM

Town	Willing-to-give/Needed-to-accept Facility	
	Compensation willing to give ($'000)	Compensation needed to accept ($'000)
A	50	200
B	10	50
C	60	3000
D	30	80
E	50	150

Study Questions

1. If you are Mr. X., the adjudicator, where would you site the facility and what compensation would you require of the other towns? Should the designated town merely get the minimum compensation that it demanded or should it get the total that the other four were willing to provide? Should compensation be divided four ways, or should it be proportional to the amounts that the towns said that they would be willing to give? What other cost-sharing schemes can you imagine—and defend?

2. Should the adjudicator believe that the values submitted by the representatives are honest? Would a town be more likely to misstate the amount it would require to serve as a site or misstate the amount it would be willing to pay another town for doing this duty? What is the effect on truth telling if the towns do not know how the adjudicator plans to use the values?

3. What if the rules are changed so that the facility will be built in the town that has the lowest CNA, and further, that it will receive as compensation the amount of the second lowest announced CNA. Raiffa states that this should induce the towns to reveal their true figures, but why? Is this still true if some of the towns can collude before making their statements?

Conclusion

Raiffa observes that the bargaining situation is made far more complex when we realize that each town is not monolithic: Citizens may have different views of what CNA and CWG their town should select; there may be several different sites within each town. The fact that development costs likely will vary from town to town and from site to site adds another complication. The state may agree to supplement the compensation. Court challenges to the site selection procedure may be pending. Many of these considerations are addressed in chapter 12 in the section on the Massachusetts Harzardous Waste Facility Siting Act.

Cross-References

Cases

The principal cases presented earlier in the book—Grayrocks Dam, South Carolina Hazardous Waste, Brown Paper, and Holston River—all contain elements that allow us to test and apply ideas about multiparty bargaining developed

TWO-PARTY VERSUS MULTIPARTY NEGOTIATION 123

in this chapter. Before you consider the following questions, take time to review these cases, particularly to identify the parties and their interests.

Study Questions

1. At first glance, the Brown Paper case (chapter 4) seems to be a pure two-party dispute between the company and the EPA; yet, there surely were other interested groups and individuals. For example, the state environmental agency had responsibility for air quality standards, and it was initially involved. Why do you suppose it dropped out? Who was aided by its nonparticipation—the company or the EPA?

2. The Brown Paper case was multiparty in another sense. Recall that at the close of negotiations, when the EPA appeared to have reached agreement with the company, the EPA's enforcement division insisted on more stringent terms and ultimately got them. The agency thus was not monolithic; there was negotiation going on within it. Catalog the ways in which intramural negotiation, that is, negotiation among groups and individual within one organization, may differ from negotiation among separate entities. (Note that in the Snoqualmie dispute, mediation had to be suspended while the environmentalists caucused to develop a unified position. Is this identical to the internal negotiation that occurred in Brown Paper?)

3. Notably absent in the Brown Paper negotiation were any environmental or citizen groups. The EPA and the company did not have to respond to any demands by others to be included at the bargaining table. Nevertheless, we saw in the West Side Highway case an attempt to invite groups to the mediation even though they had not yet been involved. Did the EPA and the company take any risk in not likewise searching out other interested parties? How would participation of groups like the Appalachian Mountain Club or the Sierra Club have affected the bargaining power of the original parties? What would have been the proper role of such groups in the negotiation?

4. The Holston River (chapter 5) case was a two-party dispute, though only in a limited sense. The EPA recognized that any waste discharge arrangement it made with Tennessee Eastman would directly affect the standards that would apply to four other companies that were polluting the river. The terms of a permit issued to Eastman would establish a precedent of sorts for the others. Moreover, if the river is seen as having a certain capacity for carrying pollutants, then the portion of that capacity assigned to one company establishes an upper limit on what can be assigned to others. With this in mind, consider the coalition possibilities in the Holston River case. What incentives might exist for any

of the other companies to try to side with Eastman? What might be the disincentives? The EPA choose to go after Eastman—the biggest polluter but also the most influential and technically sophisticated company—first. What could be said for a strategy in which the EPA met with all the companies collectively? Why do you suppose this strategy was not adopted?

5. The South Carolina hazardous waste problem (chapter 3) was clearly multiparty, and it provides an interesting parallel to the West Side Highway and Snoqualmie cases in this chapter. In all three, individuals or groups were invited to negotiate. What criteria should be applied in such a selection process? Some people have contended, for example, that hazardous waste treatment is a highly technical matter; hence, those involved in setting important policy should have particular expertise? Do you agree that technical credentials are relevant? Who should be responsible for making the selection? What recourse, if any, should a party have if he or she is not invited? Is the invitation method equally appropriate for all environmental disputes or is the scale of the conflict a relevant factor? Does it matter whether the dispute is site specific, as in the case of the proposed dam in Snoqualmie, or is about general policy, as in South Carolina?

6. The Grayrocks Dam case was another multiparty conflict; farmers, environmentalists, a power company, even state governments were among those involved. Both the farmers and the environmentalists were worried that water consumption of the proposed dam would hurt their interests. Is the fact that they shared a common concern sufficient to explain why they cooperated? Can you imagine circumstances such that, in spite of this common concern, the two groups would view each other as opponents?

7. When a dispute is of interstate magnitude, problems of coordination can be substantial obstacles to resolution. How did the parties overcome this obstacle and facilitate meetings that led to settlement? Are face-to-face meetings necessary, or could such a negotiation be conducted in other ways. What is lost when parties have to rely on mail and telephone communication? Is anything gained? Closed circuit and cable technologies are increasingly available. Do they solve the logistical problem when parties are hundreds of miles apart? Should we welcome the day when people come not to the bargaining table, but to the bargaining video monitor?

READING REFERENCES

There are many multibargaining issues that, though beyond the scope of this book, are important. For example, Howard Raiffa deals at length with problems of fair division and cost allocation in his book *Negotiation Analysis*

(Cambridge: Harvard University Press, 1982). He also illustrates strategies of voting when many parties have to choose among alternatives; voting one's true preferences is not necessarily the best route to one's goals.

The Logic of Collective Action by Mancur Olson, Jr. (Cambridge: Harvard University Press, 1965) remains a classic examination of the ways in which a group's actions may conflict with the priorities of its constituents. Olson deals at length with the question of representation, large- versus small-group behavior, and the lobbying power of special interest groups—issues that are all pertinent to environmental disputes. Thomas Schelling's *Micromotives and Macrobehavior.* (New York: W. W. Norton, 1977), is an important complement to the Olson book. Particularly relevant here is Schelling's illumination of problems of coordination that come about not so much from logistical obstacles but from diverging individual and collective incentives.

Finally, Timothy John Sullivan's doctoral dissertation *Negotiation-Based Review Processes for Facility Siting* (Kennedy School of Government, Harvard University, 1979), draws revealing parallels between multiparty bargaining in labor and other fields and environmental disputes. His dissertation is a forthcoming book from Plenum Publishing Company.

7

Prospects for Compliance

Introduction

Problems of ensuring compliance are central to the negotiation process. Disputes do not necessarily end when the parties first reach agreement. New issues may arise, and settlements may come unglued. Similarly, ambiguity in the language of the agreement may also frustrate implementation. Just as negotiation often begins before the parties sit down to bargain, so may it continue after they leave the table. The period of implementation tests the imagination and energy of the parties; if their agreement proves inadequate, the parties must renegotiate.

Potential compliance problems may also keep the parties away from the bargaining table in the first place. Savvy negotiators always are looking one step ahead. A chemical company may refuse to invest in expensive pollution-abatement technology as part of a negotiated resolution of an enforcement action if it fears that the government will impose even stricter requirements at a later date. Similarly, the government may not be willing to drop its enforcement action if it doubts the company's willingness to actually make the necessary investments.

From a negotiator's point of view, the problem is not merely whether to trust the other side but how to appear trustworthy oneself. Hostage negotiations provide a revealing example of this principle. Those seeking freedom for the hostages may be more than willing to meet the terms of the captors but may be incapable of convincing the captors of their sincerity; inevitably the captors suspect a trap. Similarly, the captors may be willing to free their hostages unharmed upon compliance with their demands, but they may be incapable of providing convincing assurances of their intentions to their negotiating counterparts. In each case, a deal may be thwarted unless the parties can find a way of ensuring compliance. In the Iranian hostage situation, this was achieved through the intervention of the Algerians as neutral third parties who temporarily took possession of both the hostages and the money tendered for their release.

Not every bargaining situation presents compliance problems. Sometimes the parties negotiate for mutual simultaneous performance, as in the case of retail transactions where money is tendered upon receipt of the purchased goods. But environmental cases rarely are so simple. Usually, they involve complex agreements that often take years to implement. Frequently, compliance problems are central.

In the case that follows, difficulty in assuring compliance nearly frustrated agreement. As you read the case, ask yourself whether there were other ways of structuring the agreement so as to remove the compliance issue from the table. This case also provides additional material on issues that have been introduced in earlier chapters. Consider, for example, the incentives that kept the parties at the bargaining table through several years of unproductive negotiations. Also, note that, as in the Brown Paper case, at least some of the parties had dealt with one another before and would certainly have to do so in the future. Finally, as in the preceding case, there were more than two parties involved in the dispute; note especially how coalitions formed and shifted over the course of negotiations.

Case Study: Jackson, Wyoming—201 Grants for Municipal Wastewater Treatment

This case study was originally prepared by Stephen Hill; it has been substantially condensed.

Introduction

Jackson Hole is a 40-mile-long flat-floored valley that flanks the east side of the Teton Mountains in northwestern Wyoming. It is an area of spectacular scenic beauty; its endowments include the Teton Mountains, Grand Teton National Park, Yellowstone National Park, Jackson Lake, and the Snake River. Fish and wildlife are plentiful, and several thousand elk winter in the valley. Each year millions of visitors flock to the area to enjoy the relatively unspoiled region.

The town of Jackson lies at the southeastern corner of Jackson Hole, within Teton County. Largely because of the region's scenic amenities, both the town and the country experienced rapid growth in the past decade; the county population nearly doubled between 1970 and 1980. This rapid growth sorely taxed Jackson's wastewater treatment facilities. Constructed in 1969, Jackson's sewage treatment plant was already receiving more sewage than its design capacity by 1971. As a result, the town was forced to look for ways to increase its wastewater treatment capacity. After exploring a number of alternatives, the town settled on a plan to construct a new plant several miles south of the town in South Park, an undeveloped area of the county. From the town's perspective, the South Park alternative was desirable for a number of reasons—a new plant would have lower operating and maintenance costs than expansion of the existing plant. It would also make it easy to accommodate future growth, which was likely to occur in South Park in any event, and, by locating the plant in one of the lower points of the valley, new growth could be serviced by a gravity-feed system, thus avoiding the need for costly backpumping (see Figure 9).

The Teton County commissioners strongly resisted the Jackson proposal. They argued that a new plant and sewer interceptor would spur intensive development in what many thought should remain a rural, agricultural area. The county commission argued for upgrading Jackson's existing plant and concentrating future growth. Ultimately, the dispute between the town and the county over wastewater planning mushroomed into a debate over the area's future growth and development, with the county urging strong growth controls and the town defending a more laissez-faire attitude. The Environmental Protection Agency became involved when it became clear that Jackson would not be able to comply with federal water pollution control standards, without some resolution of the town–county dispute. Moreover, because EPA was the source of grants for construction of new wastewater treatment facilities, it had substantial leverage in dealing with both the town and the county.

Jackson's Wastewater Planning

In the early 1970s, Jackson's town council was composed entirely of longtime residents. Dismayed at how quickly the current plant had reached its capacity, they sought a more lasting solution to the town's wastewater treatment problem. The existing plant was on a small, 5-acre site, a location that they thought limited options for plant expansion. Also, the plant could not serve future development in neighboring South Park, a 4-mile-long valley south of Jackson without costly backpumping.

The council sought a new site at the lower end of South Park that could serve all future development in South Park by gravity flow. Early in 1973, then-Mayor Lester May proposed to the Wyoming Game and Fish Commission that Jackson be allowed to construct a new treatment plant on 20 acres of the South Park Feedground, which the commission managed. The site was visually shielded by trees, and Jackson proposed exchanging a parcel of private land for a long-term lease for the site. The council was reluctant to locate the plant on private land, a scarce commodity in Teton County.

The Game and Fish Commission was not receptive to Jackson's proposal. In 1973, they voted to deny the town's request, noting that the proposal conflicted with their wildlife management objectives and that it would undoubtedly draw strong opposition from environmental and outdoor groups. The commission also doubted whether it legally could sell or lease the proposed sites; it had

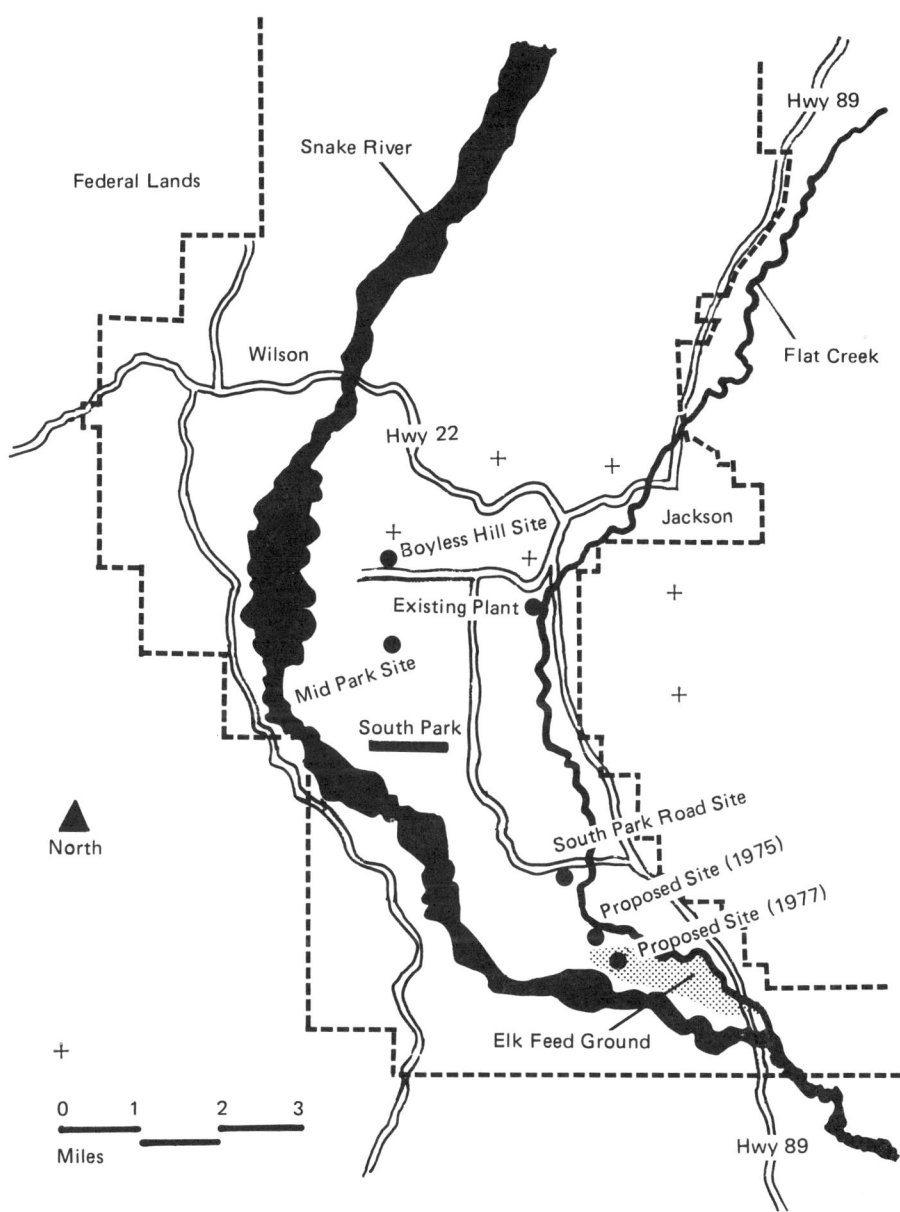

FIGURE 9. South Park and vicinity.

used federal funds to acquire the Elk Feedground, and federal regulations prohibited such transfers. In addition, the EPA Region B informed the council that if the town selected the Elk Refuge site, an environmental impact statement would have to be prepared to comply with the provisions of NEPA. Undeterred, however, the town council continued to pursue the idea for several years.

Teton County's Perspective

Beginning in 1972, the majority of county commissioners favored a planned approach to future county growth. Their position was at odds with the more laissez-faire attitude found in Jackson's town council. The chairman of the commission was William Ashly, owner of a building supply store and a respected community leader. His support of county planning efforts was not shared by the majority of town residents. Notwithstanding the fact that more than 55% of county voters were also town residents, official town and county attitudes toward growth differed markedly.

In May 1976, well after Jackson had announced its South Park proposal, the county's planning department published a summary report on alternatives for growth and development. The report described alternative development patterns, included a questionnaire, and was distributed to all county residents. Questionnaire responses indicated widespread concern among county residents about excessive growth and development in unincorporated areas.

Later that fall, the county distributed to its residents a report summarizing a proposed comprehensive plan that had been prepared pursuant to a Wyoming statute that obliged counties to engage in land-use planning. The plan consisted of a land-use element, to be implemented primarily through regulation, and a scenic preservation element, to be achieved primarily through purchase. At the same time, the county commission was supporting bills in Congress to create a federally funded purchase program for scenic easement. Once again, the great majority of questionnaire responses supported the proposed plan goals, policies, land-use proposals and scenic preservation priorities.

Several elements of the plan proposal were at odds with Jackson's South Park proposal and with the development that the plant was expected to induce. Plan goals stated that "new urban development should be compact rather than scattered in order to minimize the cost of providing public facilities and services to avoid environmental damage and to preserve agricultural, scenic, and wildlife values," and that "the pace of growth should be limited so as to avoid excessive costs to the County of providing public facilities and services, as well as to preserve the local lifestyle." One plan policy stated specifically that "sewage treatment plants and other public facilities should be located where they will not foster scattered development and where they will not cause environmental or visual damage." The land-use element called for residential development in

South Park at densities of between 3 and 6 acres per unit, depending on groundwater height. Implicit in the proposed plan was the assumption that South Park developments would be served by on-site septic systems, the constraints of which would necessitate low-density development.

The split between Jackson and the county commission widened after the November 1976 elections in which Ralph Gill was elected mayor of Jackson and Mary "Muffy" Moore was elected to the county commission. Gill, a major South Park landowner and former commission member, was firmly convinced that the town's new sewage plant belonged in South Park. Gill also enjoyed considerable influence. As one city official noted,

> Without being mayor, he's probably the second or third most powerful man in the county, because he is in control of so much property, because he has been involved for so long, and because he is interested.

Gill's election further solidified the town council's support of the South Park proposal.

Moore was elected to the county commission on a strong growth-control platform. She replaced a slightly less enthusiastic advocate of planning. She and Ashley were supported by a clear majority on the commission. They worked hard to obtain a strong comprehensive plan that was capable of controlling growth in the unincorporated areas of Teton County. In July 1977, the county planning department circulated a final draft of the comprehensive plan, incorporating virtually all of the elements of the earlier draft and stressing low-density development that would be compatible with agriculture in South Park. The plan identified 670 acres of developable land east of Boyles Hill, an area large enough to accommodate up to 3,300 residential units and close enough to Jackson to be served by the town's existing sewage plant.

At this point, both the town council and the county commission were aware of each other's positions and divergent views. Yet, neither group made an effort to discuss the impending dispute with the other. Teton County's Section 208 planning program, the obvious forum for resolution of such a disagreement, avoided the politically sensitive issue of sewage plant siting almost completely.

The Draft Environmental Impact Statement (DEIS)

Meanwhile, EPA Region 8 hired a consulting engineering firm—J. M. Montgomery—to assist with preparing of a draft environmental impact statement for the Jackson wastewater management plan. (An impact statement was required under NEPA because any new development would be funded by an EPA grant.) Like the county, EPA favored expansion of the existing facility. The staff of the Environmental Evaluation Branch of EPA Region VIII were cognizant of the severe pressure on growth that sewage treatment facilities could create and were sympathetic to the county's efforts to manage future growth. Regional Administrator Green, a Republican appointee in a new Democratic administration, wanted to avoid conflict with the EPA in Washington, D.C., and he supported his staff.

Green released the DEIS for the Jackson wastewater system proposal in May 1977. The DEIS concluded that

> the South Park Elk Feedground option would pose serious legal difficulties in acquiring the land in the feedground and adversely affect the elk here; that the proposed site was located in the one hundred-year floodplain and would conflict with the goals established by the Wild and Scenic River Study; and that the proposal would open up the entire rural South Park area to urban/suburban level development.

The DEIS examined several alternative projects, including three at Boyles Hill, one in mid-South Park, and one on South Park Road; it found several options to be more cost-effective than Jackson's proposed option.

Public comments on the DEIS were generally subdued because EPA Region VIII had already announced that it did not intend to fund a new plant in South Park through its section 201 grant program, which made money available to municipalities for construction of new wastewater treatment facilities. Several South Park landowners spoke in favor of the Elk Feedground site. Criticism of this option and support for the EPA's action came from the U.S. Department of the Interior, which opposed the municipal use of a refuge area acquired with federal funds; from County Planner Robert Ablondi, who noted conflicts in the county's comprehensive plan; and from two environmental groups—the National Wildlife Federation and the Wilderness Society—that concurred with the EPA's objections to the Jackson proposal on grounds of land use. No substantive comments came from Jackson, Teton County, the Section 208 agency, or the Wyoming Department of Envrionmental Quality (DEQ).

Environmental interests were to have very little involvement in the debate beyond their DEIS comments. Jackson was and is an isolated area. Environmental groups tend not to be well organized; instead, they depend on individual volunteer efforts. At the time, environmentalists were preoccupied with the comprehensive plan and the scenic easement proposal. Sierra Club member Phil Hocker headed a coalition, Citizens for the Plan, that fought to obtain a strong program for control of county growth. Environmentalists saw the issue of the sewage treatment plant site as peripheral and felt that the County Commission was doing a good job of representing their interests.

The Town–County Dispute Emerges

In July 1977, when the county released its final draft of the comprehensive plan, the Jackson town council announced its approval of the South Park site for the wastewater treatment plant: the new site was on private lands immediately north of the Elk Feedground and the former site. The council no longer was unanimous in its preference for a South Park site. Councilman Paul Bruun, elected in 1976, was concerned that a South Park plant would evolve into a county facility. He and two other council members, Norm Mellor and Man McCain, also worried about the distribution of costs and feared that town taxpayers ultimately would subsidize sewer service for county residents.

The majority of the council apparently gave little consideration of the alternative sites suggested in the DEIS, arguing that, in order to meet the town's long-term needs efficiently, the plant had to be as far south as possible and that all future development in South Park could be served by gravity-flow interceptor lines. Mayor Gill and the council also worried about obtaining a site and interceptor right-of-way for use of privately owned lands. They knew from earlier consultation with the respective landowners that they could obtain the new South Park site and easements at little or no cost. This was not necessarily the case for the alternative sites. Gill and others wanted very much to avoid condemnation proceedings or abandoning the project, which they thought would be costly both politically and financially.

Although Gill and most members of the town council apparently were not concerned about their differences with the county, the Teton County commissioners initiated a joint meeting of the two bodies early in August 1977 to discuss the wastewater treatment issue. The commission expressed concern about the pressures on growth that would be created by an interceptor line running the length of South Park. Commissioners Moore and Ashley knew they would be adopting the comprehensive plan in a few months and wanted some reassurance from Jackson that it would abide by the terms of the plan and its regulations. Moore suggested that the town council make new sewage hookups contingent on compliance with the pending comprehensive plan.

Jackson already was under pressure from EPA Region 8 to resolve its differences with the county. At a meeting earlier that summer, EPA Project Officer Wes Wilson told town officials that Jackson's wastewater treatment grant would be jeopardized if the two parties could not agree on a project. Neither meeting produced concessions or compromise; if anything, the positions of the town and the county hardened.

Early in September 1977, the county commission announced its opposition to a South Park wastewater treatment plant (its first formal statement to that effect), noting that such a project conflicted with the goals and policies of the proposed comprehensive plan because it would spur scattered growth and sprawl. Both the treatment plant site and interceptor route needed a county development permit, and the commission made clear its intention to withhold approval. In response to the county's action, Mayor Gill stated, "They should face it [the sewage problem] rather than saying they're against our plant." Accordingly,

Jackson placed pressure on the county by adopting a moratorium on all new county sewage taps (or hookups) and on the dumping of county septic wastes at the Jackson plant. The mayor took the commission's opposition as a personal affront, and he became even more determined not to modify the South Park proposal.

Later that month, attorneys representing both parties wrote to EPA Region 8, each asserting their authority in the plant siting matter. Dave Larson, the Jackson town attorney, stated that the town's authority to provide municipal services took precedence over the county's regulatory powers over land use. Deputy County Attorney Hank Phibbs cited two conflicting Wyoming statutes dealing with town and county powers, noting that Wyoming law did not make clear whether the town or the county should prevail.

The EPA Suggests a Compromise

During the summer of 1977, the EPA Region 8 staff became increasingly aware of the depth of the division between Jackson and Teton County. Early efforts by Wilson to persuade the town and the county to discuss the issue jointly had failed; town officials considered Wilson to be biased toward the county's position.

The EPA's role in the dispute changed markedly with the appointment of Alan Merson as regional administrator at the EPA Region 8 office in September of 1977. Former Administrator Green had been following staff advice up to that point, which was to disapprove any South Park proposal and threaten to withdraw government construction funds if Jackson did not agree to expand its existing plant. Merson came from Denver Law School with an extensive background in land-use law. He also was familiar with the growth problems of Aspen, Colorado, and felt that the lessons learned from that ski resort community could be applied to the Jackson dispute. He was reluctant to impose a solution without first exploring other possibilities with town and county officials. Also, Merson sensed that an EPA-imposed solution would only delay compliance, and he realized the political costs to EPA of a noncomplying plant in a scenic, heavily visited area such as Jackson Hole.

On October 11, 1977, Merson and other members of the EPA Region 8 staff met with town and county officials in Jackson to discuss the impasse and possible solutions. Wilson proposed two options. One was to upgrade the existing plant and set a waste-load allocation for Flat Creek (which would limit the number of small satellite treatment plants that could be located in South Park). The other option was a lagoon system at the South Park site, coupled with a tap restriction intended to limit the number or rate of hookups to the system. In his comments on the second option, Merson proposed that the EPA take an active role in enforcing some sort of tap restriction to protect such a restriction from later legal challenge.

Town and county officials had mixed reactions to Merson's South Park compromise. Both parties were enthusiastic about a compromise in principle—"they were looking for a way out"—but both were uncomfortable with the specific terms. County Commission Chairman Ashley agreed that a tap restriction addressed their prime concern about a South Park site, but he and Deputy County Attorney Phibbs worried about the legal and political vulnerability of such restrictions, contending that economic and political pressure eventually might overcome any legally binding tap restriction. Phibbs also cited the lack of state case law or statutes concerning the legality of such a restriction. Town Council members Bruun, McCain, and Mellor said they would accept tap restrictions as a price for the South Park site, but Mayor Gill pointed out that such restrictions probably would make South Park landowners unwilling to donate an interceptor right of way or plant site.

In what was to have a major effect on the dispute, the county commission, later in October, gave its preliminary approval of a major subdivision in the South Park—the Rafter J Project, with 500 potential units on 400 to 500 acres. The project included a package wastewater treatment plant and residential densities that were far in excess of those proposed in the comprehensive plan. It represented the first subdivision of any significant size in South Park. Lacking the regulatory grounds for disapproval, the commission grudgingly gave its approval. This was a major victory for Mayor Gill and the town; approval for the Rafter J Project came shortly before approval of the comprehensive plan and could now be used as evidence of the need for central sewage treatment in South Park.

Regional Administrator Merson preferred the South Park option over upgrading the town's existing site; the Rafter J Project was a major factor in this preference. Package treatment plants had a poor reputation at the EPA, and Merson feared that Flat Creek's water quality would suffer if South Park development was served by package plants and individual septic systems. The Rafter J Project indicated to Merson that South Park would develop rapidly despite the county's comprehensive plan. He was not too troubled by that prospect because, from a land-use perspective, South Park seemed like the best direction for Jackson to expand. He viewed the interceptor as a positive tool for controlling growth. The best way for the county to avoid uncontrolled sprawl was to get control of the thing that would most greatly influence the nature of growth. Although stating that EPA Region 8 would go along with either of the two options outlined earlier by the EPA, Merson made it clear that he would only approve a South Park site.

In a letter to Mayor Gill dated October 25, 1977, Merson stated that EPA Region 8 still preferred the existing site on grounds that it was compatible with the county's proposed comprehensive plan and entailed lower capital costs. He did not rule out a South Park site, however, and noted that EPA Region 8 would approve this option if the town and county could agree on tap restrictions. The county commission interpreted Merson's letter to mean that the county could dictate the terms of the tap restrictions with the EPA's full support.

Mayor Gill and the town council were furious. "If [the sewage plant] is not going to be built in the right place," said Gill, "we're just not going to spend the money." The town council agreed and threatened to spend a minimal amount of town funds to bring the plant up to state and federal standards if EPA did not relent. Gill charged that EPA was allowing the county to use the sewage plant siting issue as its primary means of controlling growth in South Park. For its part, the county commission discounted the idea of a town–county agreement. Chairman Ashley reiterated the commission's concern that future court decisions or political pressures would overturn any restriction, agreement, or contract. Ashley said, "The next move is up to the town."

EPA Initiates Mediation

In fact, EPA Region 8 made the next move in mid-November, pressuring both sides to consider a South Park compromise. Responding to Mayor Gill, Wilson repeated EPA's intention of awaiting an agreement on a South Park option. He noted that if there was no compromise by November 28, EPA would drop consideration of South Park and only make federal funds available for upgrading the existing facility. Wilson added, however, that EPA would finance an outside consulting firm or lawyer to assist with a compromise if the two parties desired. Conversely, Merson let it be known to county officials that their refusal to consider a South Park option could leave EPA no choice but to approve the town's proposal *without* restrictions.

EPA's threats were at least partially successful. The town accepted EPA's offer of a mediator and the county commission reversed itself and agreed to consider a South Park compromise. In return, EPA lifted its November 28 deadline.

The commission declared that any approval of a South Park site would be contingent on several conditions: (1) that all landowners along the interceptor route defer development for 5 years to allow Congress to act on the scenic easement issue; (2) that the town and county agree on the total number of taps along the interceptor line; (3) that all landowners agree to abide by the density restrictions of the proposed comprehensive plan; and (4) that all conditions be legally binding on all concerned parties for the life of the project (that is, until 1995). In the eyes of Commission Chairman Ashley, these conditions represented a major concession on the part of the county; no additional bargaining would be necessary because the town would have no other option but to accept the conditions if it wished to pursue the South Park site.

In fact, the town wrote back that it could not guarantee conditions 1 and 4—the most critical stipulations from the county's perspective. In his letter to EPA and the county, Mayor Gill noted that the compromise agreement probably could not be made legally binding on future town councils and county commissioners. Ultimately, binding the parties would prove to be the major obstacle to reaching an agreement.

Recognizing that serious differences still remained between the two parties, Merson in mid-December hired Andrew Briscoe, a planner and former director of public works for Boulder, Colorado, to mediate the dispute. EPA's procurement request clearly spelled out the constraints under which Briscoe would work. He was to obtain an agreement binding upon EPA, the town, and the county that (1) required county (or joint town–county) approval of new out-of-town taps; (2) specified maximum annual out-of-town taps; (3) required new-tap development to comply with the county comprehensive plan; and (4) was enforceable by EPA or any citizen of Teton County. Briscoe was instructed further by EPA officials that upgrading the existing site was no longer a viable option.

When Briscoe met with town and county officials in late December, he sensed that both sides were aware of the impasse and were willing to negotiate. Briscoe met separately with town and county officials, listened to their arguments, and gauged their responses to different proposals. At their suggestion, he also met with environmentalists, representatives of the building trades, and South Park landowners. Both the town and the county respected Briscoe's neutrality and were reasonably forthright about what they wanted and what they could accept in a compromise.

From these responses and EPA's previous constraints, Briscoe independently developed a framework for a three-way compromise agreement. He cleared this with the EPA office in Denver and presented it to the town council and county commission in a series of private meetings in mid-January 1978. His 12 proposals followed closely and elaborated upon EPA's earlier compromise package. Briscoe proposed that a maximum of 25% of the new plant's capacity be devoted to out-of-town sewage taps; this translated into a limit of 51 such taps per year. He also proposed that Jackson develop a priority system for out-of-town taps, according to their proximity to town and current impacts on water quality. Both the town and the county thus would have controlled new sewage taps in the unincorporated area—in effect, a double veto for new out-of-town taps.

The county commission was not convinced by Briscoe's proposals. Although they accepted the 51-tap-per-year restriction, Commissioner Moore feared that aggressive annexations by Jackson could render such a limit meaningless. More important, Briscoe's proposals provided no further assurances that the agreement would be binding. County officials feared that such a restriction might succumb to a legal challenge from one of two sources: future town councils might argue that they should not be bound by the actions of previous councils, and South Park landowners might claim that Jackson was legally obliged to provide them with sewer service. In a legal memorandum addressed to local officials, acting EPA Regional Counsel Raisch expressed doubt that such a challenge would succeed. Raisch distinguished between government and municipal (i.e., proprietary) functions, arguing that the provision of sewer service is a proprietary function: when a town acts in a proprietary capacity, it is bound by a contract just as a private individual would be. Therefore, the tap limitation agreement would be enforceable against future town councils as a contract. Raisch also noted that case law and statutes in most states, including Wyoming, support the view that municipalities are not obliged to provide sewer service beyond their boundaries. Thus, in Raisch's opinion, individuals residing outside of Jackson's limits had no legal right to connect to the town's sewer system.

County officials were not convinced. Deputy County Attorney Phibbs found the memorandum "specious and disconnected from reality" and doubted that a Wyoming judge would uphold the agreement if the alternative was a substandard, on-site septic system. He noted the lack of Wyoming case law concerning land-use planning to guide a judge in ruling on the validity of the agreement. Other county officials apparently shared Phibbs's feelings.

Jackson Applies Pressure

The town council also had serious problems accepting the proposed compromise. It did not address the problem of obtaining interceptor easements from South Park landowners, who probably would not be willing to donate them under the proposed tap restrictions. In a letter to Merson in late January, Mayor Gill objected to making the agreement binding for 15 years and challenged the 6%/year growth rate on which the tap limit was based. Although Gill did not fault the remainder of Briscoe's proposals, he and the town council fundamentally disagreed with many of them. They began searching for ways to circumvent the negotiation process or to force EPA to act.

Briscoe's presentation of his 12 proposals marked the end of his involvement in the case and the end of the mediation effort as such. The EPA regional office was satisfied with his efforts but apparently preferred to take the lead in further negotiations.

The next major event in the dispute occurred in early February when a number of South Park landowners along the proposed interceptor route wrote to the town council, asking the town to annex their land along the interceptor route and the 40-acre plant site. In return, they agreed to donate the necessary easements and site. Mayor Gill described their proposal in a letter to Merson, saying that he and the town council were enthusiastic about the plan and noting that the letters were unsolicited. The legality of such an annexation was questionable, but many observers thought Wyoming law would have sustained the action. In any case, Raisch countered by stating that EPA funds would be withdrawn if Jackson proceeded with annexation plans.

Nevertheless, the annexation threat had the desired effect. At a joint meeting of the town council and county commission later in February, the commission indicated for the first time its willingness to sign a three-way agreement based on Briscoe's proposals. Chairman Ashley emphasized that South Park landowners would get no quid pro quo in return for supplying easements. He also asked for clarification of several questions including whether the Rafter J Project would be exempt from tap restrictions. At this point, the county was sensitive to pressure. The construction industry portrayed the commission as the cause of all of Jackson's economic and fiscal woes, and the political mood of the county was shifting away from support of restrictions on growth and development.

Early in March 1978, Raisch sent town and county officials a draft agreement based on Briscoe's proposals. After reviewing Raisch's draft, the town council proposed two revisions: that the three landowners along the interceptor route would get sewage taps for all development allowable under the comprehensive plan and that the 6% growth rate on which EPA based total plant capacity by increased substantially. The town made further demands at a joint town–county meeting on March 14 to discuss Raisch's draft. Town Attorney Larson proposed revisions that would weaken the agreement greatly. These included omission of any limits on the number of annual out-of-town taps and the percentage of total plant capacity devoted to county taps. Mayor Gill also announced plans to visit EPA officials in Washington to resolve the impasse.

Gill and Larson met with Merson, John Rhett (deputy assistant administrator for Water Program Operation, EPA, Washington), and Senator Clifford Hansen (R.-Wyo.) in Hansen's Washington office on March 16, 1978. Gill and other town officials thought that EPA regional officials were making unreasonable demands on Jackson and hoped to persuade the next administrative level at EPA of this. Senator Hansen had been under pressure from town officials for several months to help Jackson resolve the impasse. After discussion, Rhett and Hansen concluded that the draft agreement usurped local responsiblity for land-use regulation and growth management. In response, Merson agreed to reconsider the terms of the agreement, in particular, those that restricted the number or rate of out-of-town taps. Subsequently, EPA Region 8 dropped its demand that the agreement specify an annual tap limit. (Merson's view is that EPA dropped the tap restriction in response to the intransigence of the county and the need for a timely settlement.)

Final Negotiations

After the March 16 meeting, local officials were not sure of EPA's position on the plant siting issue and of the status of Raisch's draft agreement. Mayor Gill reported that both Rhett and Merson appeared anxious to settle the dispute, but the meeting had not produced a decision. At the meeting, Merson reportedly had expressed sympathy for the town's position, noting that if Teton County would not cooperate in the negotiation, he would approve the town's preferred option with no restrictions. In subsequent weeks, Merson reportedly described EPA's position differently to town and county officials, telling each what they wanted to hear (i.e., telling county officials that he would withdraw EPA funding if Jackson proved to be inflexible and telling city officials he would give them an unconditional grant if the county refused to cooperate.) Merson was sympathetic to the positions of both sides and apparently was anxious to avoid alienating either one.

Merson and others at EPA Region 8 were under pressure to reach an agreement. The formal dispute was more than 2 years old, and Jackson's plant had been out of compliance for more than seven years. The EPA in Washington and, indirectly, members of Congress were urging Merson to solve the dispute. He decided to send Acting Regional Counsel Raisch to Jackson to negotiate an agreement.

There were several reasons why this approach seemed promising. Raisch had written the draft agreement, EPA's most recent attempt at compromise, and he was generally liked and trusted by both town and county officials. Also, Raisch was about to leave EPA to start a private law practice, and local officials did not want to have to start over with a new EPA contact.

On April 12, 1978, Raisch met privately with town and county officials in Jackson. He noted that Merson had described EPA's position differently to each side, emphasizing that EPA Region 8 could still decide to fund either the South Park option or the existing site unconditionally. Raisch suggested that both parties would risk less if they could agree upon a compromise and that a compromise was needed very soon.

Raisch had brought a copy of EPA's draft agreement to the meeting, and this became the starting point for negotiations. Jackson officials pressed for a larger plant capacity, arguing that EPA's estimate of the growth rate was conservative. They also argued that rapid utilization of plant capacity was not a problem. Under the full-cost pricing program that the town envisioned, sewer fees for hookups would cover operating and maintenance, and capital costs for the existing and new sewage plants. Unconvinced, Raisch maintained the 6% growth estimate.

Regarding the crucial issue of limiting annual taps, Raisch said there was no specific limitation that he preferred and noted that the county would have 2 years (during project construction) to adopt such a plan. Deputy County Attorney Phibbs proposed a restriction of 51 taps per year on an interim basis until the county could adopt a plan for growth management. Town Attorney Larson objected to specification of any numerical restriction on taps in the agreement.

Raisch proposed two alternatives: the 51-tap-per-year restriction form Briscoe's proposal or letting the county set a tap limit in its comprehensive plan. Jackson had no intention of accepting the first option, and the second option made the members of the county commission nervous. Although incorporating a tap limitation in the comprehensive plan solved the problem of subsequent legal challenges, it also would expose the commission and, indirectly, the comprehensive plan to considerable political pressure. If nothing else, Briscoe's proposal was obligingly inflexible. It would not allow future county commissions the opportunity to increase the allowable rate of sewage taps.

The commission felt vulnerable to political pressure. Commissioners Ashley and Moore had helped to forge a strong comprehensive plan that would limit development density substantially in South Park. They felt the public mood shifting away from controlled growth, however, a perception that would prove accurate in light of the November 1978 election results. They feared that a change in the majority on the commission could render EPA's proposed plan meaningless by simply amending the comprehensive plan. In the end, Ashley reluctantly agreed to incorporate the tap limitations in the plan but only on the condition that this alternative was supported by a two-to-one majority on the commission. The parties adjourned the April 12 meeting with an oral agreement and agreed to reconvene the next morning to ratify it formally.

At the April 13 meeting, Raisch reviewed the draft agreement from the previous day. With a minimum of discussion, Mayor Gill, County Commissioner Max May, and Raisch all signed the agreement. It specified plant size and design capacity based on an estimated yearly growth rate of 6% until 1995. It empowered (and required) the county to limit annual out-of-town taps as part of the comprehensive plan. It also required the town to develop a policy for distribution of out-of-town taps based on a specified set of customer classes as well as a policy for sewage hookup fees based on a full-cost pricing system. Both of the town's policies were subject to EPA review and approval. The agreement also specified sanctions that were available to EPA to enforce the terms of the agreement. These included withholding construction grants, limiting future EPA grants, and petitioning the courts to force compliance or recover EPA funds that had been expended on the project.

The reaction of the three parties varied, but all were glad that the dispute was settled. EPA and the town officials were pleased with the outcome. County reactions were mixed: Chairman Ashley thought the county got a reasonable deal: "All three parties had to give up something—the result was truly a compromise." Commissioner Moore and Deputy County Attorney Phibbs were bitter and felt that EPA had abandoned the county in the end. Both felt that local pressures against development and the EPA's change in position left the county with no option but to go along with the agreement. "Our backs were against the wall." By accepting the agreement (which authorized the county to restrict taps), both sides were effectively deferring the issue until the November 1978 elections at which time each side hoped to command a majority on the county commission.

The Aftermath

After the April 1978 agreement was reached, the Jackson town council hired C. E. Maguire, Inc. to update its 1975 facilities plan. Maguire considered two sites in the Lower South Park area (approximately the same as the one chosen by the council in 1977) and two treatment methods—a mechanical oxidation plant and a nondischarging lagoon/rapid infiltration system.

In September, Maguire presented an update for its draft facilities plan. The town council selected an aerated lagoon/rapid infiltration system at the Lower Bench site in South Park. The firm completed the update and submitted it to EPA Region 8 and the Wyoming Department of Environmental Quality (DEQ) in October. In February 1979, Merson released a final EIS on Jackson's wastewater treatment system. In its report, EPA Region 8 formally approved the alternative selected and described mitigation measures to be followed during construction. In mid-1979, the DEQ issued a new national pollution discharge elimination system permit to Jackson for construction of the plant.

As town officials had hoped, landowners donated the interceptor easement and the plant site. Construction of the interceptor line began in the fall of 1979, and Jackson accepted bids on the treatment plant in the spring of 1980. The full system was completed and started operating in December 1980. There has been moderate construction activity in South Park since the April 1978 agreement was signed, including the Rafter J subdivision. The town council did not issue any new out-of-town taps until completion of the new plant. Subsequently, 100 new taps have been issued to the Rafter J development.

A significant shift in county growth policy took place in 1978 after the November elections. Commission Chairman Ashley did not run for reelection. His seat was won by Jerry Tracey, head of the Building Trades Association and a vigorous opponent of the comprehensive plan and county efforts toward growth control. The election highlighted two different philosophies toward growth in the county. Commissioners Tracey and May constituted a progrowth majority.

As the earlier county commission feared, the new commission was slow to implement the terms of the agreement. The commission did, in September 1980, establish an annual limit of 150 out-of-town sewer taps in cooperation with the city. The 150-tap level was considerably greater than the 51-tap-per-year limit discussed during mediation efforts. This action was taken largely to avoid jeopardizing the remaining EPA construction grants. The city had spent nearly 80% of the construction grant funds, and the three-party agreement had set the

80% expenditure level as the deadline for implementation of the agreement's terms. Also in September, the town council adopted a "full-cost pricing" policy that established a one-time hook-up fee of $1,200 for new users of the system. Observers do not expect the annual limit on out-of-town taps to have any effect on the pace of growth in South Park.

General Study Questions

1. What were the interests of each of the three major parties at the bargaining table: Jackson, Teton County, and EPA? What constraints did each operate under? What was the source of bargaining power for each? Were there other parties with an interest in this dispute and some influence over the outcome who were not at the bargaining table?

2. As between the Jackson town council and the Teton county commissioners, should the dispute be characterized as a zero-sum or nonzero-sum dispute? What kinds of compensation or side payments might have been introduced to facilitate compromise?

3. At the outset, the EPA appeared to be in a position to impose a solution on the contesting parties. It carried the stick of finding Jackson's present treatment system in noncompliance and offered the carrot of a construction grant for a replacement system; yet ultimately, it retreated from its opposition to the South Park location. Why was EPA unable to exert its power? What factors account for the shift in its position?

4. As in the preceding cases, the parties disagreed on some basic facts. In this case, however, the technical dispute was not over the environmental impact of the proposed activity; rather, it involved the estimation of the cost of a new treatment system. Why did Jackson's estimate differ not only from EPA's but also from that of its own consultant. To the extent that Jackson favored South Park for cost reasons, what leverage did EPA, Teton County, and others have over them? How might the structure of the EPA grant program for construction of new facilities discourage recipients from making the least costly choice?

5. Early in the dispute, the state's Department of Environmental Quality (DEQ) threatened to ban any new hookups to Jackson's existing plant on the ground that the town was not acting quickly enough to cure its noncompliance. Had the threat been carried out, what would have been the effect on the negotiations? What might have kept the DEQ from carrying out its threat?

6. After protracted negotiations, the county commission finally agreed to abandon its total opposition to a South Park site, so long as certain growth

control conditions proposed by the EPA were implemented. This was a major concession on the county's part. Nevertheless, the town refused to accept the two most important conditions, and the county had to back down still further. In many negotiations, concessions are traded back and forth. What accounts for the fact that for the county there appears to have been much more give than take?

7. When mediation was initiated, EPA sent quite different messages to the town and to the county. The town was told that unless it could reach an accord with the county, the construction grant for a new treatment plant would be withdrawn. By contrast, county officials were told that unless they softened their opposition to a South Park site, EPA would approve the town's plan without any growth restrictions. What are the pros and cons of such a tactic?

8. Still later in the negotiations, the EPA administrator himself delivered two quite different messages.

> In subsequent weeks, Merson reportedly described EPA's position differently to town and county officials, telling each what they wanted to hear [i.e., telling county officials that he would withdraw EPA funds if Jackson were inflexible and telling city officials he would make them an unconditional grant if the county refused to cooperate].

9. Is it correct to characterize the efforts of Briscoe as mediation? Did Merson err by prescribing an agenda for Briscoe to pursue? Would the outcome have been any different if the mediator had been a truly neutral third party attempting to broker an agreement between the town, the county, and EPA?

10. The county balked at signing EPA's first compromise agreement on the grounds that the town's commitment to a tap restriction was not legally binding. Did the town's inability to commit itself weaken or strengthen its bargaining position?

11. In the course of refining the tap restriction proposal, EPA suggested that, instead of giving both the county and the town vetoes over new taps, power should rest solely with the county. Although this modification would have given the county greater control over growth within its boundaries, the commissioners viewed the unilateral veto as a mixed blessing at best. Why?

12. Ordinarily if one party to a dispute has difficulty binding himself to a proposed agreement, it is the other side who is nervous about coming to terms. By contrast, the doubts about the county's ability to freeze the number of taps for the life of the facility was of greater concern to itself than anyone else. Indeed, Raisch issued a legal memorandum arguing that a contractual limitation would be legally binding. Similarly, the town council did not seem worried about the county's ability to honor this aspect of its commitment. What factors account for these peculiar dynamics?

13. Some state courts render advisory opinions on unresolved questions of law, and in other jurisdictions where this option is not available, the same end may be accomplished by a declaratory judgment action. Given the uncertainty about whether tap limitations could be made irrevocable for the life of the project, why did the parties fail to seek judicial clarification?

14. Would the participation of another entity, perhaps a citizens' environmental group, in the settlement process have affected this issue of binding the parties?

15. Thomas Schelling notes in his book, *Strategy of Conflict* (1960, pp. 21–52) that one way of making a commitment binding on yourself is to subject yourself to a large penalty if you default. Were there any ways in which the county could have locked itself into the proposed tap restrictions?

Hypothetical Problem

Fairly late in the negotiation, owners of the South Park land that would be used for the treatment plant and the interceptor line to it proposed that their property be annexed by the town of Jackson. Such annexations are not unusual in the West. Town-and-city boundaries expand as it becomes desirable to provide municipal services to outlying areas. In some states, annexation can take place even over the objection of the owners and residents of the appropriated territory. Here, where the area was sparsely settled and owned by a handful of people who actively desired annexation, there apparently was nothing to stop the town of Jackson if it had been determined to proceed with this plan.

What advantages would the town have realized through this annexation? On what basis could the EPA condition its constuction grant on the town's promise not to annex? Why did the town forgo this alternative?

Now, consider what might have happened if this possibility had arisen earlier. Specifically, how would the issues have been different if, at the outset, Jackson had simply proceeded with annexation? For the purpose of this problem, assume that they could have done so in spite of any objection raised by the county.

The Compliance Issue

Binding Parties to Negotiated Agreements

In arguing that the town council was legally capable of restricting future sewage taps through a contract with EPA and the county, Acting EPA Regional

Counsel Raisch drew an important distinction between proprietary acts of a municipality and governmental functions. The general rule in the United States is that municipalities cannot contract away their policymaking powers absent specific statutory authorization. That is, they cannot agree to exercise their policymaking powers in the future in a particular way in return for a legal consideration. (A legal consideration may be the payment of money or something else of value, the forbearance of a legal claim, or a promise offered in return.) But to say that municipalities cannot contract away their policymaking powers is not to say that they cannot enter into contracts. When municipalities act not as governmental entities, but rather, as mere corporations, they may bind themselves through contract at will. For example, there is little question that the Jackson town council could enter into a contract for construction of a new sewage plant and that the contract would be legally binding upon the town. Such an action is considered proprietary because the town is acting for the *private* advantage of the inhabitants of the city and town in much the same way that an individual landowner acts for private advantage when contracting for the construction of a septic system on his or her land. To deny municipalities the power to bind themselves for proprietary acts would work a severe hardship and would likely seriously impair the day-to-day functioning of government.

There are a number of rationales that courts often cite in striking down efforts of governmental bodies to contract away their policymaking powers. First, such deals often give the appearance that governmenal policy decisions are for sale. Second, most policymaking authority derives from the police power, the inherent power of government to act to promote the public health, welfare, safety, and morals. To the extent that regulatory actions are taken in response to inducements offered by the benefiting party, the municipality is not acting in the public interest—at least in the eyes of the courts. And third, the prevailing view is that governments should not bind themselves in ways that prevent them from exercising their policymaking authority in the future in response to changing conditions. For example, an agreement never to rezone land to permit multi-family housing would be viewed with disfavor by a reviewing court because it may needlessly restrict the municipality from acting to promote the public good in the future when housing is in short supply and in great demand.

Contracts to exercise policymaking powers are not illegal *per se*, they are just unenforceable against the municipality that enters into them. As a practical matter, this creates negotiating problems of the type encountered in the Jackson case; the county rejected the first version of the EPA agreement because it was not certain that the agreement would be binding on future town councils. And although the county had no greater legal power to bind itself, shifting to it the responsibility for setting tap restrictions solved the negotiating problem because the town did not really care if the tap restrictions were enforced at all.

Because the law distinguishes between contracts governing proprietary activities and those that limit governmental functions, it is important to be able to distinguish the former from the latter. Unfortunately, such distinctions are often difficult to make in practice as the Jackson case attests.

Under what circumstances should the courts take a more tolerant view of contracts that limit the future exercise of policymaking powers? If the Jackson town council had contractually agreed to tap restrictions, should this agreement be binding on future town councils? Would the policies that underlie the rule against contracting away nonproprietary functions be served by not enforcing such a contract? For a more thorough discussion of the legal problems associated with binding communities to agreements see David Kretzmer's *Binding Communities to Compensation Agreements for Energy Facilities*, (Cambridge: M.I.T. Laboratory for Architecture and Planning, 1979).

Municipalities are not the only institutions that may have difficulty binding themselves to agreements negotiated in environmental disputes. Large organizations with diffuse memberships also may run into difficulties. Although an organization such as the Sierra Club can legally bind itself to abide by a negotiated agreement, it cannot also bind its individual members. Consequently, although the leadership of a large group may agree as part of a consensual agreement not to challenge a regulatory decision in court, the agreement is not binding on individual members, who legally may resign their membership, form a new organization, and sue to enjoin the regulatory decision. In fact, at least one major environmental organization formed in this fashion, splitting off from its parent organization because it believed that the parent was not pursuing a particular case vigorously enough.

The difficulties in binding the membership of diffuse organizations create problems for both their leaders and for people who seek to negotiate with them. As Thomas Schelling would predict, the inability to bind sometimes weakens the hands of the leaders who cannot give assurances that the deal struck at the negotiating table will be supported by the rank and file. On the other hand, savvy leaders often turn this weakness to their advantage by explaining to their negotiating counterparts that they cannot accept any agreement that would cause a coalition within the organization to bolt the group. Thus, the inability to bind everyone in an organization may actually sometimes strengthen the hand of the group's leader.

Problems of binding parties do not arise in labor/management negotiations because the law that governs labor/management relations provides for the creation of a formal bargaining unit that represents workers. Representatives of the bargaining unit are elected and negotiate on behalf of the unit's members. Once agreement has been reached, it must be ratified by a majority vote of the rank and file, and upon such a vote, it becomes binding on all parties. In his book *Resolving Development Disputes through Negotiations* (New York: Plenum, 1984), Timothy Sullivan suggests that a popular referendum may provide a means for ratifying agreements in environmental disputes and binding the parties. Sullivan envisions a process in which the final agreement is submitted for ratification by the electorate. If the negotiations fail to produce an agreement, the final offer that has been made to the community would be submitted to the electorate for acceptance or rejection. Although designed specifically for site-specific disputes (e.g., a dispute over the location of a new oil refinery in a town that has reservations about hosting it), this process could have broader application. How might the prospect of popular ratification of a negotiated agreement affect bargaining strength in cases where municipalities are parties to negotiations? Would the requirement that the last offer to the community be submitted to the electorate strengthen or weaken the hand of those negotiating on behalf of the electorate?

A Conceptual Look at Compliance Problems

Not every negotiation gets bogged down in questions of compliance. In fact, in some cases noncompliance is close to unthinkable. Why is compliance an issue in some cases and not in others? To answer this question we must take a closer look at exactly what is being bargained over by the parties.

Mutual Simultaneous Performance Concurrent with Agreement. Many simple negotiations culminate in mutual simultaneous performances. For example, although there is lots of haggling back and forth at a flea market, once the buyer and seller have agreed upon a price, the deal is closed swiftly. The buyer discharges his obligations by handing over the money and the seller his by tendering the goods. Typically, the buyer takes the goods "as is" but subject to the time honored principle of caveat emptor. Whenever all of the responsibilities of the respective parties are discharged immediately upon reaching agreement, compliance is not an issue. Because performance is rendered at the same time the deal is closed, failure to perform is simply equivalent to failure to reach agreement.

An Exchange of Promises for Mutual Simultaneous Performance. More complicated is the situation in which the parties agree to perform their obligations sometime in the future. For example, even though the buyer and seller of a used car may agree on a price at the time of their first meeting, rarely is the deal closed at this time. Usually, the need to obtain registration, insurance, and financing necessitate performance sometime in the future. Whenever performance and agreement are separated in time, a potential compliance problem exists. The buyer of the used car will want assurances that the seller will not sell to someone else in the interim. Similarly, the seller will want assurances that the buyer will actually come up with the cash on the closing date. A deposit typically assuages the seller's concerns. Buyers must usually be satisfied with the seller's good faith desire to actually be rid of his car. If he is defrauded by the seller, the buyer has contractual remedies at common law for recovery of his deposit and other consequential damages that he may establish. If the parties are careful in how they structure their agreement, they can also provide for such contingencies as accidental destruction of the car prior to closure of the deal.

Exchange of Promise for Performance. Frequently, situations arise in which simultaneous performance is not possible. In such cases, one party exchanges performance on his part in return for a promise of future performance by the other. The Jackson case provides a good example. Stated simply, the county offered its immediate acquiescence in the construction of the South Park plant in return for a promise that sewer taps would be limited in the future. This agreement made the county commissioners nervous because they realized that once the plant was built in South Park, they would have little leverage left to ensure compliance with the tap limitation. They were reluctant to perform (i.e., acquiesce) as long as there was doubt that the promise they received in return (i.e., the tap limitation) was of dubious reliability. Thus, EPA had to work hard to convince the county that the promise was actually enforceable. Just about anytime a promise is exchanged for performance, the promisor is faced with the task of convincing the party to whom the promise is given that the promise will actually be kept. We discuss below a variety of techniques for accomplishing this task.

Continuing versus One-Time Negotiations. Compliance problems arise less often when the parties have a continuing negotiating relationship than when they do not. Intentional noncompliance is seldom a problem in labor negotiations, for example, because the members of each side know that they will have to face the other over the same set of issues when the next collective bargaining agreement is negotiated. Similarly, compliance was not an issue in the Brown Paper case because of the company's continuing relationship with EPA. Occasionally, it is possible for the parties to create a continuing relationship where none previously existed by breaking down the issues to be negotiated into a sequence of smaller, individual negotiations that occur over time as the issues unfold. This is a risky strategy because sequencing the negotiations may foreclose the opportunity to strike deals over issues that overlap the sequenced negotiations.

Noncompliance, A Causal Taxonomy

Throughout this book we have argued that parties negotiate when they have an incentive to do so. But negotiation does not end when an agreement is reached. It often continues well into the implementation phase, and sophisticated negotiators spend as much time worrying about the incentives to comply as they do about the incentives to negotiate. In this section we distinguish three types of noncompliance with negotiated agreements: (1) intentional noncompliance or deliberate repudiation of an agreement by a party who is capable of performing; (2) unavoidable noncompliance that occurs when a party wants to perform but finds that doing so is impossible; and (3) unintentional noncompliance that occurs as a result of a mistake or ambiguity in the agreement. A negotiator who can differentiate the various causes of noncompliance is better situated to choose which mechanisms are most appropriate for the particular problem.

Intentional Noncompliance. Why would a party to a voluntary agreement later repudiate it? Just as a person may initially calculate costs and benefits so as to conclude that settlement is better than nonagreement, so it is possible to decide that the costs of breaching the agreement are exceeded by the benefits of doing so. It is not necessary that a deal be bad for a party to be tempted to break it—only that it is less attractive than the possibility of another deal. For example, someone who has just agreed to sell his car for $5,000 may be sorely tempted to back out of the agreement if a subsequent buyer should come along and offer him $6,000.

The costs of breaking a deal once it is made, however, may be far more substantial than were the costs of not entering into it in the first place. In addition to the possibility of being assessed damages and the expense of litigation, someone contemplating breach must consider the cost of damaged relations with the other parties and a tarnished reputation. For some people a deliberate breach would also extract a considerable cost in self-esteem.

Rarely will people know in advance the precise consequences of noncompliance; even after the fact, they may not be able to quantify its costs. Acting with imperfect information and perhaps limited analysis, they may make serious miscalculations. Nevertheless, they are likely to engage in some sort of rough calculus of costs and benefits before making the decision not to comply.

Repudiation may occur at various points in the implementation phase, and for quite different reasons. It is conceivable that a party may wish to disavow an agreement before it has been implemented in any way. Changing circumstances may have tilted the cost-benefit scales. More attractive opportunities may have arisen suddenly.

A party may also be induced to break an agreement after it has started to take effect if he has already received most of the expected benefits and has yet to incur the bulk of the costs. Again, this is what the county feared in the Jackson case, and we term it *selective repudiation.* Devious parties sometimes contemplate selective repudiation from the start.

Still another kind of repudiation occurs when a person deliberately violates an agreement—in whole or in part—but hopes the other signatories will not detect the breach. For example, a developer who has negotiated the terms of approval of a subdivision from a planning board may try to get away with building a cheaper roadbed than is called for by the permit. One need not be absolutely sure of escaping detection in order to be tempted to breach.

Finally, there are cases in which a party deliberately breaches an agreement he actually would like to see honored. This can occur if he believes that the other side has breached in whole or in part. Even if the breach is partial, this kind of retaliation can invite a similar response, and the entire agreement can unravel.

Unavoidable Breaches. Deliberate breaches, whether they are flagrant or secretive, whole or partial, offensive or defensive, should be distinguished from breaches that arise from a party's inability to comply. This sort of noncompliance may be unwelcome to all parties to the agreement.

Inability to comply may arise from a variety of causes. A developer who has agreed to make compensation payments to a community may find himself unable to do so because he has failed to turn a profit in the last few quarters. By contrast, the obstacle to compliance may have been there from the start. A community may be incapable of entering into some types of agreements without first obtaining statutory dispensation from the state legislature, which may be unwilling to grant it.

When one party claims incapacity to comply, of course, others may dispute it. The lines between inconvenience and hardship, and hardship and impossibility are often blurry.

Unintentional Noncompliance. An agreement may be breached, and seriously so, even though no one intends it. Implementation of a negotiated agreement is subject to the same pitfalls to which the management of other human activities are prone. Unintended noncompliance may also arise from poor communication or bad drafting. The unwitting blunder is not tainted with the malice of the double cross, though it may have the same consequences. One side's unintended breach of a provision of the agreement may be misread as a deliberate repudiation by the other, and the whole settlement may be jeopardized.

Enforcement Mechanisms

Having established a simple model that distinguishes the causes of noncompliance, we turn next to different devices that can be used to increase the likelihood that a negotiated agreement will be honored. Some of these devices are intended to be self-enforcing, whereas others require the involvement of third parties or institutions, should a problem of noncompliance later arise. It is important to recognize, of course, that parties may disagree about the meaning of a provision that was intended to be self-enforcing and thus go to court for an authoritative interpretation. By the same token, the fact that the parties have provided for some sort of independent arbitration of future disputes may well encourage them to work out such problems on their own.

As you consider the following catalog of compliance devices, evaluate their advantages and drawbacks. Are some suited to certain types of noncompliance but not to others?

Grievance and Arbitration Provisions. Rare is the agreement that is so clearly written and so prescient that issues of interpretation never arise. Wise negotiators anticipate disagreements over interpretation and provide for their resolution. In the labor–management field, sophisticated grievance procedures are common. In the negotiation of environmental disputes, it is not uncommon for the parties to designate a third party to resolve disputes over interpretation. The Massachusetts Hazardous Waste Facility Siting Act, described in detail in chapter 12, requires that developers of hazardous waste facilities and communities make provisions for the arbitration of disputes that may arise over the implementation of the siting agreement that is negotiated between the developer and the host community. The intent is to avoid litigation of the agreement itself.

Structured Implementation. Even if the parties stand to benefit from the substance of a proposed settlement, it may be essential for them to structure implementation in such a way that each has a continuing interest in having the contract fulfilled. A clever ordering of carrots and sticks may insure compliance, whereas a lack of foresight may encourage breach midway through implementation.

Negotiators who can break a large problem down into smaller components accomplish two things: First, they allow the parties to take measured unilateral steps without exposing themselves to high risk; and second, by creating a string of future negotiations, they create greater incentives to comply today. In effect, by structuring performance so that each action by one side earns a response by the other, the negotiators can create a continuing relationship where none previously existed. In fact, this was the strategy pursued by Raisch in the Jackson case when he tried to assuage the fears of the county that future town councils could repudiate the tap limitation; he tried to create a continuing relationship whereby the town and the county would jointly approve new hookups. Ultimately, this strategy failed for other reasons—specifically, that the county commissioners did not trust their successors to abide by the plan. Otherwise it probably would have worked.

Contingent Agreements Coupled with an Escrow Account. Unfortunately, environmental disputes are frequently marked by mutual distrust of the parties.

Communities are often skeptical about claims by developers that their projects will not have large impacts on the environment. In many cases, it is impossible for the developer to convince the doubters that, in fact, he deserves their trust. To solve this problem, the parties might negotiate a contingent agreement in which the developer spells out the corrective measures he will take in the event that specified contingencies occur. For example, a common type of environmental dispute is one in which the parties disagree over the likely future consequences of a proposed project. Usually opponents believe the consequences will be severe and thus argue for costly control measures. The developer typically disputes the estimates of the severity of the problem and is unwilling to agree to expensive capital expenditures that may ultimately prove unnecessary. The way out of this deadlock is a contingent agreement to install the control measures if conditions warrant it, backed by money set aside in an escrow account to cover the cost of the installation. If the developer truly believes that the control measures will prove unnecessary, this is a relatively cheap concession to make. Because the community is assured that money is available for installation, it does not have to trust the developer to come up with it later.

Monitoring Devices. Closely related to this strategy is the use of a monitoring device. If the parties enter into a contingent agreement, some procedure must be devised for determining when the contingency occurs. The parties may monitor themselves or engage the services of a neutral third party to monitor. Monitoring may be limited to specified contingencies or it may encompass compliance with the entire agreement. In some cases, comprehensive monitoring may be prohibitively expensive. (The McArthur Foundation, which awards no-strings-attached grants, seems to have concluded that it makes more sense to take care in selecting worthy recipients than in monitoring their subsequent work.)

Performance Bonds. Commercial and real estate contracts often contain performance bonds intended to encourage compliance and to provide a remedy in the case of breach. For example, planning boards often require some type of bonding when they approve subdivisions. The fear is that having received approval for the carving up of a large parcel of land, the developer will simply sell off the lots and disappear without constructing the required roads and services, leaving the city responsible for these costs. Planning boards use a number of techniques to avoid this problem, including requiring the developer to post a bond or placing liens on a few lots that cannot be sold until the roads and services have been provided. In each case, the size of the bond and the value of the lots held back is supposed to be large enough to cover the cost of the improvements if the developer should default. These mechanisms can be applied in other contexts as well. In a sense, a performance bond is a means of manipulating a party's costs and benefits so that he or she will continue to have an incentive to comply with the agreement.

Penalty Clauses. Penalty clauses function in the same way as performance bonds except that the money is not set aside before performance. Again, it is necessary to specify precisely the circumstances under which the penalty can be collected so as to avoid litigation over the penalty itself. Although courts generally uphold clauses that provide reasonable damages, they disfavor provisions that award far more than the actual damages suffered.

Consent Decrees. As we have noted many times before, some times disputes are settled after the parties have started litigation but before it has run its full course. In such cases, the sitting judge has the discretion to incorporate the agreement as part of a consent decree. The effect of such an action is to make available to the parties the coercive powers of the court to enforce the agreement; a party violating its terms, may be found in contempt. Judges are sometimes reluctant to embrace the private agreements of the parties in consent decrees, preferring instead that the parties pursue normal contractual remedies to ensure enforcement. We will encounter such a judge in the Foothills case in chapter 9.

A Final Comment on Trust. The problem of obtaining compliance is often viewed in terms of trust. Parties ask themselves whether they can "trust" their negotiating partners to comply. To the extent that trust and honor prevail, the mechanisms noted previously may be unnecessary. (Indeed, in the New York wholesale uncut diamond market, million dollar deals are closed with only a handshake, the parties relying upon trust and honor as their exclusive enforcement mechanisms.) Although trust is nice to have if you can get it, it is also ephemeral. In many cases, especially where the parties are unknown to each other and suspicious of each others' intentions, it may be impossible to come by. Rather than rely upon trust or honor as an inducement to compliance, it is often easier to create more tangible incentives for compliance. As a veteran labor negotiator once remarked, "You can negotiate with the devil if you can make sure he can't get out of the deal."

Cross-Reference: A Look Ahead

The principal case raises some issues that are considered more fully in connection with the Foothills case that is presented in chapter 9. In the Jackson case, for example, there was an attempt to use a mediator. It is revealing to consider why the mediator's role here was not as significant as it was in the Foothills case.

The cases are similar also in that they illustrate problems that can arise in intraorganizational bargaining. There were conflicting points of view within both the town and county governments. For example, the county proponents of growth control had to contend not only with the town but with dissenters within their own ranks. This intraorganizational dimension was particularly apparent with respect to the EPA. The political context in which the agency operated was felt from the start. The first regional administrator, a holdover Republican in a Democratic government, did not want to take a position at odds with his staff. His successor enjoyed more latitude for a time but appears to have been chastened in a meeting with a senator in Washington.

Finally, the Foothills case provides an additional look at this chapter's central issue: the problem of binding the parties. In that case, the parties attempted unsuccessfully to get judicial ratification of their settlement through a consent decree. Nevertheless, the parties still came to terms. That they were not unduly hobbled by the compliance issue provides an interesting contrast to the Jackson case.

8

MEDIATION TECHNIQUES

INTRODUCTION

In the principal cases described so far, the disputing parties undertook negotiation on their own. Increasingly, however, major environmental disputes are being negotiated with the assistance of a mediator, a practice borrowed from labor–management disputes and international conflicts. This chapter considers what the mediator brings to the bargaining table that the parties cannot provide on their own.

Mediators are sometimes called *intervenors, neutrals,* or *third parties,* but all of these terms have possible misleading connotations. A mediator, for example, may intervene in a dispute on his or her own initiative or may be invited to participate by the parties. In theory, mediators may be neutral as to the substantive issues at stake, but in practice this is not always true. In any event, the mediator has some degree of interest in the procedures used to resolve the issues. In some instances, the concerns of the mediator may conflict to a degree with those of the parties. The term *third party,* though common, likewise can be anomalous, at least in disputes where there already are many parties.

Mediators should not be confused with arbitrators and other kinds of neutrals. John McCrory, in his "Environmental Mediation—Another Piece for the Puzzle," 6 Vt. L. Rev. 49, 52-56, 1981, summarizes the distinctions among various dispute resolution processes:

> Mediation is one of several mechanisms available to disputants who wish to use a neutral to assist in achieving settlement. These mechanisms are differentiated by their degree of procedural formality and the role played by the neutral in influencing the final settlement. Mediation is relatively informal. Fact-finding and arbitration are the most prominent of the more structured mechanisms.
>
> Fact-finding involves a hearing before a neutral whose function is to

make a written report containing recommendations for resolution of the issues in dispute. A fact-finding hearing is normally informal, but provides the parties with an opportunity to present evidence and argument in support of their positions. Post-hearing briefs may also be submitted. The recommendations contained in the written fact-finding report are not binding and may be rejected by one or all of the parties.

Arbitration involves a more significant role for the neutral. A hearing is held at which the parties may submit evidence and make oral statements of position. Posthearing briefs may be submitted at the option of the parties or at the request of the arbitrator. The arbitrator renders a written decision which is normally, by statute or agreement between the parties, a binding resolution of the dispute.

Defined in most general terms, mediation is the "intervention between conflicting parties or viewpoints to promote reconciliation, settlement, compromise, or understanding." Like the more formal fact-finding and arbitration procedures, it is a mechanism for facilitating agreement in a negotiation process. It involves the intervention of a person who does not have a stake in the dispute which is the subject of negotiations. The process is voluntary because the mediator does not have the power to impose a settlement on the disputants. . . .

The function of a mediator "is to assist the parties by being creative and innovative in finding areas of agreement and compromise to reach a final resolution of the impasse." The methods employed by individual mediators to achieve this objective will vary. . . .

Mediators have procedural flexibility not available to judges or to decisionmakers who function in a quasi-judicial capacity. They need not be concerned with prohibitions against ex parte communications, with supervising the formation of a record or with other formalities which would prohibit or impair confidential relationships with the parties and would inhibit settlement efforts. A mediator may adopt procedures or methods of operation which meet the needs of each situation, and may alter those procedures if the need arises. This procedural flexibility, which includes the freedom to communicate confidentially with the parties, coupled with the mediator's ability to make timely substantive suggestions for resolution, have been cornerstones for the success of mediation. The absence of a precise or uniform format for mediation caused Professor Lon Fuller to observe: "For of mediation one is tempted to say it is all process and no structure."

In short, mediators perform distinctly different functions from arbitrators and other adjudicators. An arbitrator is essentially a private judge, chosen by parties who cannot settle a dispute themselves to render a decision. If Solomon had sliced the baby in two, that would have been an arbitrated result. A mediator, by contrast, has no authority to impose a resolution on the contending parties; instead he or she attempts to guide them to an outcome which all can accept. He or she may have influence through general reputation, familiarity with negotiation practices, technical knowledge, and control of communication, but ultimately the mediator serves at the pleasure of the negotiating parties. As we have seen, negotiation may take place without mediation, but mediation never occurs without negotiation.

The principal case study in this chapter describes a successful environmental mediation. As you read it, consider what sorts of disputes are particularly appropriate for mediation; also, what qualifications should mediators have. Mediation ethics are considered at length in Chapter 10, but you should already begin to ponder whether a mediator has any obligations beyond leading the parties to agreement: Does it matter, for example, if the mediator personally believes that a settlement that is satisfactory to the parties is against the public interest? You should also begin to think about another issue that is explored at greater length later—institutionalizing mediation. Who pays for the mediation when, as often is true, the parties are in unequal positions economically? Should mediators operate privately or under government sponsorship? Finally, do not forget that although the focus of the principal case is on mediation, it is rich in other negotiation issues, such as multiparty bargaining, joint problem solving, and data negotiation.

CASE STUDY: MEDIATION AND THE BRAYTON POINT COAL CONVERSION

This case study was originally prepared by Douglas Smith. It has been substantially edited.

Introduction

Following the OPEC oil embargo in 1973, American public policy was torn between two apparently contradictory goals: energy independence and environmental protection. If energy independence was to be achieved, America would have to turn away from her heavy reliance on relatively clean imported oil to other—often dirtier—sources of energy.

The most readily available alternative was coal. It was cheaper than oil and abundant within the United States. Yet a massive return to coal had heavy environmental costs: more mines, more air pollution, more water pollution, and more solid waste disposal problems. Continued dependence on OPEC oil entailed great cost and risk as well. Congress passed several bills intended to decrease our reliance on imported oil and to substitute domestically available energy sources. One of these acts was ESECA—the Energy Supply and Environmental Coordination Act of 1974. [PL 93-319, Section 2]

ESECA gave the Federal Energy Administration (now the Department of Energy) the authority to prohibit the use of oil or natural gas in facilities capable of burning coal. The act also required that conversions comply with existing air pollution regulations. This required cooperation between the FEA, the EPA, and state air pollution control agencies.

These agencies held widely differing priorities and goals. The FEA was primarily interested in decreasing consumption of foreign oil by bringing about conversion as rapidly as possible. The EPA and state environmental protection agencies, on the other hand, were concerned about the environmental effects of such conversion and wanted to proceed slowly and carefully.

The response of the affected utilities was mixed. Many individuals were willing to attempt or consider coal conversion because the price of oil had risen sharply and because the oil supply problems had greatly increased. Most were opposed, however, to the idea of spending millions of dollars on new pollution control equipment that would be required to meet existing air pollution regulations that cover the burning of coal (e.g., scrubbers to remove sulfur dioxide and electrostatic precipitators to remove particulates). Thus, a multidimensional conflict was created among several federal and state-level energy and environmental protection agencies and the numerous utilities affected by coal conversion legislation. After ESECA was passed, the FEA began to compile a list of power plants currently burning oil or natural gas but capable of burning coal as a primary energy source. In June 1975, thirty-two power plants (74 separate units) were issued "prohibition orders" requiring conversion to coal; 11 more generating stations (including 18 units) were issued prohibition orders 2 years later. The Brayton Point Generating Station in Somerset, Massachusetts, was among the 1977 recipients of an ESECA prohibition order.

The Brayton Point Station is located in southeastern Massachusetts at the confluence of the Taunton and Lee rivers; it is 15 miles southeast of Providence, Rhode Island, and 1 mile northwest of Fall River, Massachusetts (see Figure 10).

The facility, which is New England's largest fossil-fueled power plant, consists of four separate generating units. with capacities of 250, 250, 650, and 450 megawatts, or 1600 megawatts total. Three of the boilers, units 1, 2, and 3, are capable of burning either oil or coal. Unit 4, the newest of the boilers, is capable of oil combustion only (see Figure 11).

Brayton Point is owned and operated by the New England Power Company, which is a wholly owned subsidiary of the New England Electric System. The New England Electric System supplies electricity to over one million customers in Massachusetts, Rhode Island, and New Hampshire. NEPCo also participates in the integrated dispatching system called the *New England Power Pool*. The utilities in the pool purchase power from the utility that can produce it at the lowest cost. Thus, Brayton Point is called upon to contribute power to the pool

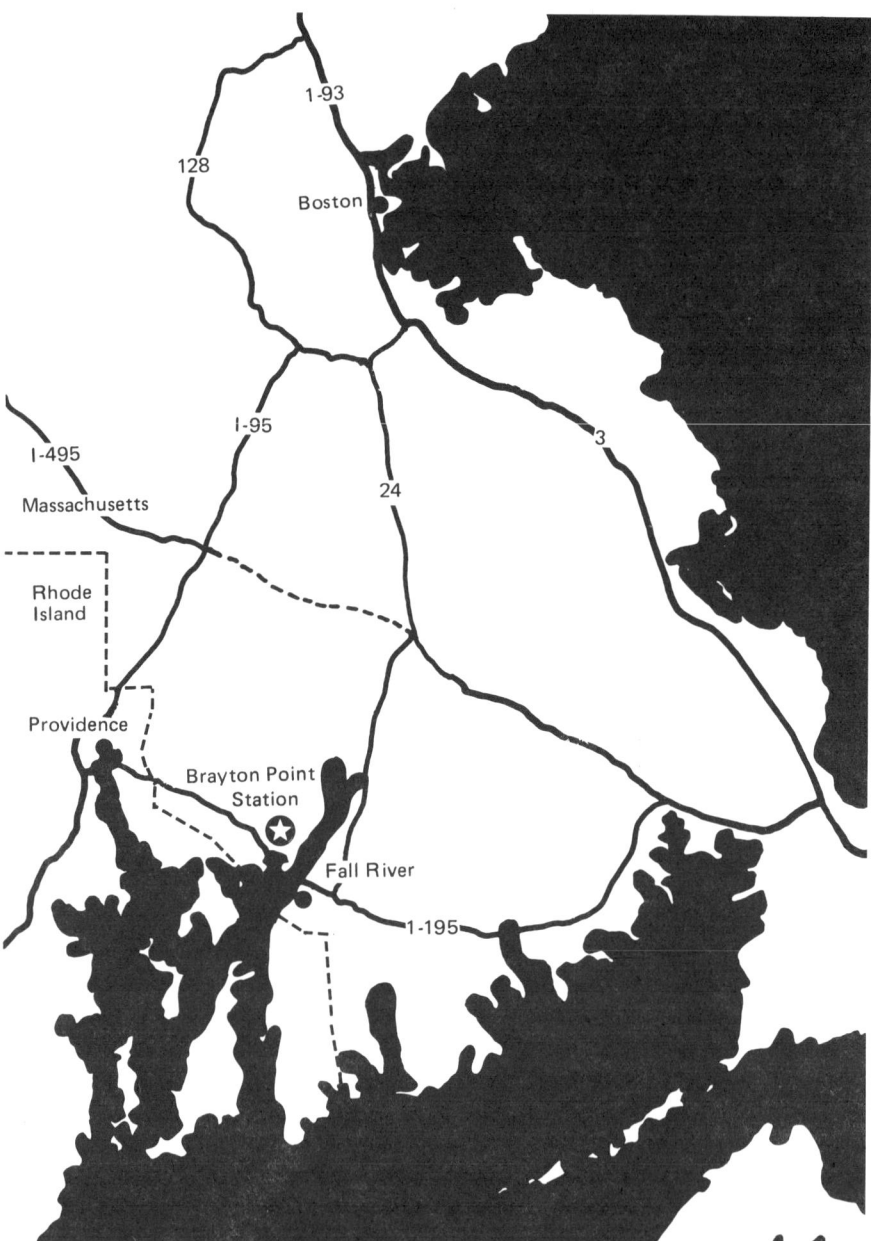

FIGURE 10. Brayton Point region. (From Final Environmental Impact Statement, Brayton Point Generating Stations 1, 2, and 3, U.S. Department of Energy, September 1979, p. 2-3.)

MEDIATION TECHNIQUES

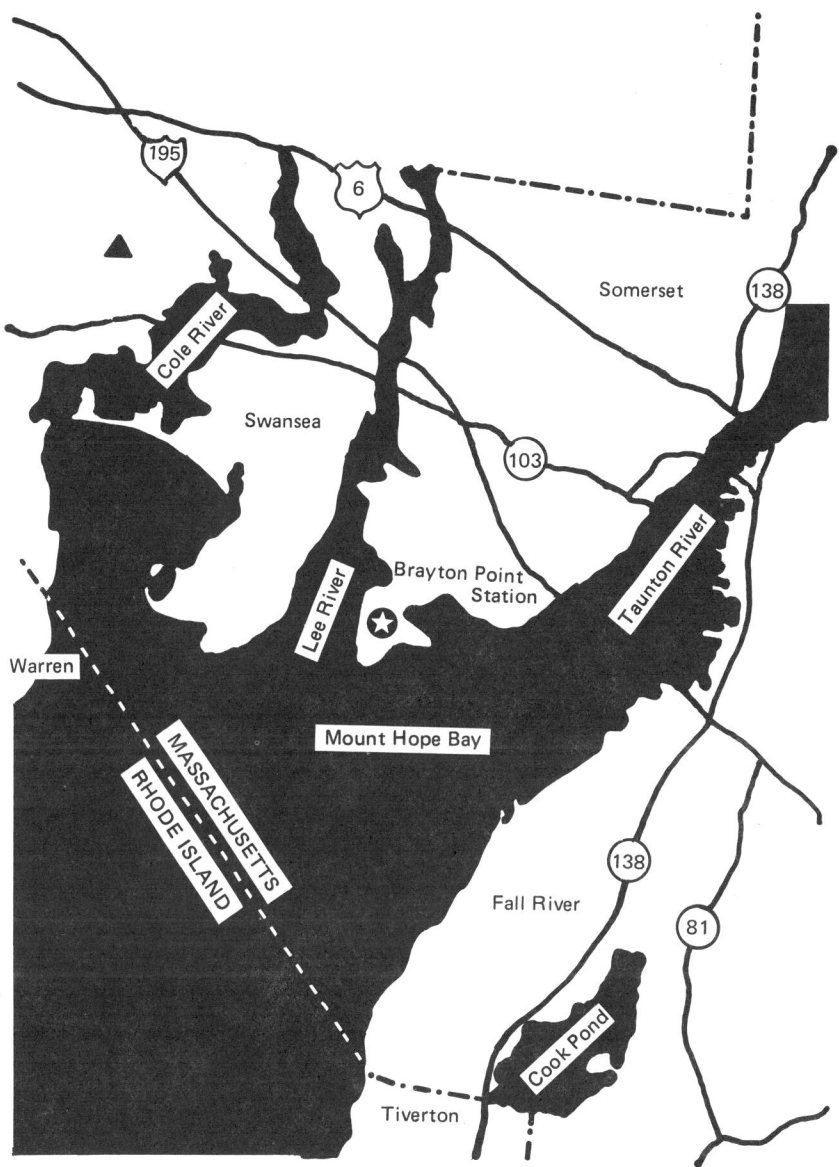

FIGURE 11. Power plant site. (From Final Environmental Impact Statement, Brayton Point Generating Stations 1, 2, and 3, U.S. Department of Energy, September 1979, p. 2-2.)

on the basis of its operating costs relative to those of other power plants in the system. Fuel cost is a large fraction of operating cost, and therefore, it is a critical factor in determining both the output of the facility and the revenues of its owners.

Before conversion, Brayton Point was a very efficient plant; consequently, it was in almost continuous operation. NEPCo management was very concerned that the plant's cost efficiency be maintained after conversion because efficiency determines its output, which, in turn, determines the amount of revenues the plant generates for its owners. NEPCo therefore opposed any added expenses in either capital equipment or fuel costs that would decrease the relative cost efficiency of the plant. Coal was attractive but only so long as it was cheaper than oil. The company did not want to be forced to convert by the FEA if the EPA was intent on imposing stern environmental controls.

The Fuel Combustion Process

In anticipation of fluctuating fuel costs, the first three boilers built at Brayton Point were designed to burn either coal or oil. Both fuels are delivered to the plant by water: oil, by tankers; coal, by seagoing colliers or barges. The oil is stored in five tanks, and the coal piled on a 9-acre storage site. The maximum coal storage capacity is 600 thousand tons, an 88-day supply. The coal is moved by conveyor belt from the coal pile to the top of the boiler house where it is ground by pulverizers, pumped into the boilers, and ignited. The heat generated in the combustion process creates high pressure steam, which is then passed through turbines to generate electricity.

Some solid matter remains after the coal is burned. "Bottom ash" falls into the bins below the boilers. "Fly ash" travels with the exhaust gases, and most of it is collected by the electrostatic precipitators. The precipitators consist of a series of electrically charged plates between which the exhaust gases travel. The particles in the exhaust are negatively charged and are attracted to the positively charged plates. The plates are rapped regularly, causing the accumulated particulate matter to drop into a collection bin. The bottom ash and fly ash are collected and transported to an approved landfill site. Although electrostatic precipitators are quite efficient, they cannot trap all the fly ash in the exhaust. It is inevitable that some of it escapes to the atmosphere as "particulate" pollution.

Gaseous emissions are also produced in the oil and gas combustion process. The pollutant of most serious concern is sulfur dioxide. Both particulates and sulfur dioxide are respiratory irritants. Sulfur dioxide also reacts chemically in the atmosphere to produce sulfates and sulfuric acid. Sulfates add to particulate problems, whereas sulfuric acid falls to the earth as "acid rain" and causes water quality deterioration and plant and property damage.

Just as particulate emissions can be controlled by burning fuel with low-ash content, sulfur emissions can be controlled by using low sulfur fuel. Low sulfur coal and oil, however, are both generally more expensive than high sulfur fuel. The alternative to using cleaner fuel is to remove the sulfur dioxide from the exhaust. The most common technology for removing sulfur dioxide from exhaust gases is flue gas desulfurization—scrubbing—in which the exhaust gases are passed through a reactive liquid or a bed of solid material, such as limestone. Chemical reactions bind the sulfur dioxide to the reactive material, thereby removing it from the exhaust before it is released to the atmosphere.

Although quite effective, flue gas desulfurization equipment is very expensive. For instance, NEPCo estimated that the cost of installing scrubbers on Brayton Point units 1, 2, and 3 would have been approximately $150 million. If the plant was to remain cost-effective, scrubbers would be a prohibitively expensive option—even accounting for the millions of dollars the company would save by burning coal instead of oil.

Air Pollution Control Regulations

The air pollution control regulations affecting Brayton Point were derived either directly or indirectly from the Federal Clean Air Act Amendments of 1970. The amendments require the EPA to establish national ambient air quality standards (NAAQS). The primary standards were intended to protect human health (including the health of those people already suffering from lung and other serious diseases). Secondary standards were set to enhance the public welfare by minimizing effects on property, human activities, and the environment. The pollutants for which NAAQS were established, which were called *criteria pollutants*, included particulate matter, sulfur dioxide, carbon monoxide, photochemical oxidants, hydrocarbons, and nitrogen dioxide. A primary standard for lead was added in 1978. In order to implement pollutant standards, states were required to adopt state implementation plans (SIPs) that specified "the manner in which the NAAQS will be achieved and maintained." In short, the states had latitude to choose the means to federally established ends.

After the 1970 Clean Air Act amendments were enacted but before the 1973 oil embargo, the Environmental Health Division of the Massachusetts Department of Public Health (the forerunner of the state's Department of Environmental Quality Engineering) submitted the Massachusetts SIP, which the EPA approved. The SIP specified an allowable fuel sulfur content of .55 lb of sulfur per million BTUs (.55 lb/Mbtu) and a total suspended particulate emissions limit of .12 lb/Mbtu for point sources in southeastern Massachusetts, including Brayton Point. (Point sources are large stationary emitters of regulated pollutants, most commonly utilities and industries.)

Although implementation of SIPs was to be handled primarily by the states, all variances, suspensions, and revisions of the SIPs were subject to review by the EPA. The agency would not allow any changes that it felt might cause violations of NAAQS. The EPA also would not generally allow any air quality deterioration in areas designated nonattainment following the 1977 Clean Air Act amendments. (Nonattainment areas are air quality control regions that did not meet NAAQS by 1977 or any other air quality control region that has two or more violations of a primary ambient air quality standard in a given year.)

Prior to the oil embargo, NEPCo could easily obtain fuel oil for Brayton Point that conformed to the SIP limits. The 1973 embargo, however, resulted in both a temporary shortage of residual fuel oil and a tremendous increase in its price. Consequently, NEPCo began to search for coal sources in the United States and abroad. In December of 1973, the company made commitments to purchase coal from American, South African, Polish, and Australian suppliers. The company simultaneously applied to the Massachusetts Envrionmental Health Division for a 5-year variance from the SIP to allow the burning of coal at Brayton Point without sulfur or ash content limits. In early January 1974, the first shipments of coal arrived at Brayton Point. Although no state action had been taken, the company anticipated that in a "crisis situation," state permission to burn coal was likely.

The state agency held hearings on the requested variance, and subsequently recommended to EPA that a variance be granted for unit 3 only, for the period from May to December 1974. The proposal contained restrictions on fuel sulfur and ash contents of 2.5% and 15%, respectively. In May, EPA approved the variance for unit 3 with a stricter sulfur content limit of 1.5% by weight. NEPCo started burning its coal later that month.

In April 1974 NEPCo again applied to the Environmental Health Division for a 5-year variance—this time for units 1 and 2. Public hearings were held in June. In July, NEPCo's request was denied because the division believed that primary standards would be violated near the plant if further variances were issued. Moreover, the regulation under which the variance was granted required that no suitable alternative to the variance be available. However, by the spring of 1974, fuel oil was again available, though at a significantly higher cost.

The Legislative Response to the Oil Embargo

In June 1974, Congress responded to the OPEC oil embargo of the previous winter by adopting the Energy Supply and Environmental Coordination Act of 1974 (ESECA). The measure was intended to cut the nation's dependence on imported oil. The law gave the Federal Energy Administration (FEA, now the Department of Energy) the authority to "prohibit any power plant [capable of conversion] . . . from burning natural gas or petroleum products as its primary energy source." ESECA specified a rather complex procedure for pursuing this mandate. FEA first cataloged the major oil consumers capable of burning coal. Brayton Point was by far the largest of the five New England power plants on FEA's list of potential converters.

Mandated conversion under ESECA required three steps: a "notice of intent," a "prohibition order," and a "notice of effectiveness." The notice of intent warned the utility that mandated conversion was being considered seriously, and it triggered public hearings. If, after receiving testimony, the FEA believed the plant met four criteria for ESECA conversion, it would then issue the prohibition order that barred the use of oil or gas in the facility at the date set by the notice of effectiveness. The four criteria for ESECA conversion were:

1. On June 22, 1974, each power plant had the capability and necessary plant equipment to burn coal.
2. The burning of coal by each power plant, in lieu of petroleum products or natural gas, is practical and consistent with the purposes of ESECA.
3. Coal and coal transportation facilities would be available during the period the prohibition orders were to be in effect.
4. The prohibition of the power plants from burning natural gas or petroleum products as their primary energy source would not impair the reliability of service in the area served by such power plants.

A prohibition order is not enforceable until the FEA issues a notice of effectiveness. Before a notice of effectiveness can be issued, the FEA must take three additional steps. First, it must prepare an environmental impact statement on the proposed conversion. Second, it must obtain the approval of the EPA administrator, indicating that by a given date the proposed conversion can take place in compliance with all applicable environmental regulations. Finally, it must get the approval of the governor of the state in which the conversion is to take place. Only when these steps are completed, can FEA issue the final order, prohibiting the burning of natural gas or petroleum products at the power plant.

ESECA also contained provisions that allowed point sources to apply directly to EPA for suspension of the state implementation plan (SIP). The law allowed suspensions to be granted until June 1975 if the source "has converted to coal and if the administrator finds that the source will be able to comply during the period of suspension with all primary standard conditions." ESECA essentially permitted EPA to exclude consideration of secondary standards in the short term.

In August 1974, only 2 months after ESECA was passed and one month after Massachusetts had denied NEPCo's latest variance request, the company applied to the EPA for an ESECA mandated temporary suspension of all fuel and emission limitations in the Massachusetts SIP. Four months later, in December 1974, the EPA approved a 6-month suspension of the SIP limitations for Brayton Point units 1, 2, and 3. This suspension was conditional, however, on

NEPCo's using only coal with a sulfur content of 1.10 lb/Mbtu (equivalent to 1.4% sulfur by weight) and ash content of .12 lb/Mbtu (approximately 15% ash by weight). The suspension also required establishment of a network of ambient air quality monitoring stations in the Fall River area, efficienct testing of the pollution control equipment, and emissions monitoring. The EPA regarded the suspension as an opportunity to gain data on the impacts of burning coal without additional pollution control equipment. This monitoring system was critical in the later debate over particulate standard violations.

The Brayton Point suspension ended in June 1975. At that time 266,000 tons of coal, worth about $9 million, remained piled in the Brayton Point storage area. The company explained their overstock as insurance against the possibility of future oil supply problems. The company also hoped that continued coal use would be allowed, but this particular supply had a very high ash content. The permissible conditions for use of the coal were the focus of much of the discussion about coal conversion at Brayton Point over the next two years.

The Massachusetts legislature also responded to the OPEC oil embargo by passing chapter 494 of the Massachusetts General Laws in 1974. This law required that state regulations, including the SIP, should be revised periodically by the Department of Environmental Quality Engineering (the DEQE was the successor to the Environmental Health Division) so that the ambient air quality standards were "achieved in an economically efficient manner." Given that sulfur levels in Massachusetts were, on the whole, significantly below the ambient sulfur dioxide standards, the practical mandate of the law was to allow the use of cheaper high-sulfur fuel.

In response to chapter 494, the DEQE proposed a 2-year (May 1976 to May 1978) revision of the SIP allowing the use of 2.2% (by weight) sulfur fuel oil, rather than the previously allowed 1% fuel oil, at major point sources throughout the state. The EPA approved the revision for some of the sources, but delayed action on several "more difficult" cases—including Brayton Point.

The Particulate and Sulfur Dioxide (SO_2) Problems

The Brayton Point case was difficult for a number of reasons. The primary problems involved possible violations of sulfur dioxide and total suspended particulates (TSP) standards if the SIP revision was approved.

An important condition of the 1975 ESECA-mandated SIP variance had been the requirement that NEPCo establish a network of air quality monitors in the Fall River region. The company contracted with Environmental Research and Technology, Inc. (ERT) to operate six monitors, and the state DEQE supplemented these with two monitors of its own. The monitors remained in operation following the conversion back to oil. Together, the machines recorded a total of 17 readings in excess of primary or secondary particulate standards during

1975 and 1976. An area is considered in violation of an ambient standard if there are two "exceedences" in one calendar year.

The EPA officially notified the DEQE of the violations in March 1977, announcing, as a result, that the Southeastern Massachusetts Air Pollution Control Region was being designated as a nonattainment region for particulates. The letter also called upon the DEQE "to evaluate whether or not a more stringent SIP for TSP should be devised in order to attain the [federal] standard." Parties supporting Brayton Point coal conversion feared that the nonattainment designation and the possibility of stricter particulate standards would be formidable obstacles.

EPA believed that burning high sulfur fuel would exacerbate the existing particulate problem—a pattern that seemed to have occurred with coal conversions in other parts of the state. Because the district was already designated as nonattainment, the EPA regarded any additional particulate emissions as unacceptable.

Modeling data also indicated that a switch to high sulfur fuel (either oil or coal) might cause a violation of the federal ambient sulfur dioxide standards. Air pollution dispersion models are imprecise. The models are computer-run mathematical simulations of pollution dispersion patterns. They are used to predict the maximum ambient pollution concentration at particular points downwind from pollution sources. The inputs to the model include meteorological variables and source variables (e.g., emission rate, stack gas velocity and temperature, and stack height). In order to predict maximum ambient concentrations, the models assume worst-case conditions. Worst-case conditions are hard to define accurately, and a model's outputs can vary greatly with a small change in any parameter.

In order to verify such uncertain predictions, the EPA used both their own models and one designed by ERT and the DEQE to investigate the sulfur dioxide problem. None of the models indicated that the use of 2.2% sulfur fuel at Brayton Point would, by itself, cause a violation of the primary sulfur dioxide standard, but the models did predict one or more exceedences when the emissions from the Brayton Point plant and those from the nearby Duro Finishing Company plant interacted. Indeed, the models suggested that Duro caused violations independently, but because these violations were modeled, not monitored, the EPA could take no action against that firm. The agency could, however, refuse to allow Duro to burn 2.2% sulfur oil, and it could do the same at Brayton Point if the power companies significantly worsened the sulfur dioxide problem overall.

The Coal Committee's Test Conversion Plan

The Federal Regional Council (FRC) of New England was created in 1972 to facilitate coordination of federal activities in New England. The group is composed of the New England regional administrators of the 10 principal grant-making federal agencies. In 1976, the New England Energy Task Force was created under the auspices of the FRC to study regional energy issues. John McGlennon, regional EPA administrator, and Robert Mitchell of the FEA were named co-chairmen. Other members of the task force included representatives from various federal agencies, the New England governors, the New England Regional Commission, and the New England River Basins Commission.

The task force split up into several committees to investigate specific issues. The coal committee focused on the problem of converting New England power plants to coal. The committee included personnel from the FEA, EPA, the Bureau of Mines, the DEQE, the Massachusetts Energy Office, NEPCo, and a nonprofit research organization called the Center for Energy Policy.

In September 1976, the committee issued its findings, *A Report on New England's Potential for Increased Use of Coal*. The document concluded that whereas coal conversion was desirable, it might be economically and environmentally costly. Consequently, they suggested that a test be conducted to determine the economics and environmental effects of burning coal without retrofitting flue gas desulfurization equipment.

Following this proposal, John Kaslow, vice president of NEPCo, offered the use of Brayton Point units 1 or 2 for the test. The company anticipated significant fuel cost savings if it was allowed to burn coal instead of oil but felt that additional pollution control equipment should not be required until it was shown to be necessary to meet ambient air quality standards. The test would give NEPCo a chance to show that no new equipment was needed. It proposed a one-year variance from the SIP requirements so that minor violations of acceptable pollutant emission levels would not cause environmental officials to halt the experiment. The company requested permission to burn off the old "dirty" coal pile to make room for a year's supply of high-quality coal.

The EPA objected to the test proposal on three grounds. First, they did not want NEPCo to burn the existing high-ash coal without controlling particulate emissions. Coal burning without emission control during the 1975 SIP suspension had caused numerous complaints from area residents about both highly opaque emissions and falling soot. Monitors had recorded one primary and one secondary standard violation in 1975. Second, the EPA believed that a test had already been done to collect exactly the same data and that the results had clearly shown that additional precipitator capacity would be needed to make coal use environmentally acceptable. Third, previous modeling and data suggested that electrostatic precipitators, and possibly scrubbers, would be required to meet ambient sulfur dioxide and participate standards. Finally, EPA suspected that the Brayton Point test was simply a NEPCo maneuver to obtain permission to burn the coal in the pile.

Before the test proposal was acted upon, the EPA notified the DEQE that the Southeastern Massachusetts Air Pollution Control District had failed to meet

NAAQS for particulates in 1975 and 1976. The EPA would not issue a variance to a SIP that would cause increases in any pollutant in an area where the standards for that pollutant were not being met. NEPCo, however, was still opposed to conversion that would require large capital outlays for new precipitators or scrubbers. In the view of David O'Connor of the Center for Energy Policy, "the prospects for conversion were as dark as they had ever been."

Nevertheless, shortly after the EPA declared southeastern Massachusetts a nonattainment area, John McGlennon (the regional administrator) sent a letter of explanation to Robert Mitchell of the FEA. The letter confirmed that EPA would not approve any test proposal if the nonattainment condition was verified, but it also included a list of conditions under which EPA would consider approval of coal use at Brayton Point. The alternatives listed included those options proposed consistently by EPA. To use coal at Brayton Point, NEPCo would have to comply with the existing SIP emission limits by improving the plant's electrostatic precipitators and either purchase low sulfur coal or install scrubbers. The letter, however, also contained a third, previously unconsidered, option—a consent decree:

> EPA is willing to consider entering into a consent decree with New England Power Company stipulating the actions to be taken to convert back to oil or penalties that will be levied if violations of regulation 2.5 (particulate emission limitations), regulation 6 (visible emissions) or regulation 5.4 (ash content of fuel) occurred during the one year test.

The enforcement branch had employed the consent decree mechanism previously in simpler two-party situations (EPA and a regulatee). When McGlennon asked the branch heads to explore all the real options for conversion, the enforcement entity suggested the consent decree.

Representatives of the FEA and NEPCo were dismayed by the letter. They believed that it signaled the end of all prospects for conversion. However, David O'Connor, who later acted as mediator, read the letter differently. He was intrigued by McGlennon's list of options that were acceptable to the EPA and considered them an invitation to negotiation. O'Connor discussed the possibility with all the parties and won their agreement to meet in April 1977.

In general, mediation (and negotiation) cannot succeed unless all key stakeholders have incentives to negotiate—otherwise some may spurn the negotiation process and try to influence the outcome through political or judicial channels. It is important, therefore, to identify the ways in which the parties to the Brayton Point conflict had a common interest in mediation.

The Parties' Incentives

New England Power Company

Because NEPCo's profit margin was dependent upon its cost efficiency, the company's management was interested in obtaining permission to burn the cheapest available fuel—coal. They did not want, however, to convert to coal if that required scrubbers or other expensive pollution control equipment.

At the time O'Connor first proposed mediation, the company knew that an ESECA notice of intent and prohibition order were likely. Because ESECA required conversion to comply with all current air pollution requirements, NEPCo also realized that expensive control equipment—new precipitators and possibly scrubbers—could be mandated. (The alternative of burning low sulfur coal was more costly than burning oil.)

The company thus was opposed to mandatory conversion under ESECA. It preferred negotiating with the EPA and the FEA to try to develop a plan for voluntary conversion that would allow them to burn the high sulfur coal without expensive control devices. Voluntary conversion might also enable NEPCo to use up the 2-year-old coal pile, which represented a dormant investment of $9 dollars. Also, conversion to low-cost coal would reduce electricity prices and thus fulfill the company's public service obligations.

The only alternative to mediation was a costly administrative and legal battle over the expected ESECA order. The FEA could not require conversion unless it was found to be "practicable." NEPCo could argue in administrative hearings and in court that it was not practicable because the cost of scrubbers was so high. The probable success of this argument, however, was unclear. Moreover, even if the company blocked mandatory conversion, it still would lack permission to burn coal on its desired terms. NEPCo thus had a strong incentive to negotiate with the various government agencies.

The Federal Energy Administration

There had been no mandatory ESECA conversions in the country by the spring of 1977. The administrative process set out in ESECA was time-consuming, cumbersome, and vulnerable to extended court battles. ESECA was enacted to reduce dependency on foreign oil, but in 3 years it had failed to produce any significant shift to coal.

In light of ESECA's poor conversion record, the FEA was willing to participate in mediation because it increased the likelihood of conversion at Brayton Point. Nevertheless, the agency insisted on concurrently pursuing the ESECA process. These dual processes were labeled *separate but parallel tracks* toward conversion. This stance allowed the FEA to increase the possibility of coal use while "remaining loyal" to the ESECA mandate. The FEA's position was not unexpected because the other parties understood the agency's need to observe its legislative mandate.

Environmental Protection Agency

The EPA had two incentives to participate in the mediation process. One was political. The national energy policy, muddled as it was, stressed the development of alternatives to continued importation of foreign oil. The EPA had been cast by the utilities and by some interest groups as a "nay-sayer" on alternative energy sources. The EPA's participation in mediation could promote its "good citizen" image. The agency wanted to be as cooperative as possible without compromising its commitment to clean air.

The other incentive was the association of two key individuals. According to John McGlennon (then Region I administrator), the regional EPA office first sat in on a meeting to discuss mediation because Robert Mitchell (his counterpart at the FEA) had personally asked him to do so. McGlennon respected Mitchell's opinion that mediation was potentially valuable, but like most of the other EPA officials, he was originally skeptical about the value of the effort. He also made clear his intention to fulfill EPA's mandate to protect air quality and ensure compliance with primary and secondary ambient air quality standards.

The Massachusetts Department of Environmental Quality Engineering

The DEQE was under a statutory obligation (chapter 494) to implement sulfur standards in an economically efficient manner. The state agency had proposed a SIP revision to the EPA that was intended to fulfill this mandate. The EPA, however, had delayed decisions on several important sources, including Brayton Point. The DEQE was concerned about the issues involved in coal conversion at that site and wanted to participate in the decision-making process. Also, if ESECA decisions were made unilaterally by federal officials, the DEQE feared that the conditions established might be unresponsive to the environmental needs of Massachusetts. The DEQE was concerned that conversion under the SIP might include inadequate control of particulates and uneconomical sulfur restriction.

Mediation also offered the DEQE a forum in which it could be responsive to the complex pressures from within the state government. Governor Dukakis was sympathetic to coal conversion, whereas Evelyn Murphy, head of the Executive Office of Environmental Affairs, predicated support for conversion on environmental acceptability. The Department of Public Utilities was concerned with minimizing the costs of providing electrical service. Mediation provided the DEQE with a forum (lacking in the ESECA administrative process) in which it could effectively address all of its concerns.

Mediation Begins

Although formal mediation did not begin until April 1977, the groundwork was laid throughout the previous year during the meetings of the New England Energy Task Force's coal committee. The task force was organized under the auspices of the Federal Regional Council, which had brought together interested parties to explore the possibilities and problems of coal conversion at Brayton Point and elsewhere. In time, committee members grew to know and trust one another; this rapport facilitated subsequent negotiations.

The parties also developed trust and respect for David O'Connor from the Center for Energy Policy, "a nonprofit organization dedicated to the peaceful resolution of energy and environmental disputes." The center had an abiding interest in the problem of coal conversion and the Brayton Point project in particular. It had participated in the coal committee meetings as an activist organization. Through his involvement, O'Connor learned more about the coal conversion problem and the specific concerns of all the parties to the debate.

O'Connor assumed a more active role in the committee after receipt of McGlennon's letter. As noted earlier, O'Connor believed that it outlined the EPA's requirements for acceptable coal conversion, whereas the FEA and NEPCo initially interpreted McGlennon's letter as precluding coal conversion. O'Connor convinced them that the letter could be the basis for a voluntary plan.

At O'Connor's urging, representatives from NEPCo, and the FEA, the EPA, the DEQE, and the Massachusetts Energy Office met in April to discuss the options for Brayton Point. The highly productive session yielded a surprising number of agreements. All the parties quickly agreed to drop the idea that Brayton Point "test" coal burning for a year without additional pollution control equipment in light of the particulate nonattainment designation. In response to suggestions from the center, the Energy Office, and the FEA, the committee proposed a joint effort to develop an environmentally and economically acceptable plan for permanent conversion to coal. The DEQE conceded that economic conditions might not allow purchase of low sulfur coal or addition of "scrubbers." They had already agreed that sulfur emissions could be raised without critical damage to air quality by assuming the chapter 494 fuel sulfur content SIP revision for Brayton Point. NEPCo, in turn, acknowledged that additional precipitators might be needed to protect the environment, given the particulate problems in the Fall River area in the previous two years.

Although the FEA would not agree to drop its ESECA conversion process, it did promise simultaneous pursuit of voluntary conversion. Continuation of the formal administrative process kept the pressure on the parties—particularly NEPCo—to negotiate a settlement.

O'Connor, who had recently completed an American Arbitration Association mediator-training program, suggested that the parties try mediation as a means of resolving their remaining differences. The group assented, on the understanding that O'Connor himself would be the mediator. John Kaslow of NEPCo later commented that O'Connor's evolutionary transition from participant to mediator was critical to the parties' acceptance of mediation. Had an outside mediator been suggested instead, Kaslow believes the proposal would have been rejected.

The participants agreed to meet 15 times over 5 months. The costs of the mediation effort—estimated to be $20,000—were divided equally among four of the parties: the FEA, EPA, NEPCo, and the U.S. Bureau of Mines (strong advocates of increased coal use).

The ESECA Process

Shortly after the April meeting, the FEA issued NEPCo a notice of intent saying that it was planning to pursue the mandatory ESECA process for the Brayton Point plant. In response, the company filed a lawsuit asking the U.S. District Court to restrain the FEA from holding hearings on the prohibition order. The suit, filed only 3 days before the date set for the first formal mediation session, charged that the agency had not followed its own guidelines in developing its conversion plans. In particular, the complaint stated that no site-specific environmental impact statement had been prepared, that the FEA had failed to provide documents needed for the company's defense, and that the FEA had failed to provide for cross-examination of witnesses at hearings.

David O'Connor contacted John Kaslow of NEPCo to find out what was going on. Kaslow indicated that NEPCo lawyers were pursuing the court action as a "procedural" matter. O'Connor relayed his feelings that such a suit would sabotage the negotiations and, he recalls, asked Kaslow to "stop it." (According to Kaslow, O'Connor did raise questions about the effect of the suit on the mediation process, but he did not specifically ask Kaslow to withdraw the legal action.) In any event, the suit was postponed pending progress in the negotiations.

The FEA then did hold hearings on the notice of intent. In its testimony, NEPCo argued strenuously against many of the FEA findings. The primary contention was that the FEA had greatly underestimated the costs of conversion. Hence, conversion was not "practicable" as the FEA had claimed. In particular, the company challenged the cost and availability of low sulfur coal. The FEA thereafter revised its estimates of coal prices but did not retreat from the stance that conversion was practicable. In late June the agency issued its prohibition order.

The coal conversion work group at Brayton Point convened in May 1977. John Kaslow, vice president at NEPCo, represented the company in the negotiations. Thomas Devine, chief of the Air Branch in Region I, represented the EPA. Duane Day, special assistant to the regional representative in Region I, participated on behalf of the FEA. Anthony Cortese, director of the Division of Air and Hazardous Materials, negotiated for the DEQE, and the Massachusetts Energy Office sent James Connelly, its deputy director. (Connelly acted largely as an observer because the Energy Office had no direct control over either the

ESECA or SIP revision process. As had been agreed earlier, David O'Connor mediated the negotiation sessions.

The first two meetings in May 1977 focused primarily on ground rules for the mediation effort. At the initial session, it was agreed that all parties would acknowledge their participation to the press but that no communication of a substantive nature would be made without prior approval by the group as a whole. The group also decided that all parties must be present in order to hold a meeting. Finally, it was agreed that O'Connor would convene meetings, take notes, prepare summaries of previous meetings, and set agendas. At the second meeting, the $20 thousand funding proposal was approved. These costs included the mediator's salary, technical assistance, and direct operating expenses.

The first substantive discussions were very broad. Although O'Connor and the parties understood that their goals were not mutually exclusive—and thus that resolution was possible—no one knew exactly what form that resolution would take nor even how discussions would proceed. Although O'Connor tried to direct the discussion to some extent, he did not have "any preconceived notion" about the order in which the issues would be tackled. He generally let the group select the agenda.

The negotiators identified key issues from the summaries of sessions written and distributed by O'Connor after each meeting. The summaries stressed areas of agreement and areas in which further decisions had to be made. According to O'Connor, these summaries were taken seriously, and they helped guide the negotiations along a productive path. They also served to involve the participants in making interim agreements. All the parties reviewed the summaries and together revised and approved them. O'Connor believed that this provided an "agreement-oriented frame of mind" and aided the process of building consensus.

O'Connor spoke up when he felt that an issue was assuming disproportionate importance and tried to guide discussion to a more fruitful area. In June, for example, the issue of the coal pile again threatened the negotiations as it had in prior discussions. Anthony Cortese (DEQE) and Tome Devine (EPA) felt that the coal contained too much ash to be burned without additional electrostatic precipitator capacity. NEPCo maintained that new equipment should not be required until proven necessary. A number of alternatives were suggested by O'Connor and others, including burning the coal at a reduced rate, mixing the "dirty" coal with other coal containing less ash, and reselling the coal. After much discussion, however, consensus could not be reached. At the suggestion of O'Connor, the parties agreed to put the issue aside until a plan for permanent conversion had been developed. Had he not intervened at this point, negotiations might have stagnated or reached an impasse.

The first major breakthrough in the negotiations came in July when Bob Thompson (EPA regional counsel for Region I) and Tom Devine (chief of the

Air Branch of Region I, EPA) initiated a study of the primary particulate standard violations around Brayton Point in 1975 and 1976. This was a major concession on the part of the EPA because the agency does not usually check air quality violations to determine their source. Rather, the EPA tries to control all the polluters contributing to the problem in the nonattainment area.

The source of the increased particulate level was of great importance to the Brayton Point negotiators, and it was a topic of considerable debate. NEPCo maintained that they were not causing the particulate violations. If they occurred at all, the company argued, they were most likely caused by local street sanding operations and general road dust reentrainment. Further, if NEPCo was not causing the problem, they contended that it was unreasonable to make them install additional electrostatic precipitators to solve the problem. At one point, they even suggested that it would be cheaper if NEPCo bought street sweepers for Fall River, rather than buying electrostatic precipitators to control their plant's emissions.

According to O'Connor, these arguments finally "got to" the EPA, which agreed to do the study to resolve the particulate dispute. Although Thompson indicated that the primary purpose of the study was to enable the EPA to defend itself if it was challenged in court, McGlennon added that as EPA personnel became more deeply involved in the Brayton Point case, they grew less cautious of NEPCo and more interested in finding an acceptable solution to the problem. Also, the complexity of the case made a detailed study necessary, according to Tom Devine. Because modeling data did not predict the violations that the monitors had recorded, a detailed analysis of the particulates' source was needed to determine how the violations could be prevented in the future and whether coal conversion was environmentally sound.

The EPA released the findings of its study on August 31, 1977. By reviewing "site location, standard operating procedures, pertinent calibrations, sample data sheets, and other documents," the EPA Surveillance and Analysis Division recalculated and verified all seventeen of the particulate exceedences, confirming the nonattainment status for the southeastern Massachusetts Air Quality Control Region.

The study, however, also confirmed NEPCo's claims that the source of the particulates was mostly street sanding and road dust reentrainment, not Brayton Point emissions. This finding was determined through a microscopic and X-ray analysis of the particulate matter in the filters. Because the two local power plants—Brayton Point and Somerset—were the only nearby consumers of high sulfur fuel, sulfer content of the particulate matter served as a good "marker" of particulate from the plants. The report showed a very weak positive correlation between sulfur and particulate emissions; hence, it concluded that fuel burning at the power plants "did not contribute significantly to the NAAQS exceedences."

The report had immediate impact on the debate over high sulfur fuel. The chapter 494 variance proposed by the DEQE to allow the use of high sulfur fuel oil at Brayton Point previously had been held up by EPA for two reasons: the anticipated detrimental impact on the ambient particulate levels and modeled predictions of possible violations of ambient sulfur standards. The report showed only a small positive correlation between ambient particulate and sulfur levels. This finding largely revived EPA concerns that the burning of high sulfur oil would exacerbate the existing particulate problems. The agency thus approved the SIP revision, though on the understanding that approval would be rescinded if subsequent monitoring showed that the interaction of Brayton Point emissions and those from the nearby Duro Finishing Company produced unacceptable levels of sulfur dioxide. NEPCo began burning high sulfur fuel oil almost immediately.

The EPA's approval of the use of high sulfur fuel oil prompted the coal conversion work group to begin discussing the use of coal with an equivalent sulfur content for permanent conversion. The experience with high sulfur oil provided important information on the effect of high sulfur fuel use on both ambient sulfur dioxide and particulate levels. Because data on ambient conditions that was gathered during the variance showed no significant problems with the particulate situation, the resistance to high sulfur fuel faded.

Interest Group and Public Participation

The inclusion of all interested parties is usually critical to the success of any negotiation. Excluded groups may challenge a negotiated settlement in court if they perceive that their interests have not been sufficiently represented. The initial selection of participants for the conversion to coal work group at Brayton Point was accomplished by including all the groups represented on the FRC coal committee that had a specific interest in the project. Because neither the Fall River community nor environmental groups were involved in the coal committee's work, they also did not participate in the initial negotiations.

In September 1977, four months into the work group's negotiations, some participants recognized that residents in the Fall River area might well be interested in the outcome. The debate turned to how the public should be involved. John Kaslow of NEPCo felt that public hearings following an agreement would allow sufficient public input. He feared that introducing new parties to the group in the middle of the process would inhibit the group's progress, given the technical nature of the discussion. By contrast, Anthony Cortese, of the DEQE, felt that it was important to avoid possible public resentment over "closed door" negotiations. Although there was some feeling that the public should be given a more active role in the process, the work group finally decided to limit further participation until after it had reached an agreement. At that point, the group would hold a public information meeting to explain the settlement and a public hearing to receive comment; the latter was required by law. In addition, Cortese agreed to ask a Fall River citizen to sit on the citizen advisory committee overlooking the DEQE's study of statewide particulate problems.

The rather low level of public participation did not arouse objections from Fall River citizens, who apparently were uninterested in the conversion debate. Neither the residents nor local officials made any attempt to assert a louder voice in the decision-making process, and at the June 1978 public hearings no significant opposition to the proposal was expressed by local residents or organizations.

Environmental advocacy groups also showed little interest in the conversion. They later offered several reasons for their inaction: lack of resources (time and money), a confidence that Cortese would protect their interests, a similar faith in the Dukakis administration, and an underestimation of potential impacts. Several groups expressed their intention to participate in future coal conversion cases.

The Particulate Emissions Debate

After resolving the public participation question, the group directed its attention to the continuing debate over particulate emissions. In late September, NEPCo informed the work group that additional precipitator capacity would be needed to meet current particulate emission standards even though the efficiency of the present precipitators would increase if NEPCo was allowed to burn high sulfur coal. Cortese felt that if NEPCo were to burn high sulfur coal at Brayton Point, the particulate emissions from the plant would have to be minimized.

In November, after reviewing NEPCo's preliminary engineering study, the DEQE asked the company to determine whether the installation of additional precipitators would allow them to comply with an even lower particulate limit. NEPCo agreed to study the costs of meeting tighter particulate emission standards, but it wanted some assurance from the DEQE and the EPA that the emission standards would not change dramatically over the life of the plant—once they were established. Without such a commitment from the regulatory agencies, the company felt that the large capital investment in a specific pollution control technology, as well as the investment in a long-term coal supply, would be unacceptably risky.

Devine resisted any long-term commitment to specific emission standards because changes in federal law, advances in technology, and new information about health effects of pollution might require that the standards be revised later. Nevertheless, in further discussions, the EPA indicated that it could approve long-term emission limit revisions if the DEQE could produce the requisite technical support for any such proposal. The DEQE agreed to try, seeing such an effort as necessary to gain the increase in pollution controls offered by the company.

There was very little debate over the actual design of the new precipitator equipment. The amount of new equipment that could be added was limited by space. The engineering study of the new equipment was discussed in the group, but the members did not share NEPCo's technical expertise and site-specific information sufficiently in order to be able to analyze the study. Because the group's mutual trust had been strengthened, the parties accepted the engineering report's accuracy largely on faith.

In December, NEPCo indicated that, with the new precipitator capacity, particulate emissions would vary between .06 and .12 lb/Mbtu (the preagreement limit) as coal sulfur content was varied between 1.21 lb/Mbtu and .55 lb/Mbtu (the higher sulfur content corresponding to the lower particulate emissions). The key aspect of these findings was that if the EPA and the DEQE agreed to raise the sulfur content limit, the particulate limit could be lowered.

By the end of 1977, the group had reached agreement on a number of points: (1) new precipitator capacity would be required to minimize particulate emissions; (2) a sulfur content limit, equivalent to the chapter 494 limit approved on a temporary basis during the negotiations, would be required to make a permanent conversion economical; (3) a particulate limit lower than that in the existing SIP might be required to make conversion environmentally acceptable; and (4) these new limits could be set for an extended period so that investments could be justified.

Final Negotiations

By January 1978, four major issues still had to be resolved: how NEPCo would comply with the new (higher) sulfur limit; what the new particulate limit would be; how long the new limits would stay in effect; and what mechanisms would be used to formalize the agreement. In order to facilitate the discussion of the remaining issues, the group as a whole authorized the DEQE and NEPCo to settle them bilaterally, with the aid of O'Connor.

According to Kaslow, the bilateral format was preferred because the remaining issues involved topics that were specifically relevant to the DEQE and NEPCo. Also, bilateral negotiation would require the time and energy of only two parties rather than six. O'Connor pointed out that the move facilitated negotiations by simplifying the communication process and minimizing the time devoted to gaining an understanding of highly technical material. The group agreed that O'Connor would keep the uninvolved parties informed of any progress and that agreements would be subject to review by the group as a whole.

Although the work group had agreed that the sulfur content limit should be raised to the equivalent of the chapter 494 limit, no method for measuring this content was prescribed. Because the sulfur content in coal varies significantly with shipments, NEPCo requested that they be allowed to average the sulfur content over a 90-day period as long as the content never exceeded 3% by weight.

Federal ambient air quality standards for sulfur dioxide take the form of both 3-hour and 24-hour averages in addition to the annual limit. Thus, the DEQE wanted a much shorter term than the company's proposed 90 days. After much debate, Cortese and Kaslow settled on a 1.21 lb/Mbtu limit over any 30-day period and a 2.31 lb/Mbtu maximum for each 24-hour period. These limits were strict enough to prevent NAAQS violations from occurring, whereas, at the same time, being lenient enough to make conversion to coal economically feasible.

The anticipated particulate emission rate for the new precipitators was .06 lb/Mbtu. However, the company wanted enough tolerance in the regulatory limit to accommodate variation in the ash and sulfur content of the coal supply. Both ash and sulfur content affect the efficiency of the precipitators, and they vary considerably from shipment to shipment. The DEQE, in turn, wanted to minimize the emission rate without being "economically punitive." After several heated debates, the particulate limit was set by compromise at .08 lb/Mbtu, which was slightly higher than the projected emission rate but considerably lower than the existing .12 lb/Mbtu particulate limit.

Terms and Form of Agreement

At the first the bilateral meeting, it was agreed that the settlement should have three components: a proposed revision of the SIP that the DEQE would submit to the EPA, a set of operating procedures to be attached to the regulation change defining compliance with the SIP (defining, for example, a sulfur averaging technique), and a memorandum of understanding to document the agreement between the state and NEPCo.

The time frame for these agreements and the form of the SIP revision generated much debate. NEPCo wanted regulation changes to be effective for the remaining lifetime of the plant because it wished to protect its investments in new equipment and long-term coal supply. Kaslow backed down slightly at the February 1978 meeting when he indicated that the company could commit itself to adding new precipitators if it could be assured that the negotiated regulatory conditions would be effective for at least 10 years. Cortese stood firm. He wanted to maintain some regulatory flexibility to accommodate unanticipated changes in federal law or advances in technology.

NEPCo and the DEQE also disgreed over the form of the SIP agreement. The company wanted the agreement to take the form of a SIP revision specific to Brayton Point so that any future regulatory change would also have to take the form of SIP revision. This would require the DEQE to hold hearings and obtain EPA approval before making changes that might threaten the value of the company's investment. The mechanism preferred by Cortese was a variance from the existing SIP, which would be effective for a predetermined period of time. The choice between a variance and a new regulation also determined whether NEPCo or the DEQE would be responsible for providing technical documentation for the action to EPA. A new regulation would be proposed by DEQE, requiring DEQE to do the technical work. By contrast, a variance would be applied for directly by the company.

The compromise reached between Kaslow and Cortese addressed the needs of both parties. It was decided that the agreement would be in the form of a new regulation, that it would be specific to Brayton Point, and that it would be effective only until November 1, 1988. NEPCo was to provide all the required documentation. Also, the DEQE was required to make a "full review of the circumstances and impacts of the conversion" prior to any "extension, modification or termination of the regulation." The compromise satisfied the DEQE's need for some long-term control over emission limits, and it provided some security for NEPCo's investment. The memorandum of understanding further stated:

> If for some reason the Regulation is not to be extended or is to be modified, DEQE will provide the Company with as much notice as possible of its intent to take such action. DEQE will encourage the Company to make any necessary changes to the facility to promote the extension of the Regulation for as long as possible. It is the hope of the Parties that the Regulation will remain in effect for a period beyond the initial ten years, preferably for the remaining useful life of these units, as long as coal burning is not precluded by new laws or new scientific information concerning the facility's ability to burn coal in an environmentally acceptable fashion.

The DEQE made as much of a commitment to coal conversion as possible, given the uncertain future.

At the request of O'Connor, the company and the DEQE agreed to sign a memorandum of understanding (MOU) to document the agreements and to demonstrate the commitment of each party to abide by them. On March 23, 1978, Cortese and Kaslow signed off on a draft MOU to be presented to the work group for review.

The Approval Process

The final negotiations had been completed by NEPCo and the DEQE alone. Their agreements now had to be approved by the entire work group and then by the EPA, the governor of Massachusetts, and the DOE (the Department of Energy, successor to the FEA.) The work group reviewed the agreement for two months and made no major changes. (Devine's approval of the agreement in the group was not a commitment by EPA to accept the SIP revision. EPA retained the right to review the revision in conjunction with the technical support data the DEQE was to provide.)

During this review period, the group raised the issue of how voluntary conversion would fit into the federal regulatory framework. The DOE was concerned that Brayton Point would be converted back to oil if oil again became cheaper than coal. The agency thus wanted to issue its ESECA order in support of the voluntary agreement. Kaslow responded that NEPCo's investment in coal conversion precluded a return to oil. NEPCo wanted the DOE to issue a notice of effectiveness (the final order to convert) so that Brayton Point could become eligible for a delayed compliance order. In support of this approach, Devine indicated that ESECA conversions would not be included as "new sources" under the prevention of significant deterioration (PSD) clause of the Clean Air Act. The DOE agreed to delay the completion of the final environmental impact statement until EPA had made a decision on the SIP revision. This decision was made to alleviate confusion that might have been aroused by studying two SIPs in the EIS.

On May 25, 1978, the DEQE sponsored a public information meeting in Fall River to answer the public's queries about the coal conversion. The purpose of the meeting was to explain the proposed regulation changes so that those interested would be able to testify at the June public hearing. About 40 residents attended and raised questions about environmental impacts, the truck traffic generated by ash disposal, job creation, the health affects, and the impact of the conversion on electric bills. (The Fall River area is served by the Fall River Electric Light Company, which is not part of the New England Electric System. Thus, ironically, the local population will not benefit directly from the fuel cost savings.)

The panel was asked if there had been any complaints or opposition to the proposal from environmental groups? "There have been expressions of interest and some questions of a technical nature," Cortese answered, "but so far no complaints or opposition that I know of." Environmental interest groups likewise presented no opposition to the proposal at the subsequent public hearing on the regulation change on June 15, 1978. According to the local newspaper, testimony at that hearing, which was attended by about 50 people, "was unanimously in favor of conversion," though in fact a representative of the Southeastern Massachusetts Lung Association expressed concern about the health effects for the surrounding population of increased sulfur levels.

In August 1978, the DEQE completed the technical support material for the SIP revision and submitted the proposal to the EPA. That month officials of the DEQE and NEPCo formally signed the Memorandum of Understanding in a ceremony in the governor's office at the "state house." The agreement was praised in the editorials of local and Boston papers.

In contrast, the November 1978 DOE hearings on a draft environmental impact statement generated some criticism. A DEQE employee expressed concern about serious errors in particulate emission estimates. A representative of the Rhode Island Department of Environmental Management noted the lack of support for the contention that Rhode Island's air quality would not be affected. A spokesman for the Greater Fall River Safe Energy Alliance charged that alternate energy sources and the potential dangers of acid rain were not given serious consideration.

Nevertheless, in March 1979, the EPA announced its proposed approval of the SIP revision in the *Federal Register*, and it allowed 30 days for comment. Approval was based on a review of the technical data. Although there was some doubt about the adequacy of the DEQE modeling of anticipated ambient sulfur conditions, EPA officials drew greater comfort from the availability of 15 months of monitoring data taken while burning high sulfur oil under the chapter 494 variance.

The EPA announced its final approval of the SIP revision in May 1979. This decision triggered the completion of the final EIS, which was issued in September 1979. The notice of effectiveness from the DOE was finally issued in June 1980, completing the formal ESECA process.

Subsequent to the EPA's approval of the SIP revision, the company filed for a delayed compliance order (DCO), which was granted in November of 1979. The DCO allowed the company to burn coal without additional pollution control equipment. On December 2, 1979, the first of the three coal-capable units at Brayton Point was back on coal. The first coal to be burned was that in the 4-year-old coal pile that had been the subject of much concern at the outset of the negotiations. NEPCo planned to install the new precipitator equipment one unit at a time over a 3-year period.

The Role of the Mediator

It is impossible to know what the outcome of the Brayton Point dispute would have been without David O'Connor's intervention. His first crucial act was to intitiate formal negotiations in April 1977. The FEA and NEPCo both believed that McGlennon's letter to Mitchell had ruled out voluntary conversion. Had O'Connor not intervened, the FEA probably would have pursued the mandatory ESECA process, and NEPCo would have challenged it in court.

Once negotiations began, O'Connor held it on a constructive course, skillfully avoiding several dangerous obstacles. For example, when NEPCo filed suit against the FEA, he managed to postpone litigation. When the parties were stalemated over the use of the old coal pile, O'Connor directed discussion toward other issues. Another crucial point came at the end of the bilateral negotiations when the parties were at an impasse over the particulate question. Ultimately, O'Connor moved the representatives to separate rooms, and, shuttling between the two, he tried to arrange an agreement. When this failed, he brought them back together and—much to everyone's surprise—he yelled at them, charging that only their lack of courage was preventing agreement. "There is nothing more I can do," he said, as he stalked out of the room. "If you want to negotiate further and make some compromises, you'll have to let me know."

This outburst shocked everyone, even O'Connor himself, who had "reached the end of his rope." Kaslow and Cortese were amazed that O'Connor, much younger than they were, was yelling at them as if they were children. "They didn't realize how much strain I was under," O'Connor reflected later. "They didn't realize how much I cared." Shortly after he left, Cortese and Kaslow called him back. "They calmed me down," he said, "and then they reached an agreement." O'Connor is unsure how decisive his anger was; consensus may have been the result of the accumulated negotiations. Still, his expression of intense feelings reenforced the parties' own commitment to find a solution.

O'Connor performed other important services throughout the negotiations. At the outset, he operated primarily as a facilitator and organizer. His first task was to obtain group approval of a set of informal procedural ground rules for setting agendas, raising issues, making proposals, dealing with the press, documenting discussions, and formalizing agreements. O'Connor was given responsibility for convening meetings, keeping written records of the meetings, and documenting areas of agreement. The session summaries he produced were instrumental in charting and propelling the negotiations, as they gave the parties a shared platform on which they could work. O'Connor also "brow-beat people to be sure they attended meetings" so that time was not lost explaining and possibly renegotiating matters that had been covered before. Continuity was crucial, he believed because it helped build group cohesion and trust. O'Connor also moderated discussions, ensuring that each party had an opportunity to be heard. He summarized technical and legal issues to help the parties understand unfamiliar material.

As a mediator, O'Connor also spent much time meeting privately with the individual parties. In these sessions, he sought to understand the concerns and technical factors underlying the parties' positions. He tried to help each party clarify their own positions by diplomatically challenging assumptions. The parties were thus encouraged to prepare a rationale for their positions; this, in turn, facilitated group presentations.

In the private meetings, O'Connor also tried to encourage considerations of creative solutions. He served as a sounding board for new positions and proposals, allowing parties some feedback on ideas without the risks inherent in presenting them to the group as a whole. He also encouraged parties to examine each other's positions and to discover potential concessions as well as areas where failure to compromise might lead to a stalemate.

Finally, O'Connor played an active role in the work group meetings. With as little bias as possible, he asked probing questions so that positions underlying a party's statements would become clear to the group. On occasion, he also presented ideas and options of his own in an effort to broaden the spectrum of possibilities under consideration.

Taken together, O'Connor's efforts provided an atmosphere that was conducive to fruitful negotiation. He did not invent the terms of agreement on his own, as an arbitrator would; instead, he helped the parties work together to develop their own agreement. He provided discipline, advice, and encouragement. Most of all, he provided the conviction that resolution was possible—and he kept the parties working together to that end.

Study Questions

1. Consider the interests of the various parties. (1) The New England Power Company had constructed three of the four generators so that they could be converted from oil to coal; in the face of rising oil prices, they had initiated conversion on their own. Why, then, did they resist the federal agency's attempt to mandate conversion under ESECA? (2) Under ESECA procedure, conversion cannot be ordered without approval by the EPA administrator. Given this veto power, why did the EPA feel compelled to attend the negotiation sessions, let alone relax its earlier air quality standards? (3) Even though area residents seemed to have the worst of both worlds—they were threatened with increased air pollution but were not going to enjoy the savings on generating costs—there was virtually no public objection to conversion. What may account for this seeming anomaly? Would public opposition have altered the course of negotiation? (4) Initially in this dispute, NEPCo was caught in the cross fire between the conflicting aims of the FEA and the EPA; yet ultimately, the company found itself in bilateral negotiations with the Massachusetts DEQE rather than with either of the federal agencies. What accounts for this transformation? What bargaining power did the DEQE have?

2. The power company committed itself to buy substantial amounts of coal even before it had received the variance from the state that was a prerequisite to its use. (1) Was this necessarily an oversight on the company's part? (2) The $9 million pile of coal that remained after the limited variance expired was a further incentive for the company to seek conversion. The company could not recoup its investment by selling the coal, because transportation would be a large component of its cost. Both state and federal authorities objected to the use of this coal at Brayton Point because of its high ash content. When the parties first met with the mediator, they were unable to resolve the coal pile issue; therefore, it was put aside pending agreement on other points. Consider the implications of structuring an agenda: What are the pros and cons of deferring basic issues?

3. The power company tried to institute coal conversion in 1974 but was unable to obtain government permission until late 1978. Was this delay unavoidable? To put the question somewhat differently, given that the 1978 settlement was not brought about by the invention of a new antipollution device or the passage of a new law, what prevented the parties from reaching the same resolution sooner?

4. On what important facts did the parties disagree? To what extent did these factual disagreements explain the dispute? How were the differences resolved? What parallels in this regard can you draw to principal cases we have already considered?

Cross-Reference

The Brayton Point case echoes an important issue introduced in the previous chapter—the problem of binding the parties. Just as in the Jackson Hole case in chapter 7, the parties here reached a point where they could envision a mutually satisfactory outcome, but settlement was held up by serious doubts about making it stick.

The Department of Energy, eager for a successful coal conversion, worried that if oil prices fell NEPCo would simply convert back to that fuel. The company spokesman responded that the substantial investment in added precipitators would make a return to oil uneconomical. (Does that necessarily follow?) As it happened, issuance of a notice of effectiveness—the final order to convert—proved to be in NEPCo's interest, as well as the agency's because it qualified the company for less stringent treatment under the Clean Air Act. Had this not been true, NEPCo officials probably would have been less willing to tie their own hands.

A more serious obstacle was the company's understandable need for some assurance that the control devices that they agreed to install would satisfy future as well as current environmental regulations. The company was reluctant to spend considerable amounts in conversion to coal if it could not be sure that it could recover the investment over the next decade. The government authorities, however, felt that they could not promise that more stringent rules would not be enacted. Further discoveries about the health impacts of air pollution, development of control technology, and changes in the political climate could all bring regulatory changes.

NEPCo ultimately had to accept less than an ironclad guarantee from the government. The parties did, however, sign a memorandum of agreement in which they expressed the "hope" that the revised regulation would remain in place for the useful life of the units. In some instances, such a pledge can carry great weight even if it is not legally enforceable. This is particularly true when parties know one another well and subscribe to a common ethical code. The negotiators here did succeed in developing substantial trust, but the agreement that they signed was intended to be in effect longer than it was likely that most of them would be involved in the case. Even while the dispute was being negotiated, the Federal Energy Administration and the state Environmental Health Division were replaced by the DOE and the DEQE, respectively. Undoubtedly, there was some continuity from each agency to its successor. Still, it is fair to question the depth of institutional memory over the years.

NEPCo, of course, did not rely entirely on the good faith of future bureaucrats to honor an unenforceable agreement made in 1978. Company officials insisted that the agreement take the form of a SIP revision specific to Brayton Point so that the burden of any change would fall on the DEQE. If the state agency later wished to impose more stringent control standards, it would have to muster the technical documentation, hold public hearings, and possibly defend its action in court. None of these steps would be impossible for the agency, but, together, they would constitute a substantial deterrent to change. This is a clear example of how negotiators can manipulate future procedures—here by incorporating the rules governing regulation revisions—to redistribute bargaining power. If, for instance, the DEQE one day did see the need for new controls, it might well seek the cooperation of NEPCo rather than press forward unilaterally and risk a protracted administrative battle.

Finally, it is worth noting that although negotiation had begun in earnest when the EPA expressed its willingness to consider a consent decree, this device was not ultimately used to anchor the settlement. The consent decree, a judicial order ratifying a negotiated agreement, sometimes is favored when one or more of the parties wants the assurance of a simplified enforcement mechanism in the event of a breach. In the Brown Paper case, described in chapter 4, the EPA insisted on a consent decree in order to prod the company there into completing its antipollution steps. A consent decree was less valuable in Brayton Point because ESECA's notice of effectiveness and the DEQE's regulation specifically revising the SIP served much the same function.

Mediation Study Questions

1. Parties who are in the midst of negotiation still may hesitate to enter mediation. Generally speaking, what are the incentives—and disincentives—to take this further step?

2. How did the mediator enter this particular dispute? Clearly, the parties' familiarity with O'Connor enhanced everyone's receptiveness to mediation. If he

had not been involved, how might mediation with someone else have been instituted?

3. O'Connor was qualified to serve as mediator in three important ways. Working for the Center for Energy Policy, he was conversant with coal conversion issues. From his early involvement on the coal committee, he knew the parties to the Brayton Point dispute well. Having been through an American Arbitration Association training program, he was familiar with mediation techniques. Obviously, all of these experiences were important to O'Connor's success, but which one do you think was most valuable? In many disputes, it may be impossible to find potential mediators who have technical knowledge, personal associations, and negotiation expertise. What are the virtues of one kind of specialty as opposed to the other? Do some disputes call for particular types of mediators?

4. One of O'Connor's initial functions was to organize and moderate meetings. At his urging, the parties adopted specific ground rules. They agreed, for example, that all parties would acknowledge their participation to the press but that no communication of a substantive nature would be made without prior approval by the group as a whole. What is the likely purpose of such a rule? What other effects might it have? How would such a rule be enforced? The parties also agreed that they all had to be present for a meeting to be held. What is the impact of such a rule? More broadly, in what ways do procedural decisions, such as these, affect the balance of bargaining power and possible outcomes?

5. Paradoxically, O'Connor was successful in opening communication among the parties, but he sometimes did so by organizing private meetings and caucuses. Consider the function of these private sessions. Why might it be in the interest of a party to reveal something privately to O'Connor that he or she would not disclose directly to the group? When does open communication among the parties inhibit settlement?

6. Reconsider the question asked in the introduction to this chapter: What is it that the mediator brings to the bargaining table that the negotiators cannot provide themselves? O'Connor had no legal or economic power that could force them to settle; yet, he appears to have played a substantial role at the outset in persuading others of the possibility of voluntary conversion, and at the end, in prodding the parties into settlement. Why would it have been difficult for NEPCo, the DEQE, DOE, and EPA to negotiate on their own?

Mediation Skills

Those who have attempted to mediate environmental disputes have understandably looked to the experiences of mediators who have practiced in other fields. Indeed, there is a significant literature on the theory and practice of mediation of both international and labor/management conflicts. Undoubtedly, some lessons can be transferred to environmental problems. As you read the following excerpts, however, consider how the science and art of mediation may have to be bent to fit the special nature of environmental conflict.

The first selection addresses the question that began this chapter: what is it that the mediator brings to the bargaining table that the parties lack themselves? The two excerpts that follow deal with the stages and techniques of mediation generally. Readings in chapter 10 raise specifical ethical questions.

Stulberg, Joseph B. "The Theory and Practice of Mediation: A Reply to Professor Susskind," 6 *Vt. L. Rev.* 85, 91-94, 1981.

> Lon Fuller, the distinguished professor and arbitrator, described the goal of the mediator in elegant fashion when he wrote: "[T]he cental quality of mediation [is] its capacity to reorient the parties towards each other, not by imposing rules on them, but by helping them achieve a new and shared perception of their relationship, a perception that will redirect their attitude and dispositions towards one another." What functions of office does the mediator have that enable him to fulfill that objective? A brief listing would include the following functions.
>
> A mediator is a catalyst. Succinctly stated, the mediator's presence affects how the parties interact. His presence should lend a constructive posture at the discussion rather than cause further misunderstanding and polarization, although there are no guarantees that the latter condition will not result. It seems elementary, but many persons equate a mediator's neutrality with his being a non-entity at the negotiations. Nothing could be further from the truth. [Stulberg takes issue with those who contend that because the mediator is concerned more with process than with outcomes, he or she should be regarded as merely having a passive role.] The active/passive distinction, however, seriously misrepresents the impact of the mediator's presence on the parties. Much as the chemical term *catalyst* connotes, the mediator's presence alone creates a special reaction between the parties. Any mediator, therefore, takes on a unique responsibility for the continued integrity of the discussions.
>
> A mediator is also an educator. He must know the desires, aspirations, working procedures, political limitations, and business constraints of the parties. He must immerse himself in the dynamics of the controversy to enable him to explain (although not necessarily justify) the reasons for a party's specific proposal or its refusal to yield in its demands. He may have to explain, for example, the meaning of certain statutory provisions that bear on the dispute, the technology of machinery that is the focus of discussion, or simply the principles by which the negotiation process goes forward.
>
> Third, the mediator must be a translator. The mediator's role is to convey each party's proposals in a language that is both faithful to the desired

objectives of the party and formulated to insure the highest degree of the receptivity by the listener. The proposal of an angry neighbor that a "young hoodlum" not play his stereo from 11:00 P.M. to 7:00 A.M. every day becomes, through the intervention and guidance of a mediator, a proposal to the youth that he be able to play his stereo on a daily basis from 7:00 A.M. to 11:00 P.M.

Fourth, the mediator may also expand the resources available to the parties. Persons are occasionally frustrated in their discussions because of a lack of information or support services. The mediator, by his personal presence and with the integrity of his office, can frequently gain access to the parties to needed personnel or data. This service can range from securing research or computer facilities to arranging meetings with the governor or president.

Fifth, the mediator often becomes the bearer of bad news. Concessions do not always come readily; parties frequently reject a proposal in whole or in part. The mediator can cushion the expected negative reaction to such a rejection by preparing the parties for it in private conversations. Negotiations are not sanitized. They can be extremely emotional. Persons can react honestly and indignantly, frequently launching personal attacks on those representatives refusing to display flexibility. Those who are the focus of such attacks will, quite understandably, react defensively. The mediator's function is to create a context in which such an emotional, cathartic response can occur without causing an escalation of hostilities or further polarization.

Sixth, the mediator is an agent of reality. Persons frequently become committed to advocating one and only one solution to a problem. There are a variety of explanations for this common phenomenon, ranging from pride of authorship in a proposal to the mistaken belief that compromising means acting without principles. The mediator is in the best position to inform a party, as directly and as candidly as possible, that its objective is simply not obtainable through those specific negotiations. He does not argue that the proposal is undersirable and therefore not obtainable. Rather, as an impartial participant in the discussions, he may suggest that the positions the party advances will not be realized, either because they are beyond the resource capacity of the other parties to fulfill or that, for reasons of administrative efficiency or matters of principle, the other parties will not concede. If the proposing party persists in its belief that the other parties will relent, the question is reduced to a perception of power. The mediator's role at that time is to force the proposing party to reassess the degree of power that it perceives it possesses.

The last function of a mediator is to be a scapegoat. None ever enters into an agreement without thinking he might have done better had he waited a little longer or demanded a little more. A party can conveniently suggest to its constituents when it presents the settlement terms that the decision was forced upon it. In the context of negotiation and mediation, that focus of blame-the-scapegoat can be the mediator.

O'Connor, David, with Charles Foster. "Founding a Center for Environmental Mediation in New England," Paper prepared for the 45th North American Wildlife and Natural Resources Conference 1980, pp. 11–13.

The Mediation Process

In the course of a formal mediation process, the mediator proceeds through four stages to assist the parties in dispute. He first undertakes a "fact-finding consultation" which has two objectives: to assess the physical dimensions of the problem, and determine why the parties have not been able to reach agreement on their own. This culminates in a recommendation on how to proceed, either with or without the assistance of a mediator.

If the parties and the mediator decide to continue working together, there follows a period during which the mediator meets separately with the parties and explores with them the interests they most want to protect. He then helps them clarify what they want the other party to do to protect these interests.

Next, there is an "inventing phase," a series of separate and joint meetings during which as many solutions as possible for each aspect of the problem are discussed and evaluated. None is formally selected or rejected until every possible alternative has been thoroughly considered.

Finally, the mediator begins to offer draft agreements to each party for review and revision. After receiving suggested changes, the mediator submits an amended version for review. This revision process is repeated as long as is necessary to reach a final agreement.

The Analogy to Mediation of International Disputes

In labor mediation, the mediator's role is to guide and catalyze the final stages of negotiation. This approach assumes that negotiation has already gone on for some time and only needs to be re-oriented. With this image in mind, it is not surprising that parties to environmental disputes associate mediation with extreme pressure, difficult decisions, hard bargaining, and painful concessions. It is no wonder that they are reluctant to retain a mediator.

Yet, environmental disputes are much more like international disputes than labor disputes. Often, there are more than two parties with a direct interest in the outcome. Questions of social and political importance compete with economic and technical issues. And, in some cases, delays in decision making work to the advantage of one or more parties. As in international disputes, environmental disputants can make use of a mediator's help from the moment a serious conflict can be identified. Even if a formal mediation process is not established, a mediator can help the parties clarify their complaints and demands, see that all of the parties have heard and understood the concerns and proposals of the others, and establish a framework within which the parties can negotiate without a mediator.

Cormick, Gerald and Knaster, Alana. "Mediation," The Mediation Institute at the University of Washington in Seattle, 1971.

A mediator cannot perform *any* role in a dispute unless he can gain the confidence of the parties or, to put it another way, reduce the parties' distrust to a level that will allow him to function. He has to get the trust of the parties (who usually don't trust each other) so that they will be willing to take some risks with him and eventually with each other. . . .

No matter where the mediator comes from or what has gone on before, trust must be established (then re-established) in each situation. Trust is more a matter of what you do than what you say. It is also very fragile, and not owned but constantly earned.

Ways of earning trust are:

1. *Explaining the mediator's presence and role while letting the parties explain the dispute.* In explaining his role, a mediator has to make it clear that he is not there as a partisan spokesman for either side. To the extent that the mediator advocates a bargaining process and a continuing agreed-upon relationship between the parties, he is obviously an advocate of change, rather than a neutral. But he is not a partisan. Once he's pegged as a partisan, his credibility and ability as a intermediary is at an end. This does not mean that a mediator doesn't or shouldn't have views on the issues. It means that his views aren't important to a resolution of the dispute. The views that are important are the views of the parties. Those are the views which should be explored, not his. He should make it clear that he is there to help the parties get what they want, not what he wants.

At the same time, he has to learn what the parties want by letting them *tell* him. It's important that a mediator put himself in the posture of being educated by the parties. This is true even if he thinks he knows what the dispute is all about and even if his pre-entry data-gathering has given him a great deal of information on the problem. . . .

2. *Reducing defensive communication.* Defensive communication, and, as a consequence, inaccurate communication, occurs when an individual is distrustful of another or feels threatened, sometimes subconsciously, by what the other says or the way he says it. There is a way to listen and a way to ask questions which minimize defensive communication and actively support communication that is open.

(a) *Description, not evaluation.* A mediator has to avoid value judgments, particularly in the early stages. Statements like "you'll never get that," "if I were you, Mr. Administrator, I'd never take that position" or "that's far out" or "unimportant" are guaranteed to stop the flow of communication and tarnish your objectivity as well. These statements may seem extreme, but it should be remembered that the thoughts behind them don't even have to be put in words to be conveyed. (There may come a time in a dispute when the parties will be interested in your opinion, but that's much later. . . .)

(b) *Problem orientation, not control.* Initial indications that a mediator is attempting to exercise control can turn a group off before communication begins. Consequently, a mediator must be open, convincing the parties that he is there to aid them in resolving their problem and that he has no hidden agenda or strategy. This is not to say that the mediator, at some point, doesn't want to manage the proceedings and have communications flow through him.

Obviously, if he can make the proceedings manageable and exercise some control of their ebb and flow, the chances for an agreement are enhanced. But this doesn't happen, if it happens at all, until trust is established and it will never happen if the mediator seeks to impose control or his view of the "desired" solution at the outset.

(c) *Empathy, not neutrality.* Being neutral can mean being detached, clinical or disinterested [sic]. If a mediator is neutral in this sense, if he exhibits a lack of concern for the welfare of a group or its positions, verbally or nonverbally indicates that the group is nothing more than an interesting object of study, he will not get very far. He cannot be clinical or disinterested [sic]. He has to convey empathy and respect for the group. This does not mean that he must agree with all they say. But he has to express an understanding of their problems and positions, and accept their emotional reactions to the situation at face value.

(d) *Equality, not superiority.* Obviously, a mediator who is there to help parties resolve a problem cannot convey superiority if he expects to be helpful. The moment he creates the impression that the dispute is beneath him or that he attaches any importance to differences in status or ability that may exist between him and the group, his usefulness is at an end.

(e) *Provisionalism, not certainty.* Those who seem to know the answers tend to put others on guard. A mediator is no exception. Not only must he be open, provisional and undogmatic, he must also constantly remember that it is the parties who will have to live with the eventual answers, not him. . . . [After analyzing the various roles of the Mediator, Cormick describes the steps in mediated settlement.]

If the parties have confidence in the mediator, he will, after these separate meetings, have a pretty clear picture of what the parties will accept. If the parties' position match, his work is virtually over. If they don't as is more usually the case, he has a number of techniques which he can use to bring an agreement into being. In using or contemplating the use of any or all of them, the mediator should be mindful of the following general principles—(1) no one makes a decision if there is any possible way to avoid it, (2) all disputes must end sometime, and (3) no settlement is entered into without doubt.

The techniques (for settlement building) are not set down in order of importance or in strict chronological sequence. Nor is there any way to "teach" when or in what circumstances they, or any one of them, can be used. That comes from experience alone.

(a) *No-risk narrowing of positions.* When the mediator is told (or accurately surmises) that a party is willing to compromise on a point, he can do one or two things. Obviously, he can simply tell the other side. If he does, that party may interpret the offer as a sign of weakness, reject it and demand further concessions as a price of settlement. However, if he communicates the offer as a hypothetical possibility, which he is yet to (but will) explore, the risk of rejection is minimized. And even if the "hypothetical possibility" is rejected, the effect of rejection is minimal. The offering party's position is not weakened, because no offer has been made. The "possibility" was just a thought of a mediator, nothing else.

(b) *The "Collection of Agreements" vs. "Boulder in the Road."* There are two ways to attack issues which divide parties—take the easy ones first or the hard ones. The "collection of agreements" approach creates a momentum, an aura where other agreements become possible. On the other hand, the momentum may be for naught if the "boulder," the matter of principle, is so huge that nothing can move it. . . .

(c) *The use of external pressure.* A mediator can sometimes generate movement by the judicious use of external pressures. A federal deadline may have to be met. A citizens group may be prepared to blast an institution. A mayor may be under pressure to use police. The media, with its well-known accuracy, may be prepared to highlight the dispute. Any number of forces external to the dispute may, by narrowing the choices and time available to the parties, bring the agreement closer.

(d) *The use of deadlines and marathon sessions.* This technique is related to the previous one in that deadlines play a part in both, but here the deadlines (or pressures) are internal. In labor–management disputes, deadlines, such as the contract expiration, are apparent. In community disputes, internal deadlines are often less delineated, but not less real. An upcoming meeting of a board, for example, the opening or closing of school, the necessity to get the institution's machinery functioning to avoid a larger conflict, an unchangeable ultimatum, etc. Such deadlines often exist and the mediator can use them. He can also use his efforts to change them if agreement is near and change is useful.

(e) *Deflating extreme positions.* In this context, whether or not a position is extreme has nothing to do with its merit. The question is one of power. Given all the factors in the dispute, is the position attainable? If the mediator is convinced that it's not, or almost certainly not, then he has to bring this point home. (Obviously, this is done in a caucus or separate session, not a joint meeting.) The party so educated may choose to forego an agreement rather than yield, but it should do so with the facts not without them.

(f) *The consequences of "no agreement."* Being deeply involved in a dispute does something to one's perspective. It often distorts it. Parties tend to weigh possible settlement terms against other possible settlement terms. This package vs. that package. Often, and particularly in community disputes, the real choice is between agreement on particular terms and no agreement. Thus, the mediator has to dwell on the consequences of not reaching agreement—what happens then, does the community group have a breather and some period of stability or not, is there a relationship, however fragile or not? In short, what are the consequences of no agreement at all, and how do they balance up?

(g) *The mediator's proposals—uses and dangers.* The mediator is often trying things on for size, floating ideas or informal suggestions, but he rarely makes proposals. The reason is obvious. Once he proposes particular settlement, he is telling the parties how he thinks the dispute should be resolved. If one party doesn't agree with his proposal, his usefulness is over. Consequently, a mediator doesn't make settlement proposals unless (1) he knows they will be accepted, or unless (2) he is at the very end of the line and believes that such proposals will aid an ultimate settlement in some other way, such as public pressure or in the hands of a mediator who follows him.

9

MEDIATING LARGE DISPUTES

INTRODUCTION

The principal case that follows draws together many of the ideas introduced earlier in the book. The central dispute was over the proposed construction of the Foothills Dam near Denver, Colorado. It involved scores of parties, raised questions about the accuracy of key data, ultimately required mediation, and almost went unresolved because of problems in binding the parties. In short, the case provides a rich opportunity for synthesizing material that has been developed throughout the book.

Indeed, the issues in the Foothills case are so complex and the details so varied that there are important comparisons that can be drawn within its own boundaries. This was a case, for example, that was litigated, negotiated, and mediated. Mediation, moreover, was attempted twice: one effort failed whereas the other succeeded. A comparison of the two attempts reveals important lessons about the timing of mediation and the choice of a mediator. The second Foothills mediator deliberately flouted commonly held precepts about the importance of neutrality and open participation. In large measure, he succeeded not in spite of these transgressions but because of them.

The Foothills case is also significant because it tests the proposition that some environmental disputes are simply too cumbersome to be negotiated. Although it certainly does not prove that all large cases can be settled out of court, it is persuasive evidence that some can be. It is true that the process was protracted and that some of the parties agreed to the settlement only with great reluctance. Yet, it is also true that whatever the shortcomings of negotiation, the parties preferred it to alternative social processes. They did, after all, agree to abandon a concurrent lawsuit. Moreover, the situation was highly politicized; had the case not been settled, the president might have been required to arbitrate differences among competing federal agencies. If Foothills illustrates the difficulty of negotiating complex problems, it also demonstrates that legal and bureaucratic machinery may be even less adequate.

The case also reveals the important distinction between specific disputes and the more general conflicts that usually underlie them. The parties here did come to agreement over the terms under which the Foothills Dam would be built, but the conflict over growth control, water conservation, and local autonomy persists. Even with specific disputes, there may be degrees of resolution. In the case of Foothills, there have been continuing skirmishes over implementation of the agreement.

The case study is followed by a series of general questions intended to raise issues about the interrelationship of the parties and their relative bargaining positions. A second series of questions deals specifically with mediation issues. The important matters of mediator ethics and accountability are the core themes of the next chapter.

CASE STUDY: THE FOOTHILLS WATER TREATMENT PROJECT

This case study is adapted from an account prepared by Heidi Burgess.

Introduction

Background on the Foothills Project

The story of the Foothills water treatment complex is filled with political and theoretical surprises. Designed and developed by the Denver, Colorado, Water Board (DWB), the Foothills complex was planned to extend the DWB's raw water treatment system. The system treats mountain water and supplies it to the city and the surrounding suburbs for residential, commercial, and industrial use.

The DWB began studies of the project in 1952 and obtained the necessary federal rights-of-way in 1967. Because projections showed a need for additional treatment capacity by 1977, the DWB asked the voters of Denver to approve the issuance of a bond in a 1972 election. The proposal was rejected. In 1973, the DWB submitted another proposal to the voters and coupled it with a heavy advertising campaign. This bond issue was approved by a solid majority, although turnout was only 13.7% of all eligible voters. The bond issue did not specify the Foothills plant in particular; however, the DWB assumed that vote represented support for the Foothills project, and they decided to begin construction in 1974.

Five years, two lawsuits, and many million dollars later, construction of the Foothills plant began. During that time, two federal agencies wrote four environmental impact statements (three drafts, one final version), and six federal agencies fought internally and with the DWB over numerous technical, economic, environmental, and aesthetic issues. Local and state agencies as well as various environmental groups joined the fray, which concerned a project that was approved by a local government, apparently supported by local voters, required no federal money, and involved only 87 acres of federal land (see Figure 12).

The extent of the conflict was striking in itself; that it was resolved by mediation out of court appears even more improbable. The success of the mediation effort defied the emerging rules of environmental mediation. The principal mediator was considered to be biased in favor of the Foothills project before the negotiations began. Instead of involving all the parties initially (as theory dictates), the mediation began with only three parties; subsequent sessions included other interested stakeholders.

The Foothills water treatment project is a raw water treatment facility designed, owned, and operated by the Denver Water Department—an independent municipal government agency. In addition to a treatment plant, the project would include a concrete diversion dam and reservoir to be located on South Platte River, approximately 25 mi southwest of Denver. The proposed dam would rise 243 ft above the existing river channel, creating a reservoir 1.7 mi long, 400 ft wide, and 240 ft deep. The elevation of the dam would be 6,002 ft above sea level.

The treatment plant would be located about 3.5 mi northeast of the diversion dam on 490 acres of low, rolling, undeveloped grassland. The plant would be linked to the reservoir by a tunnel, emerging as a conduit a few hundred feet before entering the plant. The plant's elevation would be 5,889 feet; the drop in altitude from the dam to the plant would allow the DWB to install a hydroelectric turbine at the treatment plant intake. The turbine would produce electricity in net amounts far greater than needed for plant operations. In addition, the filtration plant's location above Denver would allow for the use of a gravity feed to distribute the treated water throughout much of the DWB service area.

The plant was designed to be built in units. The first unit was to be capable of treating 125 million gallons of water per day (mgd) with expansion capability for three additional 125-mgd units on the same site. Because the capacity of the existing DWB treatment system is only 520 mgd, the first unit represents almost a 25% increase, and the ultimate 500-mgd facility would represent almost a 100% increase in the plant's capacity for raw water treatment. At a treatment level of 125 mgd, the system uses only existing water supplies. The project would expand Denver's treatment facilities but not its raw water supply system.

If the new system's capacity expands to 500 mgd, the total capacity of DWB's treatment system (1,020 mgd) would exceed the existing raw water supplies. New supplies of raw water would eventually have to be developed to meet peak summer and winter needs. Because Denver (and much of the western

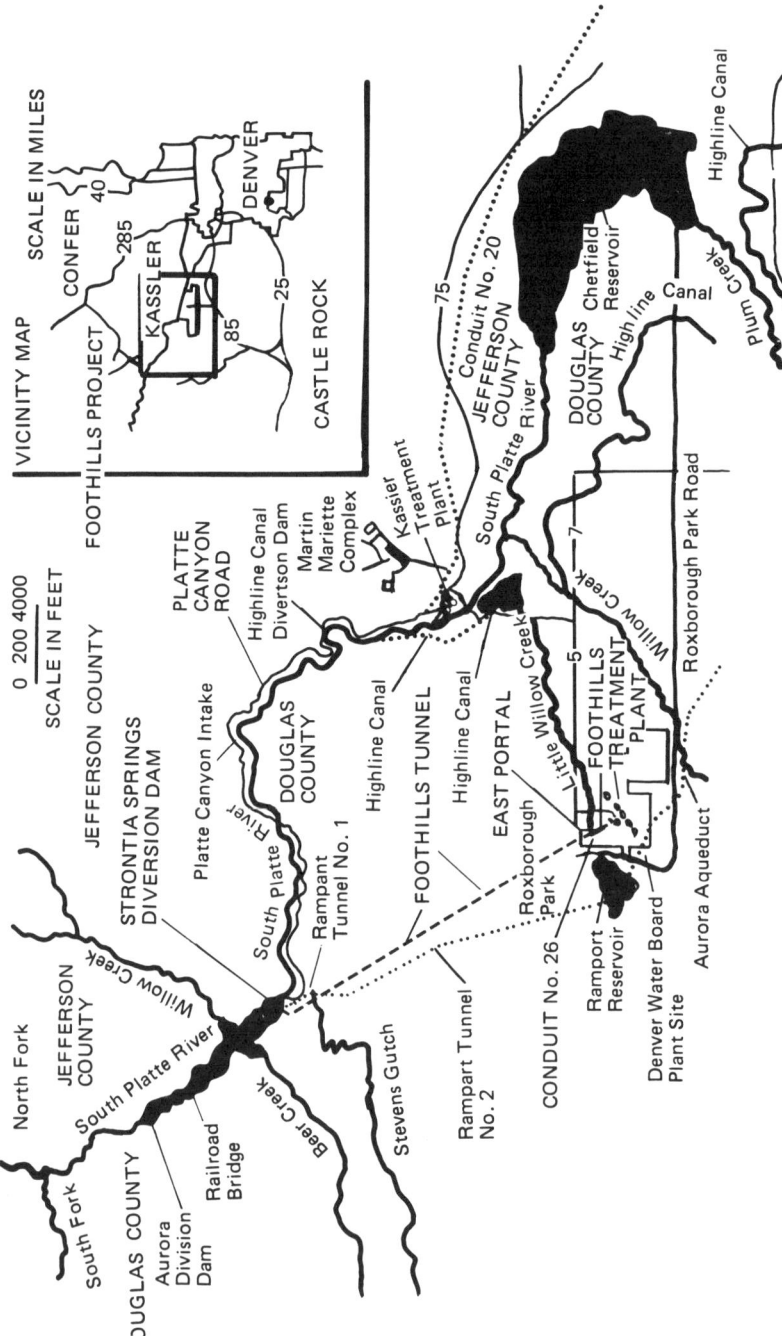

FIGURE 12. Proposed Foothills plant.

United States) is situated in a semiarid region, additional water supplies are not easily obtained. Much of Denver's water now comes from the western slope of the Rocky Mountains (Denver is on the eastern slope) because the western slope is much wetter. This requires extensive transmountain diversion of water through 10- to 20-mile-long tunnels underneath the Continental Divide. However, western slope water is scarce when it is compared with the water supplies in the eastern part of the county, and residents of the western slope are increasing their demands that western slope water be saved for their area. Although the DWB claims that it has the right (through very complicated Colorado water laws) to divert much more water to Denver from the western slope, many environmentalists and residents of the western slope disagree. Further water diversion has become a major political and legal issue.

Because the Foothills dam, tunnel, and conduit systems were designed to accommodate the final 500-mgd plant, many opponents of the project are sure that the plant will inevitably be expanded to that size. This would mean the development of additional systems for transmountain water diversion. For this reason, opponents viewed even the initial 125-mgd plant as a potential threat to western slope water supplies. They have fought construction of the plant on these and other grounds.

Problems in Obtaining Federal Permits

The Foothills Project was funded entirely by the Denver Water Department, which raised the necessary money through general obligation municipal bonds; federal money was not involved. Nevertheless, the federal government got deeply embroiled in the Foothills controversy because the dam and reservoir were to be located on 27 acres of national forestland and on 22 acres of land administered by the Bureau of Land Management (BLM). For this reason, the DWB was required to obtain right-of-way permits from these two agencies. Although these permits originally were issued in the 1960s, they had expired and had to be reissued in the early 1970s. By that time, the National Environmental Policy Act had been passed, and the project required completion of an environmental impact statement (EIS).

The EIS process lasted 5 years; the lead agency designation was contested and the extent of the EIS necessary was debated. Opponents of the project tried to block approval by arguing that the three draft statements and the final impact statements failed to deal adequately with either direct or indirect effects on the environment and did not consider sufficient alternatives to the proposal.

The EPA was one of the many agencies, groups, and individuals to denounce the impact statements. After the BLM issued the final environmental impact statement (FEIS), the EPA used its statutory authority under section 309 of the Clear Air Act to request the Council on Environmental Quality (CEQ) to intervene in the Bureau of Land Management (BLM) and United States Forest Service (USFS) right-of-way decisions. Although the EPA's complaint was lengthy, its objections were summarized initially in a letter from Douglas Costle, EPA Administrator, to Secretary of Interior Cecil Andrus:

> 1. Construction of the Foothills Project would make the attainment and maintenance of national ambient air quality standards in Denver more difficult and perhaps impossible. (Because providing water would encourage growth and urban sprawl.) 2. Construction of the Foothills Project would result in significant environmental degradation to a unique aquatic wildlife and recreational resource that could be avoided by other practicable alternatives.

The EPA also contended that there was virtually no analysis of secondary impacts as required by CEQ guidelines.

After quick deliberation, the CEQ concurred with the EPA's findings and recommended that the federal permits for the construction of the Foothills Project at the 500-mgd and 125-mgd levels be denied, or alternatively, that the department withdraw the EIS as inadequate under the National Environmental Policy Act for failure to analyze air quality impacts and water conservation and other reasonable alternatives. The CEQ statement concluded,

> The analysis in the impact statement indicates that raw water supply will not be a limiting factor until 1990. We believe that there is time available to develop the new analysis needed to reach an informed decision on the Foothills project and on its relationship to federal resources, water conservation, air quality, and to the development of an overall water supply and treatment policy for the growing Denver region.

Despite the CEQ's findings, the BLM and USFS chose to issue the permits on the grounds that issues of growth, air pollution, and land use were state and local issues and thus outside the authority of the BLM and the USFS. The CEQ did not exercise its only remaining option (to refer the issue to the president), although the permits were challenged in court by both the DWB (which did not like the conditions that were imposed) and by environmentalists who wanted the project stopped altogether.

In addition, because a dam involves fill materials, the Denver Water Department had to obtain a 404 dredge-and-fill permit from the United States Army Corps of Engineers. The Corps of Engineers is the primary agency responsible for reviewing 404 permits, which are required under the terms of the Clean Water Act. The construction process also was expected to deposit additional fill material into the South Platte River, which is a navigable stream. This meant that an additional 404 permit was necessary.

The corps is required to consult with EPA to be sure that projects under consideration are not excessively damaging to the environment. If the EPA finds a project unacceptable from an environmental standpoint, it may object to issuance of a 404 permit. The matter is then taken up by the federal hierarchy to make "every effort . . . to resolve differences at the Division Engineer level before referring the matter to higher authority." If the differences between the corps and the EPA remain unresolved, the division engineer must refer the case to the chief of engineers in Washington who must try to resolve remaining differences with the administration of the EPA. If the differences still cannot be resolved, the EPA administration has the final authority to deny the permit under section 404c of the Clean Water Act. Although the Foothills dispute between the EPA and the corps was not referred to the chief of engineers by the EPA administrator, it was referred to the division engineer; it could have been sent to Washington had the mediation process failed.

The United States Fish and Wildlife Service (FWS) also became involved as an opposition group through the Fish and Wildlife Coordination Act, which requires the coordination of federal decisions on permits affecting fish and wildlife with the FWS. In addition, regulations of the Corps of Engineers prohibit the district engineer from issuing a permit over the unresolved objections of another federal agency in the event that the agency requests that the application be referred to a higher level of authority for review. Under these conditions, therefore, the FWS has the authority to prevent approval of the permit at the district or division level.

In addition to these permits, numerous other permits were required from additional federal, state, and local agencies. Those permits were easily obtained because the local and state governments favored the project. The other federal permit (actually a "checkoff" on the BLM and USFS rights-of-way permits) from the Federal Energy Regulatory Commission was also granted easily after the other disputes ended. Thus, the primary federal agencies in the dispute were the BLM, the FWS, the USFS, the Corps, and the EPA.

Issues and Disputants

Introduction

Many factors transformed this simple, locally supported, and locally funded project into a complex intergovernmental controversy. In its application and throughout the controversy, the DWB contended that the current capacity of the Denver Water Department's water treatment facility would be inadequate by 1977 due to increasing population growth and increasing per capita water use. The DWB wished to incorporate the Foothills Project into the system by 1977 to assure continued unlimited supplies of treated water to the metropolitan area. If the project was not built, the DWB contended that it would not be able to meet peak summer demands for water in 1977. Although the shortage predicted for

1977 was not severe in its extent and duration, future shortages of greater impact were anticipated (because the DWB assumed both continuous population growth and a continued increase in per capita consumption).

Nevertheless, the DWB's analysis was disputed by many environmentalists, the EPA, and, to a lesser extent, by the Corps of Engineers. The environmentalists charged that the DWB's and the BLM's projections of future water needs were excessively high, thus rendering the entire environmental impact statement erroneous. Both environmentalists and the EPA charged that additional treatment capacity provided by the Foothills Project "is needed only to allow unlimited lawn watering through the year 1988" (when the supply of raw water, not treated water, would become the limiting factor). This contention was supported in the FES, which states that

> the principal purpose of the project is to enable the Denver Water Department to meet projected max-day demands in order that Denver Water Board customers can irrigate horticulture without restriction during the hot summer months.

According to opponents of the project, if watering restrictions similar to those instituted in 1977 were imposed, the need for the Foothills system could be delayed until 1988 or later, when additional supplies of raw water also would be needed. In fact, the 1977 restrictions on watering reduced the max-day 331 million gallons per day, or 64% of existing capacity. If these restrictions were applied in the future (taking into account expected population growth), the Corps of Engineers estimated that the treatment capacity would not be exceeded until the year 2000.

Thus, according to the analyses of the EPA and the environmentalists, the Foothills Project could easily be replaced by a program of conservation or water rationing, or both. The environmentalists contended that either measure would have significantly fewer harmful effects and be much less expensive than the proposed Foothills Project. The EIS pointed out that in 1977, restricted watering increased the health of lawns. The argument over the Foothills Project included, among other issues, debate about proper lawn care.

Growth, Sprawl, and Air Pollution

A major concern underlying the Foothills controversy was control over growth, urban sprawl, and air pollution. Denver is one of the fastest-growing urban areas in the United States and ranks second to Los Angeles in the severity of its air pollution problem. The pattern of growth in Denver has been characterized by low-density urban sprawl, resulting in high per capita use of the automobile—the source of most of Denver's air pollution. It has been suggested that the best way to control Denver's air pollution would be to reduce its rate of growth and to direct growth toward existing centers of urban activity. Often there is a strong link between the provision of public services and settlement patterns; hence, many viewed control of water distribution (in this case, by preventing construction of the Foothills Project) as the first step in a major effort to control growth and air pollution.

Proponents of the Foothills Project disagreed with this analysis for several reasons. First, they argued, growth would occur with or without the Foothills Project. Citing the experiences of several other rapidly growing cities in arid regions (among them, Salt Lake City, Las Vegas, Phoenix, and Los Angeles), both project proponents and the BLM (in its FES) contended that the Foothills Project would neither suppress nor encourage population growth in the Denver area. The only party that disagreed with this position was the EPA. The agency pointed out that the cities just cited have "gone out of their way to provide water for growth," implying a relationship between the provision of water and the influx of population.

Although the parties debated the potential of water supply planning to influence the rate of growth, almost all agreed that water planning would affect development patterns. The DWB is required by law to fulfill all requests for service (water taps) within the city limits. Limitation of taps would favor inner-city growth over urban sprawl because the limitation would apply only to the suburbs. The existing limits on new taps, according to the *Denver Post*, "spurred a dramatic increase in the number of single family homes being built or planned within Denver." Thus, the EPA and many environmentalists contended that not constructing Foothills would promote higher densities by encouraging housing construction in the city of Denver where water taps would be more readily available. According to the EPA analysis, this situation would decrease reliance on the automobile and thus lessen air pollution.

The EPA also maintained that "construction of the Foothills project would make the attainment and maintenance of National Ambient Air Quality Standards in Denver more difficult and perhaps impossible." The EPA concluded that the project was "unsatisfactory from the standpoint of public health, welfare and environmental quality" and therefore subject to CEQ intervention. In addition, the EPA warned that, as required by the Clean Air Act, it would impose costly economic sanctions if the national ambient air quality standards (NAAQS) were not met by 1987. These economic sanctions could make Foothills Project prohibitively expensive.

Foothills Project proponents—especially the DWB—contended that limit on water service would increase urban sprawl. They argued that such limits encouraged the development of separate, independent water districts during a limitation on new water development in 1951. According to the DWB, "without Foothills, the limitation on new water taps must continue, and leapfrog development . . . is bound to occur. EPA's action, if successful, will only serve to create and stimulate the very kind of development which it abhors."

In addition to the arguments over impacts on growth, the DWB and others contested the federal permitting agencies' right to make land-use decisions for the city of Denver and the surrounding region. According to DWB, such decisions should be made by local residents and their elected representatives, not by the agencies required to supply water to the area or by federal officials.

In the DWB's view, local residents made their decision in the 1973 bond election, and it was clear that the duty of the board was to carry out this choice. According to the DWB,

> As the public interest group, Historic Denver, has pointed out, the citizens of Denver over 100 years ago chose an esthetically pleasing environment patterned over the English Garden style of the East. The people have continued this lifestyle, the people have voted water works improvements with this in mind and neither narrowly based special interest groups nor federal officials have the right to reject the people's choice and destroy this Historic Value.

The BLM and USFS shared this view, maintaining that,

> the issues involved in the Foothills project are based on questions of land use and population distribution in the Denver area, and, thus, should be dealt with by state and local governments, not federal land management agencies.

On this basis, the BLM chose to issue the right-of-way permits over objections from the EPA and the CEQ.

The EPA continued to believe that such matters were its appropriate concern under its mandate to protect the environment. Thus, it actively opposed the project. The debate widened to include not only matters of land use and pollution but also the issue of local, state, and federal control of land use.

It was also feared that the Foothills Project would have other adverse environmental impacts, particularly in Waterton Canyon, where the Strontia Springs dam was to be built. Waterton Canyon is the only relatively undeveloped, steep-walled, narrow canyon near Denver, with a rapidly descending, free-flowing river coursing its length. In addition to providing excellent white-water kayaking, the canyon offers excellent opportunities for hiking, fishing, bicycling, and wildlife observation near Denver.

Development of Future Raw Water Supplies

The fourth major issue concerned the future development of additional supplies of raw water for the Denver system. As was indicated previously, without new supplies Denver would face a shortage of raw water beginning in 1988. As a result, severe summer watering restrictions would be needed, whether or not Foothills was built. The plans to expand the Foothills Project to 500 million gallons per day depended on the development of additional supplies of raw water. The source of such supplies was a matter of deep concern.

One means of increasing future water supplies would be to construct an additional large dam and reservoir on the South Platte River, which was upstream from the Foothills treatment plant. The DWB had been planning this project, called Two Forks Dam and Reservoir, for many years, although it had not set dates for construction nor applied for the necessary permits. Two Forks was highly controversial because of its probable local environmental impacts (which were substantially greater than those of Strontia Spring because Two Forks would be 19 miles long, whereas Strontia would be only 1.7 miles long) and because it would depend on further diversion of western-slope water to the eastern slope.

Sensitive to the diversion controversy, the DWB insisted that Foothills and Two Forks were separate projects and that Foothills was necessary regardless of whether Two Forks was built. However, they conceded that additional water supplies would have to be developed for Foothills to function at its full 500-mgd capacity.

Other Issues

Competition between the city and the suburbs over water, population, land use, and political power surfaced throughout the controversy. The intensity of the debate about the need for lawn watering revealed how tightly attitudes were tied to differing life-styles and conflicting visions of the future. To an outsider, unlimited lawn-watering capacity might seem like a trivial issue, but project proponents saw it as a potent rallying point, claiming it was essential to the preservation of historic values. The Corps of Engineers concurred in this analysis, citing a passage from an encyclopedia of gardening:

> In Denver some statisticians estimated that as many as 86 million man-hours were being devoted to lawn care during the five-month season; calculating the labor at $1.60 an hour, they concluded that lawn tending was the largest single industry in the area.

Environmentalist John Bermingham countered that he observed changes in life-styles that would actually decrease the demand for water. Among the changes he noted was the back-to-the-city trend, which he believed was a response to problems of energy and home financing. This trend, he predicted, would lead to a decrease in water use because "apartment dwellers do not water lawns." In addition, he felt water use could be controlled by the emerging "conservation ethic" that could be fostered through increasing public awareness.

Thus, Bermingham concluded, prevailing life-styles would not suffer if the Foothills Project was abandoned.

Although the presence of a multiplicity of issues can aid negotiations by providing the parties with ample grounds for bargaining, the issues must be separable and manageable. The number and complexity of issues in the Foothills controversy, the relationships among them, and the uncertainties of projected demands and impacts greatly complicated the negotiation process.

The Disputants

Most of the major disputants have been introduced. The project sponsor was the DWB—an independent government agency of the city and county of Denver responsible for supply water to Denver and many of the surrounding suburbs.

Federal agencies involved included the USFS and the BLM, which worked on environmental impact statements and collaborated in receiving and issuing their respective rights-of-way permits. Although the EPA and the Corps of Engineers were supposed to collaborate on the 404 permit, they fought at length before they reached a settlement. The CEQ was involved briefly, as were other federal actors, including the Federal Energy Regulatory Commission and, to a lesser extent, the FWS.

A number of state agencies were involved in the dispute, including the governor's office, the Department of Natural Resources, and the Colorado Department of Health. These three agencies supported the Foothills Project at the 125-mgd level, and participated actively in the dispute either by issuing permits (the Department of Health) or by taking part in the negotiations and mediation between the DWB board and the federal agencies.

A multitude of interest groups were aligned with the major disputants. The Denver Regional Council of Governments and numerous local governments supported the project and conducted studies, wrote press releases, lobbied, and generally tried to influence the public and the government agencies in favor of the Foothills Project. Local businesses also became involved, as did a nonprofit group known as Water for Colorado, which lobbied with the businesses and the DWB in favor of the Foothills Project.

Aligned with Foothills Project opponents were numerous local, state regional, and national environmental groups. These include the Water Users Alliance (Bermingham's group), the Colorado Open Space Council, the Sierra Club Legal Defense Fund, the Rocky Mountain Bighorn Society, the Colorado Mountain Club, the Colorado White Water Association, the National Wildlife Federation, the American Rivers Conservation Council, the Environmental Policy Center, the Concerned Citizens for Upper South Platte, the Environmental Defense Fund, Trout Unlimited, Friends of the Earth Foundation, the

Wilderness Society, the Foothills Coalition, and the League of Women Voters. These groups were plaintiffs and defendants in two suits linked to the dispute. One was brought by the DWB in Denver, whereas the other was instituted by environmentalists in Washington. Many more individuals and environmentalists were involved to a lesser degree.

Major political figures in Colorado were involved, including the governor, the senators, and the congressional representatives. With few exceptions, all of them perceived the Foothills Project to be a very dangerous political battle and steered clear of taking sides, particularly as the 1978 election drew near. At the same time, all were under pressure from their constituents, as well as from the DWB and the EPA, either to support or oppose the project. In fact, Alan Merson, the EPA regional administrator, stated that if only one major political representative, senator, congressman, or the governor had supported the EPA's stance, the agency would have opposed the project steadfastly and probably would have prevailed on its terms. As time went on, however, more and more politicians publicly voiced approval of the DWB, which was a crucial factor in Merson's decision to drop his opposition first to the project and, later, to the Strontia Springs dam site.

Obstacles to Mediation

Polarization

In addition to the controversy's size and complexity, there were other obstacles to successful mediation. Most important was the DWB's unwillingness to compromise and its hostility to the idea of mediation. Although delay was increasing costs daily, the DWB was firm in its insistence that Foothills Project be built as designed. They maintained that the dam at Strontia Springs was the only technically satisfactory alternative. Because the environmentalists and the EPA were opposed to the Strontia Springs site (and many were against the project altogether), the DWB saw no benefit in pursuing mediation.

The polarization of the conflict grew over time, and the personal animosity between several of the major actors was intense. For instance, Merson received fierce public attack from the DWB, the *Denver Post* (which was strongly pro-Foothills), local politicians, and numerous other Foothills supporters. They claimed that both he and the EPA staff, which supported him, were incompetent, irresponsible, and politically motivated. In their comments to the corps in regard to the 404 permit. the DWB said:

> The EPA has simplistically characterized "the principal purpose of the project" as permitting the Denver area unrestricted summer lawn watering. This provides an excellent example of the danger inherent in any attempt to oversimplfy a complex issue. . . .

> Both the Board of Water Commissioners and the Environmental Protection Agency are charged with the responsibility of ensuring that the people of Denver are provided with high quality drinking water. It is submitted that EPA, in taking its recent action, has shirked that statutory responsibility. The temptation of social meddling to further political philosophies was apparently too great.
>
> The "true fact" of the matter is that opponents of Foothills are using this as a tool to advance their political philosophies; political philosophies which are not held by those whom the people have elected to public office.

Although none of these statements mentioned Merson personally, many similar comments in the *Denver Post* did. The editorial writers at the paper were particularly hostile:

> The decision is broadly troubling for the simple reason that one man, not elected by the people, has worked with his Washington superiors [also not elected] to overturn a decision on a water project made in 1973 by the voters of Denver. . . . Like other transplanted Coloradans, Merson wants "quality" growth now that he has established roots here.

A subsequent *Denver Post* artcle reported that

> Merson is well aware of the accusations that having lost three attempts to win elected office, he now is trying to impose his personal political philosophy on an unwilling public.

Merson did not respond to the personal attacks in kind. However, until the mediation process began, he took a firm anti-DWB stance and, like his opponents, made extensive use of media to wage his war against the project. After the USFS and the BLM had issued the permits and the case had been sent to the Corps of Engineers, Merson flatly told the *Denver Post* that "the fact is Strontia Springs is not going to be built. I think the Denver Water Board should realize that and modify their proposal." Such strong public statements were characteristic of Merson's strategy and made backing down or negotiating much more difficult for the other parties. This strategy heightened the DWB's distrust of his motives, making the initial stages of mediation very difficult.

In addition, John Bermingham, a leading anti-Foothills environmentalist, became embroiled in a similar public debate in which he tried to destroy the DWB's credibility and image. Bermingham issued a lengthy report entitled *Foothills $135,000,000—In the Wrong Place* that charged the DWB with misleading the public, inaccurately assessing the need for additional water treatment, and withholding the figures that the public and government must consider if they were to make an informed decision. Bermingham even asked the Denver district attorney to file criminal charges against DWB manager James Ogilvie on the grounds that he refused to release financial information about construction costs for the project and the cost of acquiring new supplies of raw water for future expansion of Foothills. If these figures were made available, Bermingham charged, the information would "show that Foothills is going to be worthless in a few years . . . [and that it would] be cheaper to put in a recycling plant than to go to the mountains for additional water supplies."

This bitterness had to be replaced by some measure of mutual respect and trust before constructive negotiations could begin. Even after the DWB agreed to negotiate, the mediation team faced a major challenge in defusing the battle and creating a climate of cooperation.

Scope of the Dispute

Foothills is particularly interesting to students of mediation because two people attempted to mediate the dispute at different times. Because the second effort succeeded after the first attempt failed, the case provides an instructive comparison of alternative methods for initiating mediation.

Representative Pat Schroeder tried to introduce mediation fairly early in the dispute (May 1977) after she had heard from a friend about the University of Washington-based Office of Environmental Mediation. Because the DWB had recently gone to court to seek an end to federal delays, Schroeder thought mediation might bring an even speedier settlement.

The DWB and other Foothills supporters promptly rejected this suggestion for a number of reasons. Most important, Schroeder was perceived to be an environmentalist. Although she was careful not to take a public stand for or against Foothills, most proponents of the project assumed from her voting record that she was against it and that her proposal was simply another tactic for delay. Also, little was known about the Office of Environmental Mediation members—what they could do, how they worked, and if they could be trusted. Those who believed in Schroeder (the environmentalists) thought the organization's participation might be helpful, whereas those who distrusted her thought that it probably consisted of more antigrowth environmentalists who simply would further delay the decision process.

The problem was exacerbated by a major strategic error. Instead of privately selling the idea of mediation to each party individually, Schroeder immediately publicized the idea, which made it appear as if agreeing to mediation was agreeing to her environmental position. In addition, it made many of the contesting parties believe that she was primarily seeking good press coverage and had little real interest in resolving the dispute.

In January, Schroeder again tried to begin mediation by holding a press conference at the mouth of Waterton Canyon. According to the *Denver Post*, she said the DWB and Foothills opponents "should work out their differences rather than have lawsuits from now to kingdom come." She further accused both sides in the dispute of ignoring the possibilities of compromise because "a lot of people think in the long term they're going to win and egos get involved."

The conference did not change the DWB's position, and they continued to rebuff Congresswoman Schroeder's efforts. Later, when asked why they opposed Schroeder's initiative, Wayne Williams, DWB general counsel, stated,

> We felt we would be dealing with a situation where there was no possibility of our coming out with our project. We would have to sacrifice one or more important features . . . In my experience, that is normally the result of mediation.

At the time, the formal position of the DWB was presented in a letter from Charles F. Brannan, then president of the DWB, to Congresswoman Schroeder. Williams read the letter in a meeting called by Schroeder to consider mediation further. In addition to the primary parties present, there were also other interested parties, including Gerald Cormick and Leah Patton from the Washington Office of Environmental Mediation, Helmut Wolff from the American Arbitration Association, and John Kennedy and Susan Carpenter from the Rocky Mountain Center on Environment. All gave presentations on the merits of mediation, but Williams had his "mind made up" and read this statement:

> Mediation and compromise are, of course, useful in many situations, but we do not believe that anything is to be gained, and much may be lost by attempting to pursue such a series of meetings at this time. I am writing this letter to give you a statement of the more important reasons for our conclusion as follows:
>
> First: The people of Denver approved the Foothills Project and authorized the bonds for its construction in the 1973 election. We do not believe that the expressed will of the people can be mediated or negotiated away.
>
> Second: The vital decisions to be made respecting the Foothills Project are to be made, according to law, by various agencies of the Federal government. Mediation has no proper place or function in the completion of this process. The agencies are obliged to make their determinations according to the standards provided by the law and without reference to which special groups are able to speak the loudest or get the most publicity. Whatever conclusions might be arrived at by mediation, they cannot take the place of review and action by the prescribed federal governmental agencies.
>
> Third: The duty of the federal agencies to act is presently involved in a pending law suit before Judge Winner, Chief Judge of the U.S. District Court, and certain dates for federal action have been established by agreement in that case. A mediation effort involving various special interest groups would simply tend to remove the controversy from the forum where it belongs, and substitute talk for proper agency action. Mediation could contribute nothing to the progress of the case now before the Court, and would probably delay resolution of the entire matter with consequent mounting expense.

Fourth: As stated in my letter to you of 4 August 1977 on this same subject, negotiation with people who have no authority in the matter and whose agreement or consent would be utterly unenforceable in any administrative or legal forum would be without purpose and be an indefensible waste of public funds.

Cormick and others from the Office of Environmental Mediation held a number of meetings before and after the DWB made this statement, but they were unable to alter the DWB's position. Other parties, including the environmentalists and the EPA, said that they supported the idea of mediation; however, mediation was meaningless without the DWB's participation.

Moreover, most observers agreed that Schroeder's efforts came too early in the process, at a time when Foothills' proponents had no incentive to negotiate. The FES had not yet been issued, and the major permits (BLM, USFS, 404) were not yet under active consideration. No mediation proposal offered the DWB an outcome that would have been superior to that that they expected from litigation. Federal district court Judge Fred Winner was believed to hold a strong antienvironmental, prodevelopmental position. Many thus expected that he would rule in favor of the DWB. The DWB also expected the litigation to proceed quickly, ending the federal delays and perhaps the conflict itself. (This optimism later proved to be unfounded; a credible mediator might have persuaded the DWB that success in a court trial could still be followed by years of appeals.)

Schroeder followed the Waterton Canyon press conference with a seminar on mediation a few days later. Representatives from the American Arbitration Association, the Rocky Mountain Center of Environment, and the Office of Environmental Mediation each gave presentations illustrating the potential usefulness of mediation in solving such disputes. Most agreed that the meeting accomplished nothing. The environmentalists who were present were receptive because they thought mediation would force the DWB to reveal information about the plans and costs related to additional diversion of western-slope water. This information, they believed, would vindicate Bermingham's assertions that the plant was unnecessary and lead to resolution that was completely in their favor.

The DWB again rejected mediation, citing its earlier letter. Thus, the dispute continued in administrative and judicial forums, and the idea of mediation was put aside.

The 404 Permit Dispute

The third draft environmental impact statement was issued in August 1977, and hearings on it were held in September. One major concern was whether the statement anticipated the impacts of the proposed 500-mgd facility. In November 1977, Judge Winner effectively removed this hurdle by requiring the DWB to limit the first phase of the plant to 125 mgd and by requiring additional EISs and permits for any future expansion. The BLM was ordered to finish the FES by February 1978, and to issue their final permit decision by the following month. Although a great deal of last-minute work was required, the BLM met these deadlines and issued the right-of-way permits despite CEQ's objections. The EPA continued to oppose the project, fortified by the power to veto the pending 404 permit and thus send the case to federal court.

None of the parties wanted the case to take this course. Foothills was widely considered a local dispute, and a major political battle was brewing over local-versus-federal control. In addition, both the EPA and the DWB were afraid of losing if the case went to federal court. The DWB estimated that the delay cost them $630,000 each month, and the project itself was at some risk because President Carter's administration was perceived to be firmly against water development.

Still, the EPA was not confident about support from Washington. EPA Administrator Costle apparently pressured Merson to settle the dispute at the regional level. Costle wanted to stay out of what he saw as a politically dangerous local fight. Merson did not report feeling such pressure, although he recalled that after the BLM and USFS had issued their permits, Costle had said to him, "Well, that's it, isn't it?" as if conceding defeat.

Others anticipated little, if any, federal opposition to the Foothills Project. The CEQ refused to take the matter to the president on the grounds that "environmentalists in the Carter Administration didn't plan further protest actions." The *Denver Post* quoted an unnamed federal official, who explained that "the federal handle was small" in attempting to block the Foothills Project. The two best arguments developed by the EPA and the CEQ—protests of environmental impact to both the Colorado western slope and the Denver area—had been turned down.

Finally, on April 7, 1978, the EPA retreated under public pressure and announced that it would not block construction of the water development project, but it added that it hoped to attach stipulations regarding water conservation before the 404 permit was granted. The BLM permit already required that the DWB institute a loosely defined plan for water conservation before permits were granted for future expansion of Foothills or other aspects of its water system.

Merson attributed his change in position to the "lack of firm political support for the EPA's position on Foothills." Still, he did not back down any further than was necessary. The battle was far from over. The EPA could still threaten to veto the 404 permit and force the DWB to make concessions in favor of the environment. Specifically, the agency hoped to generate sufficient opposition against the Strontia Springs dam to force the DWB to build a smaller, less damaging dam in another part of the canyon.

To gather such support, the EPA planned another round of Foothills hearings in April 1978 on both the eastern and western slopes. The DWB objected strenuously to further delay, maintaining that the prior BLM hearings made them unnecessary. The earlier sessions had focused on the right-of-way decisions. However, hearings had not been held on the 404 permit and water conservation issues the EPA wished to emphasize. A *Denver Post* article quoted Merson as saying, "We've lost all of our leverage, quite honestly, if we issue the permit, and then talk about water conservation." The article went on to report that Merson was personally leaning toward approval of the water board plans for a 245-foot high dam and reservoir at Strontia Springs. "We'll never smile on Foothills," he said, but he would accept it "if mitigating measures can be applied."

The DWB also objected to the EPA holding any hearing on the western slope, contending this was beyond the EPA's authority. The EPA justified the action by saying that many still feared that Foothills was linked to future projects for western slope water diversion and that residents of the western slope deserved a chance to voice their concerns on this issue. Although lawyers for the DWB attended the western slope hearings, they refused to answer questions, saying that the DWB's presentation would be given in the public hearing to be held in Denver the next evening.

More than 300 people attended the EPA hearing in Denver. Although most of them opposed the project, a number of important statements were made in favor of it. A spokesman for the mayor of Denver pointed out that in 1973 Denver voters had approved a DWB bond issue by a martin "approaching a landslide. . . . Now EPA, here today, is in a very devious way trying to circumvent these decisions made by the administration through the guise of the 404 permit program." The mayor's spokesman also accused the EPA of "attempting to change the lifestyle of Denver residents, superimposing its fanatical water conservation ethic on our area."

DWB Manager Ogilvie also attacked the EPA, saying it had no legal authority to require water conservation as a condition of the 404 permit and that the DWB could not accept such conditions. Ogilvie asserted that the DWB's "record on water conservation is exemplary and unsurpassed by major utilities in the area." He went on to say that whereas the board

> intend[s] to be at the forefront on conservation, Denver will have no part and will not permit the EPA to extend its federal tentacles beyond the lawful jurisdiction provided that agency by Congress. Denver views this procedure [the hearing] as a baldfaced attempt by the regional administrator of the EPA to interject himself in a matter of local concern and local prerogative for which he has no legal authority.

Despite the anti-EPA comments, the bulk of the testimony at the hearings strongly favored water conservation and opposed the destruction of the natural ecosystem in Waterton Canyon. As the EPA had hoped, the hearings supported strict conservation and impact mitigation measures. As a result, on May 25, 1978, Merson announced that the EPA had recommended to the Corps of Engineers that it deny the 404 permit for Foothills on the grounds that the Strontia Springs dam would have serious adverse impacts on the environment of Waterton Canyon and that alternatives less damaging to the environment warranted further examination.

The EPA's recommendations were then forwarded to the corps district office, which sent them to the division level along with the district engineer's findings. Most observers expected the division engineer to route the dispute to Washington to be resolved by top EPA and corps officials. Because the 404 permit allows the EPA final veto power, Costle eventually would make the permit decision. According to Merson, "Mr. Costle is aware of my position and has expressed his concurrence with it." Nevertheless, Merson refused to predict whether Costle would veto the corps if it chose to issue a 404 permit without additional mitigation measures.

Wirth Initiates Mediation

Tim Wirth, the other congressional representative from the Denver area, entered the controversy at this stage. According to his aides, Wirth could see that the dispute was going nowhere and he considered a stalement unacceptable. He also believed that the dispute was local and should be resolved locally—not at the federal level. Although he could have remained neutral, as did the rest of the Colorado congressional delegation, Wirth had voiced his support of the 125-mgd project, a stance that sparked criticism from his environmental constituents. At the same time, the pro-Foothills forces were demanding he do more to "get the Feds off Denver's back," and James Kenney, DWB president, had specifically requested Wirth's help in attaining final federal approval of the project.

Wirth's seat in Congress had never been secure. He realized that neutrality would anger both sides; action would alienate, at worst, half of his constituents. If he was careful, he might achieve a politically advantageous resolution that would satisfy most of the parties. Neither a continued stalemate nor a federally imposed solution was desirable to any of the disputants. Neither side was certain of a satisfactory outcome. The situation was ripe for the intervention of a mediator.

Timing was not Wirth's only important advantage. Contrary to theory, Wirth's success is also due to his widely perceived, long-standing support of the 125-mgd project. Specifically, when he suggested mediation, the DWB expected it to work to their benefit. In addition, Wirth began the process by persuading the Corps of Engineers to complete what he hoped would be the definitive study of the remaining issues. Because both the corps and the DWB favored water development and structural solutions to water-related problems, most DWB members believed the cards were stacked in their favor. Getting the EPA to concur was more difficult, but Wirth fashioned his proposal so as to leave them little choice.

Unlike Congresswoman Schroeder, Wirth started the negotiation process privately and carefully, First, he talked with Merson, who was also a friend. Merson, already under heavy political pressure, had reduced his opposition to the project. Yet, he still objected to issuance of the 404 permit because, as he told Wirth, he felt that the alternatives to the Strontia Springs dam had not been adequately considered. Wirth thus concluded that if he could arrange a study of such alternatives, Merson would have to agree to the process. Because the corps already was involved in the dispute and its expertise in dam design was widely acknowledged, a corps study of alternatives seemed appropriate.

Wirth then consulted another friend, Clifford Alexander, secretary of the army, as well as several members of the corps. He suggested that Merson (and thus the EPA) probably would acquiesce if the corps was to expand its normal review process for 404 permits to review thoroughly the alternatives to the Strontia Springs dam. After brief discussion, the corps agreed to the expanded review; the DWB also agreed to this because they expected the corps to agree with their analysis and rule in their favor.

Although the EPA objected to Wirth's choice of the corps as the reviewing agency (some said it was like letting the "fox look after the chickens"), no one could dispute the corps' expertise in dam design or contest its statutory authority to review the dam and proposed alternatives. Because Merson had already publicly and privately defined the two remaining issues as the dam's location and size, he had no political or legal grounds for blocking the corps' analysis.

But Merson could make his approval of the mediation process contingent upon the adequacy of the corps's review. He told Wirth and the press that he would not abdicate his decision-making responsibilities. Thus, he could not agree to binding arbitration before knowing the quality of corps review. However, he asserted, if both he and the EPA staff were satisfied with the corps' procedures and results, "we would find that very persuasive." Although the EPA did not make a commitment to accept the corps' study, Merson acknowledged that "he would find it very difficult to ignore."

The DWB also refrained from publicly accepting the corps' review as binding, although they came very close by saying that "if we're wrong, then these are the people to tell us we're wrong. If we're right, then they are the ones to tell the public." When Kenney was asked what the DWB would do if the corps' study "went against the Water Board," he replied, "I just don't know. I guess we'd go to the people of Denver . . . I just don't know."

Neither party explicitly agreed publicly to be bound by the corps' review, but they agreed to pursue the study with the hope of reaching a settlement. This agreement was announced to the public in a press release issued by Wirth on June 16, 1978:

> Wirth today reached agreement with the Denver Water Board, the Environmental Protection Agency, and the Army Corps of Engineers on a proposal for a final, full, and complete consideration of the most significant outstanding controversy still surrounding Foothills, the Strontia Springs Dam in Waterton Canyon.
>
> Under the terms of the Wirth proposal, the Army Corps of Engineers will go beyond its normal review procedures to conduct a major study of the Strontia Springs Dam and alternatives to it. Both the Water Board and the EPA have agreed that they would have no problem with the Corps' decision as long as the procedure established and carried out is fair and thorough.
>
> Failure to consider the alternatives to the dam was the ground on which EPA based its recommendation to the Corps that the necessary federal permit not be granted. Under the regulations governing such matters, a ruling on whether or not to grant the Water Board a permit to construct a dam in Waterton Canyon now rests with the Corps of Engineers, but normal Corps procedures do not call for a review in the depth contemplated by the Wirth proposal. Under the proposal, the Army's review would not consist solely of approval or disapproval or a permit for the Strontia Springs Dam, but would be expanded into a full-scale review of the dam and its alternatives.
>
> At present, should the Corps differ with EPA's recommendation, EPA holds a veto power over the granting of the permit. Under the Wirth proposal, EPA and the Water Board said they should have no problem with the final result if it was reached after a full and equitable study by the Corps.

The press release did not refer to the corps' review as mediation; however, the participants perceived it as such and hoped the review would result in a settlement. When the *Denver Post* asked Wirth about the implications of the agreement, he replied, "We think we've got this thing sorted out," and he said that "he was hopeful the study recommendation would be binding" to the DWB and the EPA.

The Corps' Review Process

The Corps' Role

Although many factors contributed to the success of Wirth's mediation, one important factor was the care with which he initiated discussions. Just as he proposed the mediation in a way that induced both the DWB and the EPA to accept, he structured the mediation process in a way that made their withdrawal difficult. As the study went on, the *Denver Post* continued to reinforce the notion of binding fact-finding, thereby increasing public pressure, and, in turn, the obstacles to withdrawal. Although the EPA and the DWB each continued to claim that they could always retire from the negotiations if they were dissatisfied with the corps' study, anyone who did so risked looking as if they had gone back on their word. Given the EPA's tenuous political position in the state at the time, the agency especially could not risk creating an adverse image.

According the Bob Drake, Wirth's aide, "we could see that the EPA wanted to throw in the towel [withdraw from the mediation process], but they could not because of the power of the press and the process itself." Press coverage put pressure on the DWB as well. On several occasions, according to David Aylward, the DWB threatened withdrawal from the negotiations but did not do so when Wirth or his aides threatened to renounce them publicly.

In addition, Wirth quickly established an interactive and cooperative procedure that created additional pressures on the parties to remain in the process until they reached a settlement. Rather than allow the corps to design and complete the review on its own, Wirth arranged a series of meetings among the corps, the EPA, and the DWB in order to clarify jointly the remaining issues in contention and determine the scope and procedures for the corps' review. This turned the process into a cooperative venture and set a precedent for the joint consideration and discussion of issues. Further, this joint procedure improved the study and increased its legitimacy. Because the EPA was so heavily involved in both the design and implementation of the study, they could not easily repudiate the corps' review.

The parties spent the entire first month of negotiation, July 1978, discussing and defining the outstanding questions and determining what information they needed in order to resolve them. Although both the corps and the DWB sought to complete the study as quickly and as simply as possible, the EPA insisted upon a thorough review. Because the study's usefulness for resolving the dispute was contingent upon its acceptance by the EPA, the agency was able to enforce its demand, despite the objections of the other parties.

The *Foothills Newsletter* was introduced as a medium for public announcements concerning the progress of the corps' study. The first issue identified three key topics:

> The need for additional water treatment, [the] existence of superior alternatives [to either the project as a whole or just the Strontia Spring dam], and [the] severity of environmental, social, and economic impacts, whether direct or indirect.

The resulting study was far more comprehensive than the corps originally had planned.

Although the EPA's veto power gave it the upper hand in the early negotiation, the agency was not to prevail throughout. Although the corps agreed to expand the study considerably, they also promised (at the DWB's insistence) to complete the study as expeditiously as possible and to limit the analysis to unresolved issues. In the Foothills Review Procedure Outline, approved by the DWB, the EPA, the corps, and Wirth on July 4, 1978, the corps asserted that its study would not "beat any drum twice. If an alternative is demonstrably infeasible from the standpoint of construction or operations, we shall not enter into time-consuming economic, social or environmental evaluations." The EPA was wary of the emphasis on speed and insisted that the study last at least five or six months. According to the same *Denver Post* report:

> The U.S. Environmental Protection Agency has served notice it will abide by the findings of a U.S. Army Corps of Engineers review of the first phase of Denver's Foothills water treatment project only if sufficient time is devoted to that review.
>
> That means a study of at least five or six months, said Roger Williams, EPA's Deputy Regional Administrator. And that effectively would put off the start of construction of Foothills for another year because it would miss this year's construction season.

By placing this demand in the newspaper, the EPA tried to strengthen its hand and maintain an escape should the study lead to conclusions they did not wish to accept. Thus, the EPA tried to counter Wirth's manipulation of the press in kind. Though it was less successful, largely because of the abiding press bias against the EPA, the agency enjoyed less success than Wirth with this tactic.

The Role of the Public

Another item of contention in the early phases of negotiation was the public's role in the corps' review process. Although the EPA strongly favored a high level of public participation, including more hearings and a public advisory committee, the corps was against further public participation and wanted their studies to remain confidential until the findings had been reviewed by the parties involved. In fact, the corps tried to conduct the first few meetings among parties confidentially and, at one meeting, ejected a reporter from the *Rocky Mountain News*. The EPA then refused to take part in the meeting, and the session was adjourned.

After this scene, the parties agreed to forego additional public hearings but to allow the public and the press to attend the meetings on the corps' review process. In addition, the corps agreed to issue additional *Foothills Newsletters*, which would review for the public the course of the study and its major findings. Dale Vodehnal of the EPA Region 8 pointed out that this procedure ultimately may have hurt the EPA because the EPA gave the corps its comments after publication of the newsletter, the public saw only the corps' point of view on some topics. Although the EPA's objections were upheld and the analysis was revised, the preliminary results already had been publicized in previous newsletters and, as Vodehnal described, "the damage had already been done." Nevertheless, the publicity allowed some public scrutiny of the process, which further pressured the corps to produce an adequate study.

The Need for the Plant

The first phase of the corps' review was an analysis of the need for the plant. The corps released its initial findings on need on August 9, 1978, very soon after the study guidelines had been established. Although the corps concluded that the FES estimates of further water use in Denver were "unrealistically high," the corps' analysis indicated that:

> Based upon a continuation of typical past consumption patterns, . . . Denver's treatment capacity will be exceeded in the summer of 1980. A permanent program of conservation comparable to that experienced in 1977 would postpone the need for more treatment another fifteen years, but not without some economic and environmental costs.

Furthermore, according to the corps' newsletter, these costs would be sufficiently high enough so that options for conservation should be considered at length before a particular plan was chosen. In addition, the corps pointed out that a substantial unmet need already existed as a result of drought and the DWB's subsequent limitation on new taps, which the board said they could not lift until Foothills was built. This limitation resulted in a two-year wait for water hookups in some areas. The corps concluded:

> The near-term need for treatment appears valid since conservation alone is not the answer. Nonetheless, the Water Board has an obligation to continue a vigorous program of water conservation as it prepares to cope with future growth. Governmental agencies in the metropolitan area which are charged with land use management and zoning activities should also bend their efforts toward this end.

The EPA and several private citizens disagreed with this analysis and contested the findings during both the August 9 meeting and in follow-up letters. The main issues of contention, as listed in the next *Foothills Newsletter* were that the corps' analysis

> did not present additional evaluations of rationing which postponed the need for treatment by more than the fifteen years [as] shown by MRD. Did not evaluate Denver's ability to utilize existing clear water storage and overload capacity of existing plants to meet peak demand without building new treatment facilities. Did not disclose that additional sources of water treatment will be needed under many assumptions of future use. Did not substantiate the economic, social, and environmental costs attributed to immediate im-

position of conservation measures sufficient to remove the need for additional treatment.

These issues were discussed at the August 9 meeting, at which time additional analyses were presented by William Ganner, a private consultant, and John Bermingham, head of the opposition group called the Water Users Alliance. Both their analyses were included in the *Foothills Newsletter* of September 27, 1978. The analyses contained different estimates of future need that were based on different assumptions about future per capita use. Neither changed the corps' overall assessment significantly. Ganter determined the per capita use to be about 10% higher than the corps' estimate, thereby confirming that Foothills, indeed, was needed immediately if unlimited per capita use were to be maintained. Bermingham pointed out an error in the corps' estimates of the impact of 1977-style rationing. This meant that if rationing patterned after 1977 rules was instituted immediately and permanently, Foothills would not be needed until the year 2,000 (the corps previously had estimated 1995). The corps readily agreed to this correction and based its subsequent conclusion on that estimate.

In addition, the newsletter contained a brief analysis of the possible effects of various conservation measures at the request of the EPA. An analysis of the actual capacity of the current Denver Water Department's system by an engineering consulting firm was also published in the newsletter.

These additions were responses to comments made by the EPA and Bermingham in the August 9 newsletter. Although the answers were not as extensive as the EPA had wanted, the dialogue continued as the corps' study moved on. The EPA and Bermingham did obtain some responses to their questions for themselves and the public.

After completing another analysis of the impacts of rationing and other water conservation measures, the corps concluded that:

> If conservation measures were used to keep average and peak (with rationing use rates to the 1977 level), need for additional treatment would not occur until the year 2000 and the need for more water would occur at about the same time . . .
>
> Efforts to predict future trends become increasingly uncertain the farther they are extended; nonetheless one conclusion is apparent from an examination of the data presented above: Denver's water dilemma has not one horn but the conventional two. In selecting a consumption pattern—whether unrestricted or constrained—the citizens of the Denver area will be establishing water supply requirements as well as water treatment requirements.

This statement confirmed the view of the EPA and the environmentalists that Foothills was linked to meeting future water needs of Denver. Because most future water would come from the western slope, they felt the issue of water diversion from the western slope should be considered in making choices about treatment or conservation. The corps maintained that state and local governments should make this choice and that it was not appropriate for federal consideration once the state and local governments had made their decisions. Therefore, the corps did not further consider western-slope water diversion versus water conservation on the eastern slope.

A Review of Plant Impacts

The focus of the study then switched to the second key item—the analysis of secondary environmental impacts. The October 7, 1978, newsletter dealt with Foothills' impact on urban sprawl, air pollution, and "the operating regime of Dillon Reservoir (the western slope reservoir feeding Foothills) and the quality of water on the western slope."

The urban and regional planning firm Llewelyn-Davies-Carson, Ltd., of Toronto, Canada, performed the analysis of growth, sprawl, and air pollution. This firm previously had conducted similar studies for the United States National Water Commission and the Bureau of Reclamation in Utah, New Mexico, and Arizona.

Although the firm's report was lengthy and polished, it was completed quickly and lacked substantial documentation. The EPA was angered by the study's conclusion that

> there is little direct relationship between water supply and pattern of growth. The Foothills Project, whether built or not, will have little effect on urban sprawl in the Denver Region and therefore cannot influence ambient air quality in the area.

Llewelyn-Davies-Carson based their conclusion on the observation that water is only one of a large number of urban services that must be planned, regulated, and coordinated in order to control urban sprawl. Unfortunately, the firm said, the "institutional framework of municipal government" in the Denver region is highly fragmented and thus is unable to engage in the necessary comprehensive planning. Therefore, they concluded, any attempt to control growth in one segment of Denver by controlling one service (e.g., water) will only divert growth and quite possibly increase urban sprawl in other areas. This, they pointed out, had happened in nearby Boulder, which has limited the number of building permits available as well. Llewelyn-Davies-Carson also asserted that:

> It has been suggested the provisions of service such as water, sanitary drainage and major roads be used as growth management tools. It is not, however, the responsibility of these service agencies to engage in land use planning. Rather it is their duty to provide the service where it is required.

The firm maintained that the responsibility for land-use planning is regional and must spring from a coordinated effort between local and regional government, utilities and the private sector.

The EPA strongly disagreed with this assessement and wrote a letter to the corps at the conclusion of the study summarizing its criticisms. After pointing out that the study had twice misquoted or misinterpreted EPA statements and positions, the letter charged:

> There is a disturbing theme throughout the report with which we must take issue. The report contends that nothing can or should be done about a given urban service since there is not a regional entity with appropriate authority to plan for and manage all urban services. Unfortunately, the many problems we face in urban areas like Denver (such as the Denver air situation) are too serious to await new institutional arrangements.

The EPA continued to believe that growth and air pollution could be controlled by water use limits and that enforced conservation would ameliorate both population density and air pollution in the Denver area. However, they did not have further data with which to support these claims. Nor did they have the time, money, or expertise to generate such data or reports. As a result, the contractor's findings were published without dispute in the October 7, 1978, *Foothills Newsletter*.

The newsletter also included a discussion of the impact of the 125-mgd Foothills unit on the supply and quality of water on the western slope. Because this area was outside the region of the corps division that was conducting the analysis (the Missouri River Division), the corps contracted the work to the Colorado district of the United States Geological Survey (USGS). The USGS presented a "preliminary technical analysis" on this issue.

According to this analysis, the Foothills Project would slightly affect the quantity and quality of water on the western slope by decreasing the level of water and increasing the concentration of dissolved solids in Dillon Reservoir. The analysis pointed out that this pattern would occur even if Foothills was not built—although more slowly. Not building Foothills, the corps concluded, would delay minor negative impacts but not prevent them. Effects on the quality and quantity of water on the western slope therefore were not regarded as sufficiently negative impacts to prevent further consideration of the Foothills plant. The EPA disputed this analysis. They held that the increase in salinity and the removal of water from the Dillon Reservoir were significant negative impacts and that the corps underestimated their importance. However, this objection was submitted after the last newsletter was published, and thus was not publicized.

A Review of Alternatives

The final segment of the corps' study and the final substantive newsletter dealt with an analysis of structural alternatives of the Strontia Springs dam. This was the corps' area of expertise, and the EPA acknowledged the quality of the corps' analysis. The analysis considered six different proposals, including the corps' original proposal for the Strontia Springs dam and four alternatives that would utilize an existing reservoir. The EIS considered only two of these alternatives; the other three arose after completion of the FES.

The most promising alternative involved a smaller dam at the mouth of Waterton Canyon. Although such a reservoir would be longer than that at Strontia Springs, its location far downstream was strongly favored by environmentalists because that section of the canyon was already heavily impacted. Flooding the lower canyon area, therefore, was not as objectionable as was flooding the wider, upstream section of the river.

In addition, use of the lower dam site had significant implications regarding land use. Because the Strontia Springs dam was considerably higher than the DWB service area, the DWB would be able to provide water to much of the southwestern Denver area by using a gravity-flow distribution system. Building the Strontia Springs dam would remove economic constraints that had previously deterred extensive urbanization. The "canyon mouth" dam was 337 ft lower than the Strontia Springs dam and would not be able to provide water to the southwestern Denver area without extremely high pumping costs. Merson and the EPA thus favored the canyon mouth dam, seeing it as "much less damaging from the urban sprawl focus, as well as with respect to direct environmental impacts."

The corps released its findings on the alternatives in the October 31, 1978, newsletter. The engineering, environmental, and economic advantages and disadvantages of each proposal were presented. The list demonstrated that the assets of the canyon-mouth dam were comparable to those of the Strontia Springs dam without stating any conclusions.

The Strontia Springs dam had several advantages. First, it could be built 1 to 3 years sooner than the canyon mouth dam because engineering designs had been completed and construction only awaited the 404 permit. In contrast, the canyon mouth dam would require extensive engineering and design work.

Second, the Strontia Spring dam was designed to produce more energy (through hydroelectric generators) than it would use. In contrast, the canyon mouth dam would not be capable of generating hydroelectric power and would consume 41 million kilowatt hours a year. The corps estimated that the annual operation and maintenance costs for the Strontia Springs dam would be one third of those at the canyon mouth dam.

In addition, the Strontia Springs dam was designed to provide water for Aurora, a Denver suburb, whereas the canyon mouth dam would not. If the canyon mouth dam was chosen, Aurora would have to replace its temporary and insufficient structure with a new dam, intake, and tunnel. According to the corps' estimates, this would add $22,770,000 to the construction costs of the canyon mouth dam. Nevertheless, the construction costs of the canyon mouth dam were significantly lower ($133 million) than those for the Strontia Springs dam ($170 million), even including the additional Aurora project.

Thus, both proposals had certain economic advantages. Although the DWB maintained that the pumping costs for the canyon mouth dam were prohibitive, the economic consultant hired by the EPA to assess the economics of the two proposals concluded that "the economic differences between the Foothills alternatives are actually very narrow and likely to be well within the errors of estimating the projected future costs." He also added that "economics should not play a determining role in deciding the issue."

The primary advantages of the canyon mouth were environmental. Although the corps' newsletter tended to deemphasize the environmental advantages of the canyon mouth site, it indicated that a "moderate" deterioration in aquatic habitat would occur at Strontia Springs as compared with a negligible impact at the canyon mouth site. In addition, the canyon mouth dam probably would have less of an impact on prehistoric artifacts than would the Strontia Springs dam.

The Study Concludes

The final newsletter stated that the next step in the process "will be the formulation of a decision on the 404 permit application." It requested that comments be sent to the division engineer by November 16, 1978. He would then make a decision and forward it to the EPA for concurrence or veto. If the EPA still disagreed with the corps' decision and chose to veto the permit, the dispute would be taken to Washington for resolution.

Because a Washington decision was still unpredictable, all of the parties preferred settlement at the regional level. Yet the corps' study was not as conclusive as the parties originally had hoped, and, in the end, the EPA still disputed major aspects of the corps' analysis. In particular, the EPA still disagreed with the corps' assessment of the viability of conservation as an alternative to Foothills:

> With respect to conservation and environmental benefits, the newsletter contained no discussion of benefits such as postponement or environmentally damaging water development projects, less fluctuation in reservoir levels, or reduced mass emissions of pollutants from wastewater treatment plants.

The EPA also disagreed strongly with the corps' lack of emphasis on the environmental advantages of the canyon mouth alternative.

Whereas the EPA continued to favor the canyon mouth location, the DWB steadfastly refused to agree to that site because it would take longer to build, and it would be expensive to operate. Although the corps apparently agreed with the DWB's assessment, it did not wish to confront the EPA directly, nor did the EPA with to refer the case to Washington. Discussion continued over the remaining issues until after the November elections when Wirth reentered the dispute.

Concurrent Suits

The DWB Files Suit

As the corps carried on its study, the controversy was escalating on various legal fronts. The DWB and the BLM were locked into a dispute over the terms of the right-of-way permit. In March 1978, the BLM issued the permit with broadly worded conditions requiring that the DWB ensure minimal streamflows below the Strontia Springs diversion dam and implement a water conservation program. The details were to be worked out between the parties in the following months. The BLM held several meetings with the corps to discuss these issues, but when the revised BLM permits were issued in July, the DWB rejected the conditions attached as unacceptable and illegal. They contended that maintenance of streamflow would require the DWB to release water over which they had legal rights.

In addition, the permits required that the water conservation program be approved by both the Denver Regional Council of Governments and the state of Colorado before the DWB could either expand Foothills beyond the 125-mgd capacity or "alter any existing federal water agreement or seek a new one that would use or affect any federal resources." The DWB charged that both stipulations forced it to "hand over some of its charter-delegated power" to regional and federal agencies. This, their lawyers argued, would "cause havoc" with their future plans for water development in the Denver area. Therefore, the DWB rejected the BLM permits and continued their lawsuit in the U.S. District Court in Denver to seek the unconditional right to begin construction on Foothills.

Presiding Judge Winner tried to broaden the case by involving all interested parties, both those for and against the projects, to participate in the proceedings. According to Winner, the invitation "was designed to combine as many legal aspects of the Foothills controversy as possible in one lawsuit." Winner anticipated "more litigation and suits . . . in connection with the project, no matter what direction it takes." He also requested that the DWB file an amended complaint because so much had changed since it filed its first pleading.

The DWB did so on August 22 as the corps completed the first phase of its study on the need for Foothills and after the DWB, Wirth, and the EPA had accepted the Wirth/corps mediation process. The complaint was expanded considerably from the earlier version and named seventeen federal officials and fourteen environmental groups as defendants. As described in the *Denver Post*:

> The suit asks that Winner assess $30 million in damages against Jack Horton, former Assistant Secretary of Interior, Dale Andrus, State Director of Interior's Bureau of Land Management, and Curt Berklund, former Director of the Bureau of Land Management. It was alleged that the three officials "conspired together and agreed to impede, delay and prevent the construction of the Foothills project" in 1975 by unlawfully expanding the scope of the original impact statement, even though federal law didn't require that expansion. . . . [The complaint also said the officials] "conspired to attach illegal conditions to permits which, if allowed to stand will cost taxpayers millions of dollars and will further delay the Foothills construction."

In addition, the lawsuit asked that three million dollars in damages be assessed against Alan Merson and Roger Williams.

It is clear that the DWB was not relying solely on the corps and Wirth to obtain approval of the project; further, they may have hoped the demand for damages would inhibit further opposition. Also, as U.S. Attorney Hank Meshorer suggested, they might have expected the charges to "galvanize the federal agencies and get them to do something fast, for once."

The charges worried most of the defendants, but they had been processing the permit application as quickly as they believed the regulatory procedures allowed. The charges simply increased their insistance on obeying every detail in the statutes and regulations. Some observers believed that the charges worried Merson more than anyone and may have influenced his final decision to accept the Strontia Springs site. Merson, however, felt that other factors were at least as important and that the hostility created by DWB's attack actually hindered the permitting process.

Even more self-damaging was the DWB's decision to name the seventeen environmentalists in the suit. Supposedly, the purpose was to involve all interested parties in the court settlement to avoid subsequent suits on the same grounds. But most observers interpreted the tactic as harassment and typical of the DWB's attempt to "steamroll over" the opposition.

Also, the complaint seeking an injuction against the environmentalists potentially violated their rights under the First Amendment. They were charged with having interests "adverse" to those of the DWB and with expressing those interests in public hearings and in meetings with personnel from federal agencies. In some respects, the DWB was suing individuals and organizations for exercising their constitutional and statutory rights to speak in opposition to the DWB's plans. If these grounds proved to be an acceptable basis for suit, many of the environmentalists and federal attorneys believed the entire NEPA process—and in fact, all public participation programs—would be threatened. The DWB was forced to make additional concessions in the later stages of the mediation process to amend these "wrongs."

The Environmentalists' Suit

One week before the DWB filed its amended complaint, the environmentalists filed a lawsuit against the Foothills Project in the United States District Court in Washington. They told the press that the grounds for the suit were that "basic Foothills decisions have been made by government officials in Washington and the interpretation of the federal statute (the Federal Lands Policy and Management Act) sought in the new suit will have national implications." In fact, the reason for the suit was that the Washington judge was proenvironmentalist and thus more likely than Judge Winner to decide in their favor.

The environmentalists planned to claim violations of both the Federal Land Policy and Management Act (FLPMA) and the NEPA and to seek a temporary restraining order against further construction by the DWB. (The DWB already had started some road construction with permission from the BLM, the USFS, and the corps.) Although the environmentalists did not have sufficient money for a protracted court battle, they were hoping to obtain quickly either a temporary restraining order or a summary judgment. They also hoped that, by developing the dispute into a landmark case over the FLPMA, they could enlist the help of national environmental groups, but the suit was eventually withdrawn.

Outcomes of the Corps' Review

The corps' study precipitated conflicts between the parties who participated in the process and those who did not. Some federal agencies did not accept Merson as an adequate representative of their views. He had already angered USFS and BLM officials by speaking too quickly and too strongly on Foothills soon after he had become EPA regional administrator and before he was fully aware of the details of the controversy. Merson acknowledged that he had "jumped into the fray" more quickly than he would have liked. "Substantively our position was correct, but we needed much more time to inform the public of our stance."

Although the corps' study did not generate sufficient agreement to enable the parties to move easily to a settlement, the process did provide an important platform from which further negotiations could be initiated. Most important, forcing the disputing parties to interact frequently helped to break down much of the previous hostility and distrust among the parties. They began to understand each other better, to trust each other more, and although they continued to disagree, to realize that their disagreements were based on legitimate differences that might be negotiated.

In time, the tone of the exchanges became more polite, though differences of opinion were still strongly expressed. Debate focused more on facts and less on personalities. With this new civility, the parties came to appreciate each others' beliefs, biases, and values. This helped both the parties and the mediators to identify possible areas of compromise that set the stage for achieving a mutually beneficial solution. The new emphasis on compromise rather than conflict facilitated achievement of a positive outcome. It was no longer acceptable to take all and give nothing, and even the DWB recognized the need to make concession. The disputants finally saw that compromise could protect most of their interests and that continued conflict would achieve less satisfactory results. This recognition probably was the most important outcome of the corps' review process, and it helped to hold the negotiations together after completion of the formal review.

Wirth Resumes Negotiations

Introduction

The structure of the negotiation had to be changed because the corps study had not led to indisputable conclusions. The corps briefly persisted in the role of primary mediator, but their study had made them advocates of the Strontia Springs dam, casting them in opposition to the EPA. Wirth, who had been recently reelected, recognized this problem and reentered the negotiations as mediator to try to break the impasse. Wirth was assisted by Aylward, his legislative director in Washington, Bob Drake, his district representative in Denver, and Hank Meshorer, the attorney representing the federal agencies in the DWB lawsuit.

Meshorer had mediated among the various federal agencies to resolve their internal and interagency differences in order to unify his clients' defense. Although Meshorer was not a formal member of the Wirth mediating team, he worked in close coordination with it.

Wirth, Aylward, Drake, and Meshorer spent the next 10 months holding meetings and making telephone calls; during this period they also drafted, negotiated, and redrafted letters, interim agreements, and settlement documents. Because few records were kept and memories have become blurred, some of the details of the negotiations have been lost. Nevertheless, the general strategies and positions of each agency and the mediator indicate the important aspects of the negotiation process.

Preliminary Work

Rather than try to settle all the issues at one time, Wirth segmented the negotiations. He devoted the initial stages to the major parties and the largest issues; he integrated other issues and parties into the later negotiations.

Wirth's first step in November and the first half of December was to determine the negotiable demands of the EPA, the DWB, and the corps—the conditions that each agency required in a settlement. Getting the parties to articulate their positions for the record was a slow and difficult process.

Most of this footwork was done by Drake and Aylward who constantly shuttled back and forth between the EPA, the DWB, and corps. They attempted to persuade each party to agree on a stated position that they thought the other parties would find acceptable.

At first they focused on the location of the dam because the canyon mouth dam appeared to be a logical compromise between the positions of the EPA and the DWB, but the DWB would not accept it because it felt that the pumping costs would be excessive. Citing the analysis done by the economic consultant for the corps' review, the EPA challenged this argument.

The EPA met in early December with the corps and Aylward in Omaha to discuss the technical and economic merits of the two sites. The technical adviser to the corps and the EPA spent the entire day arguing about technical issues. According to Aylward, by the end of the meeting, the EPA's argument was destroyed. "Merson and I both went in there open-minded . . . but the EPA just didn't have a case that would stand up." Although Merson did not formally surrender the canyon mouth dam at that time, he was "visibly deflated," according to Aylward. This meant that a major element of the controversy was resolved, though at the cost of a promising compromise solution. Aylward and the mediation team had to devise another compromise that the DWB could make that would persuade the EPA to drop its opposition to issuance of the 404 permit. Merson indicated that a good program of water conservation together with substantial impact mitigation would meet his goal.

Although some members of the EPA staff were firmly opposed to Merson's concession, Merson had decided that relinquishing this point was almost inevitable from a political standpoint. As he pointed out later, in December he still believed that barring a change in the EPA administration, he had support from Washington to prevent construction of the Strontia Springs dam. Because the fight in Washington would take a long time, Merson thought it impossible that the Washington office would change its position. If this were to occur, Region VII would lose all its control and easily emerge without any gains. Merson felt that the best strategy was to use the leverage over the dam to require a good program of water conservation.

A water conservation program had been one of the EPA's major goals all along because a good conservation program would eliminate the need for expensive and environmentally harmful projects for subsequent water diversion and treatment. The loss of the natural ecosystem in Waterton Canyon was a lamentable price to pay, but Merson decided that mitigation and conservation would produce long-term environmental benefits for the Denver region.

The DWB was not pleased with the idea of water conservation, especially if it was to be administered by a federal agency. But Wirth and the other mediators thought the board would comply if they were allowed to build the Strontia Springs dam. By the middle of December, Kenney and Merson were close to an agreement, although members of their staffs and other DWB members still opposed the concession. Thinking that one more concerted effort was all that was needed for the parties to agree to terms for a settlement, Wirth scheduled a negotiation session for the evening of December 15, 1978.

The First Negotiation Session

The December 15 meeting was preceded by drinks and dinner, during which time the participants did not discuss Foothills. Instead, Kenney talked about his childhood, and Merson told stories about his days in Alaska. This relaxed the participants and strengthened the friendliness and trust that had been developing since June. Negotiations began in earnest at 10 P.M. and continued (at Wirth's insistence) until the parties reached an agreement. At the beginning of the meeting, each party summarized its positions.

The corps contended that there was need for a measurable conservation program, for minimal streamflow provision, and for a habitat improvement program to mitigate the adverse environment impacts of the project.

The DWB insisted that Strontia Springs was the only dam site that the DWB would consider but acknowledged the need for water conservation. The DWB also indicated its willingness to agree to minimal streamflows and mitigation if the released water could be recaptured for DWB use. (This could be done by allowing the DWB to store the water in the corps-owned Chatfield Reservoir.)

The EPA continued to state that the farther down the canyon the dam was placed, the more environmentally sound it would be. They also stated that water conservation would be required with any permit. Wirth concluded that water conservation was essential to any agreement. He urged everyone to exert maximum effort to solve the remaining issues.

Because the parties had reached a clear consensus on the need for water conservation, the discussion focused on the design of an acceptable conservation program. The EPA wanted a much stricter program than did the DWB.

Discussion proceeded slowly, lasting into the night. Wirth intended to break down the resistance of the disputants through marathon negotiation sessions. As Aylward pointed out, "People fight harder at 10 A.M. than at 10 P.M.—and by 5 A.M. resistance was nil." The strain also increased the sense of common purpose and cooperation; the goal became not only to reach an agreement but simply to go home. Although some of the participants disapproved of these tactics, all stayed until they reached an agreement. They feared that, if they left, it would appear as though they cared less about a settlement than did the others.

The psychology seemed to work and, finally, at 5 A.M. a Memorandum of Understanding was approved by all the parties.

This memorandum, signed by Wirth, Merson, Kenney, and Selleck, established the basic outline of a settlement. In essence, it said that the corps would issue a 404 permit for the Foothills Project and Strontia Springs dam without EPA opposition but that the permit would contain conditions that required water conservation and impact mitigation. All the federal parties involved were required to agree to these terms. These programs were to be worked out in detail over the next few weeks, and the parties agreed to meet again no later than January 8, 1979, to reach a more definitive agreement.

The memorandum was accompanied by a statement by Mr. Kenney that confirmed the DWB's agreement with the EPA and the corps concerning water conservation. According to this agreement, the DWB would design a conservation program by March 15, 1979, that would reduce the average per capita consumption of water by 3% by January 1, 1982, and by 5% by Janauary 1, 1984. At the end of each period, the corps would evaluate the program and its progress and, after the first 5 years, determine a goal for the next 5 years that would fall in the range of 3% to 5%. Similarly, the corps would then evaluate the program after 10 years and set the goal for the next 10 years in the range of 5% to 10%. The DWB alone would determine the manner in which these reductions would be accomplished. The board could use any measures it thought best as long as the program was credible and the goals were reached.

This agreement differed somewhat from conditions concerning water conservation in the UFS and BLM permits. The BLM and USFS did not set numerical goals or dates as did the EPA/DWB agreement. But they did require the support of the water conservation plan by the city and county of Denver, the Denver Regional Council of Governments, and the state of Colorado before the DWB could either increase the capacity of Foothills to more than 125 mgd or apply for future permits or grants from the federal government. The BLM felt that such local participation and approval were very important and, in fact, should prevail over any federal determination of the adequacy of any conservation plan. Consequently, the BLM was very concerned about the difference between the EPA/DWB agreement and the conditions of their own permit. In addition, according to some sources, the BLM was irked at being left out of the discussions on water conservation.

Yet, the BLM and USFS had not participated in the corps' review process, which was designed specifically to resolve the EPA/DWB dispute. Wirth had felt that the negotiations would be facilitated if the basic EPA/DWB agreement was outlined before the other parties became involved. Therefore, he had risked the displeasure of the BLM and USFS by keeping them out of the negotiations until after the December 15, 1978, session. Only then were the other parties invited to participate in the discussion, which turned to the question of impact mitigation.

The BLM continued to express concern about the discrepancies between the two water conservation plans, despite assurances from Wirth and the other mediators that the differences were insignificant. According to Wirth, the DWB acted in good faith, and the water conservation plan would satisify all the parties involved as well as the city and county of Denver, the Denver Regional Council of Governments, and the state of Colorado. No further efforts were made to resolve the discrepancy, which still exits in the signed settlement agreement that is now in effect.

Negotiation over Impact Mitigation

Wirth, the EPA, and the corps agreed that concurrence of the BLM, the USFS, and the FWS on the impact mitigation program was very important. Consequently, Wirth invited the three agencies to join the negotiations at this point. Although the agencies were disturbed by their initial exclusion from the sessions, all three readily accepted the invitation and participated fully in the subsequent negotiations.

In order to help coordinate the newly broadened negotiations, Wirth asked Meshorer if he would be willing to speak for and help coordinate the federal parties. Meshorer supported the mediation effort and willingly linked his efforts with Wirth's after the December 15 meeting.

In order to appraise the remaining differences, Wirth asked Meshorer to prepare a draft settlement agreement that would be acceptable to all the federal parties. In the interest of time, Meshorer, in turn, asked each agency to prepare separate draft settlement agreements, which he then revised and combined into one document. Meshorer gave the document to Aylward, who combined it with a similar draft from the DWB to produce a negotiation text that provided the basis for subsequent discussions.

The most controversial BLM and USFS impact mitigation condition had been the requirement that the DWB ensure minimal streamflows below the dam to protect fish and other aquatic wildlife. The DWB had turned down the BLM and USFS permits, charging that this condition would force them to relinquish water to which they held legal rights. Once it became clear that they could recapture the water in Chatfield Reservoir, the issue of water rights ceased to be a problem, and the DWB agreed to maintain the minimal streamflows. They also agreed to a program of habitat improvement that was intended to ensure good trout fishing below the dam in compensation for the trout habitat destroyed by the reservoir. None of these conditions provoked controversy, although formulating the details to the satisfaction of all parties was a slow process.

These negotiations settled the disputes and the litigation between the DWB and the federal agencies. However, the environmentalists' dispute with the DWB remained unresolved. Although some environmentalists still favored carrying on the case, the cost of the Washington litigation was mounting rapidly, and the case was proceeding much more slowly than the group had hoped. The environmentalists also had doubts about their chances of winning because their lawyers disagreed considerably about the strength of their case.

Therefore, when Aylward and the DWB approached the environmentalists about a settlement, most of the leaders were very receptive. Bermingham especially was interested in settlement because he had spent $20,000 of his own money on the case and saw no end in sight. Others too had expended much time and effort, but thus far they had won little and future prospects were unclear. Now that the federal parties were reaching a settlement, the environmentalists felt that their case was further weakened and that their chances of stopping the project were very slim. Therefore, they decided that negotiation was their best option and were determined to bargain for as much as they could obtain.

The environmentalists met independently and frequently to generate a set of demands to present to the DWB through Wirth. Included among their demands were the following:

1. The DWB should implement an open planning process through which its short and long-range planning would be carried out with extensive public consultation. Such a process should include a commitment to complete a systemwide EIS for any further project (including extensions of Foothills) before they are built.
2. The DWB should form and fund a Citizens' Advisory Committee to help in the open planning process and to otherwise advise the board on its future activities.
3. The DWB should provide compensation for the recreational losses on Waterton Canyon by improving similar sectional facilities elsewhere. For instance, they should construct a white-water kayak chute on the South Platte River downstream from the dam and buy enough land (now privately owned) to provide recreational access.
4. The DWB should admit in the settlement agreement that joining the environmentalists in their lawsuit was illegal and that the environmentalists were "not proper parties defendant."
5. The DWB should agree to pay the environmentalists' attorney fees.
6. The agreement between the environmentalists and the DWB should be awarded so as to be enforceable by the public—that is, by making a third-party beneficiary proviso.

Some of these demands were tactical, establishing ground that the environmentalists intended to yield, such as a kayak chute in Denver. The group held some terms to be nonnegotiable: most important among these was that the DWB should admit that the environmentalists were not proper parties to the Denver suit. The environmentalists felt that, without such a provision, the NEPA would be weakened permanently because the DWB would appear to have succeeded in its strategy to force unwilling parties to settle or be sued for simply voicing their objections to a project.

The issue of attorney fees also was very important. Although several environmentalists considered this condition nonnegotiable, and most of the environmentalists thought the fee demand was legitimate, others thought that the attorneys' fees played far too great a role in the decision to settle. Several environmentalists agreed with their opponents that the fee payment had bribed the most influential environmentalists—those with the greatest amount of money at stake—to settle for terms that they would not have otherwise accepted. When Bermingham and his associates agreed to settle, the others had to settle as well because they could not press the case without the leaders' financial support.

The DWB negotiated the environmentalists' demands over the next few weeks while they also carried on negotiations with the federal agencies. The DWB's negotiations with the environmentalists were kept separate from those with the federal agencies, but Wirth, Drake, and Aylward provided mediation for both.

Although the negotiations might have succeeded if the mediators had not provided this service, the mediators' credibility and fairness facilitated compromises that otherwise might not have been reached. When Aylward or Meshorer said to the DWB, "You are not going to get a settlement if you don't agree to the following . . . ," the DWB was more likely to agree to the demands that had been communicated directly by the federal agencies or by negotiators for the environmentalists. The mediators also were able to talk freely enough with both sides to identify grounds for compromise. Negotiations continued sporadically, both directly (among the parties) and through go-betweens (Aylward and Meshorer) throughout the following two weeks until the parties seemed to approach agreement.

Settlement

Terms

Wirth then scheduled a set of meetings for January 3, 1979, to formulate the final principles of the settlement. The first meeting was between the environmentalists and the DWB. It was held at the DWB offices in the afternoon. The environmentalists (represented by Robert Golten of the National Wildlife Federation, John Bermingham of the Water Users Alliance, and Robert Weaver of Trout Unlimited) presented the DWB (through Aylward) with a revised list of demands.

At Wirth's request, Aylward then joined a meeting between the DWB and the federal representatives. The environmentalists moved their meeting to this location. The environmentalists met with the DWB attorneys in one room, while Kenney and other DWB attorneys met with federal attorneys and principals in another room. Aylward mediated in both meetings until the parties reached a settlement. Late into the night the federal representatives and the DWB signed a document entitled *Principles of Agreement* and the environmentalists and the DWB signed a *Memorandum of Understanding*.

The *Principles of Agreement* was an elaboration of the document signed earlier in December and contained few significant changes. The agreement listed a number of conditions that were to form the basis of formal legal documents (the settlement agreement and the consent decree) to be drafted in the following weeks. The conditions were the following:

1. The Corps will issue a 404 permit for Foothills and the Strontia Springs dam without objection by the parties to the document.
2. The DWB will comply with all the conditions of the permits granted previously by the BLM and the USFS.
3. The DWB agrees to accept BLM and USFS permits without objection.
4. The DWB may use Chatfield Reservoir for storage of water released downstream in order to comply with the minimal streamflow requirement.
5. The DWB will undertake a program of water conservation that incorporates the goals of the December 16 agreement but assigns the primary role of monitoring to the EPA, not the Corps of Engineers.
6. The DWB agrees to increase public participation in its decision making.
7. All parties will cooperate in developing the South Platte River as a recreational resource for the entire community.
8. The federal agencies will do a systemwide environmental analysis of all DWB projects currently under construction and/or any future water project "to determine site specific and cumulative effects of these projects." (This was requested by the environmental parties, which took this to mean that the federal government would prepare a system wide EIS for the next DWB project.)
9. The Foothills suits in Denver and Washington will both be dismissed with prejudice.
10. Finally, the parties pledge to continue to cooperate for the purpose of avoiding future litigation over similar issues.

Nothing in this agreement differed from the earlier tentative agreements, except that the primary responsibility for enforcing the minimal streamflows and habitat improvement programs was transferred from the corps to the BLM and USFS.

Although some of the environmentalists' demands were included in this document, the environmentalists and the DWB also signed a Memorandum of Understanding. This document specified that the DWB was to include the public in their decision making by establishing a citizens' advisory committee to participate in long- and short-range planning. The agreement also stated that the

DWB would pay the environmentalists' attorney fees up to a total of $47,000. In exchange, the environmentalists agreed not to initiate litigation or otherwise try to prevent the construction of Foothills and to

> use good faith efforts to discourage other individuals and environmental organizations from undertaking litigation or claims, or administrative proceedings to challenge, contest, disrupt, interfere with, or prevent the construction of the currently-permitted Foothills project.

The only major demands that the DWB refused to accept immediately were the statement of wrongdoing regarding the lawsuit, a citizens' enforceability provision, and a provision that they prepare a systemwide EIS on their next project for raw-water development (the Williams Fork River on the western slope). The DWB's refusal on the last point was not considered to be very important because the federal agencies already had agreed to make an environmental assessment of current or future DWB projects. Though the stipulation did not specifically say that this assessment should include an EIS, it did say that the agencies would follow CWQ regulations which require "the integration of the NEPA process with other planning."

Discussions continued for several weeks while the parties reviewed the draft settlement agreement. Finally, the DWB conceded and agreed to include in the settlement agreement a statement that said,

> The plaintiffs hereby recognize that the environmental defendants have asserted their opposition to the construction of the Foothill project, and related facilities, in good faith and within their Constitutional and statutory rights. The Denver Water Board and its members now recognize that in light of the affidavits of the environmental defendants and other facts, the environmental defendants are not proper parties to this litigation.

The environmentalists also won an important concession concerning future DWB projects. This agreement stated that

> when any such [environmental mitigation] measures are lawfully required, their cost shall be considered as part of the cost of developing and implementing those projects and should be borne by the DWB.

Thus, the DWB committed itself to internalizing the costs of environmental mitigation in all of its future projects. The environmentalists found these terms acceptable.

The settlement agreements were signed by all the parties on February 14, 1980. The documents included the Stipulation to Dismiss and the Settlement Agreement, which settled both the DWB and environmentalists' litigation with prejudice. A proposed Consent Decree, which included the same items, was then presented to Judge Winer for his approval and signature.

Enforceability

Much to the surprise and dismay of all the parties, the judge refused to sign the Consent Decree because he felt the document raised several serious legal problems. Instead, he wrote a three-page addendum to the Consent Decree in which he explained his reasons for not signing:

> 1. I think it quite possible that all the parties have done is to enter into an agreement. I do approve the agreement and I do sign the "Consent Decree" with the limiting comments.
> 2. My signature does not indicate any opinion as to whether a contempt remedy would lie for breach of the provisions of the document. . . . Nor do I rule on whether I retain jurisdiction to change the "Consent Decree" because of change of conditions.
> 3. Under no circumstances should this "Consent Decree" be interpreted as being a ruling by me on the intent or meaning of the Blue River Decree [part of Colorado law] or a ruling on any question of state or federal law.
> 4. I do not approve the payment by plaintiff of any attorneys' fees to attorneys not employed by plaintiffs. I do not rule on the legality of such payment and I do not rule on the ethical propriety of the acceptance of such payment. Indeed, I have serious doubts as to both the legality and the ethical propriety.
>
> I sign this "Consent Decree" with this addendum because the public, the plaintiff, and the court are under unbelievable pressure to get on with the building of the project. From an academic standpoint it would be nice if the questions I have raised could be finally answered in advance but inflation being what it is, I cannot indulge the luxury of a briefing schedule, an opinion, and an appeal. . . . The public can't afford the thousands of dollars a day it would cost to engage in litigation over counsels' legal or moral right to demand payment of their fees as a condition of settlement of a case they essentially lost. The public's pocket isn't that deep. The project can be built and the legal and ethical questions can be answered in due time.

All the parties were disturbed by this action because it made the enforceability of the settlement very unclear, particularly in view of the judge's stated doubts about the propriety of a contempt remedy should the agreement be violated. Yet, like any contract, the agreement still could be enforced through a new lawsuit. Moreover, it was partially self-enforcing because payment of the environmentalists' attorney fees depended on the environmentalists first having met their obligations in the settlement agreement. (The funds were placed in escrow immediatly after the agreement was signed but were not to be released until a year later, and then, only if the environmentalists had kept their pledge not to obstruct the Foothills Project.)

The environmentalists and the federal parties, however, had no reciprocal assurances, and instead, they had to rely on public and government pressure to assure DWB compliance. The federal representatives and Wirth made it clear that they would monitor the DWB's compliance carefully and exert whatever legal and administrative pressure they could if the DWB violated the terms of the settlement. Meshorer, now acting as a federal attorney, asserted that he would *drag* the DWB back into court if they violated the agreement. The other federal representatives indicated that the DWB's ability to obtain future permits would depend on their implementation of the water conservation plan. Although other conditions of the permits were not explicitly linked to future DWB projects, these too were enforceable through normal provisions for permit enforcement. Therefore, the DWB had strong incentives to abide by the settlement, as did the federal parties who now wanted to end the dispute. Judge Winner's unwillingness to endorse the terms of the agreement was considered unfortunate, but that did not foreclose implementation.

The Aftermath

The environmentalists and the EPA regretted the construction of Foothills, but they realized that this settlement was the best they were likely to obtain. They were generally satisfied with the agreement and optimistic about its potential influence on future DWB plans and planning procedures. Bermingham was quoted in the *Denver Post* as saying, "We may have lost the battle [the dam] but we think we've won the war." Robert Golten, a lawyer for the National Wildlife Federation, agreed: "I think we got much more than we would have achieved had we continued the litigation."

Merson, too, was pleased with the settlement, although some EPA staff members remained disappointed and perhaps, somewhat bitter. They believed that they could have stopped construction of the dam if Merson had not "sold out" under pressure. Some environmentalists also were upset that Bermingham and others "sold out" for the attorney fees and would have preferred to press the litigation. Bermingham had been paying most of the legal fees, however, making it difficult for the other environmentalists to carry on without his support. Even those who were least happy with the settlement realized that it had its benefits. Ben Harding was harshly critical of Wirth and the mediation process, commenting, "It wasn't mediation—we were beat over the head." But he thought the environmentalists could use the settlement as a legal base to influence future DWB projects. Aside from this, however, he felt that the DWB "managed to beat us down on everything . . . we got our money back, nothing else."

Although the outcome was not satisfying to all parties, most acknowledged that it was probably better than what they would have achieved without negotiations. Each party backed the settlement in the hope that, despite the judge's reservations, the agreement would be binding.

Although the DWB and the federal agencies apparently are satisfied with the implementation of the settlement, the environmentalists have become increasingly disappointed with the DWB. They contend that the DWB has violated first the spirit and then the letter of the settlement. Weaver and Harding did not become members of the advisory committee. Bermingham was head of the committee in 1979 and 1980; and he is the only environmental or conservation organization representative on the permanent advisory body. According to Golten:

> John Bermingham for over a year has been urging restraint on some of us most concerned about post-settlement developments. John felt that, through his position in the Advisory Committee, with time he would be able to incline the Board and the Water Department to the position we thought we had achieved in February 1979. I think John now concedes that he was mistaken. Some of the rest of us are convinced of that—and that the Citizens' Advisory Committee (on which he appears to be the only environmental/conservation organization representative) is ineffective as presently constituted for assuring that the Water Board will be accountable to the "public" (as opposed to special) interests.

These remarks appeared in a letter from Golten to DWB General Counsel Williams. Enclosed with the letter was a list of seven alleged DWB violations of the Settlement Agreement. These included:

> 1. Violation of the commitment to full public participation in future decision making by the Denver Water Board.
> 2. Violation of the DWB commitment to provide staffing and technical support for the Advisory Committee.
> 3. Violation of the DWB commitment to "develop" a "program" making available to the Citizens Advisory Committee and to the public its present and future plans.
> 4. Violation of the DWB commitment to keep the Advisory Committee regularly appraised of the status of its planning for water diversion and storage facilities and structures.
> 5. Violation of the commitment to coordinated planning and coordinated implementation of planning.
> 6. Violation of the commitment to provide advance notice of the pending proposals well before they are considered for adoption.
> 7. Violation of the commitment to achieve timely consultation with affected property owners concerning management of flows of the North Fork of the South Platte River.

Golten concluded his letter by saying,

> What is more important than fussing over these matters through the mail, however, is to sit down and review the implementation of the Foothills settlement.

We think it time, before further thought is given to other approaches, to meet with you and one or two other representatives of Tim Wirth's office. We suggest Congressman Wirth's involvement since he has asked to be kept informed and he also assured us of his continuing interest.

The DWB has agreed to meet though the meeting has not yet taken place.

General Study Questions

1. There were many agencies, groups, and individuals that had a clear stake in the Foothills controversy. What impact did the number and diversity of these interests have on the course of negotiation?

2. Unlike some of the other cases we have examined, the Foothills dispute was played out in the public eye from start to finish. Indeed, some influential news organs—most notably, the *Denver Post*—not only reported developments but actively supported one side. In a proper analysis of this dispute, should the *Post* be regarded as a party? (Other aspects of the role of the press are discussed in a section that follows these questions.)

3. In several other cases, there were divisions within one or another group; this was true in Brown Paper presented in chapter 4. In Foothills, there was division within the EPA and among the environmentalists. First, consider the EPA. We have seen it in a number of roles in earlier cases: as a permitting agency in Holston River, as an enforcer in Brown Paper, and as a grant maker in Jackson Hole. (In chapter 11 we shall see the agency engaged in administrative rule making.) What was its role here? In what ways were its priorities conflicting? How did divisions within the agency affect the course of negotiation? In contrast, consider the situation of the environmentalists: Did division among them weaken or strengthen their hand?

4. In this case, perhaps more than in the others we have looked at, many of the principal parties were represented by legal counsel. They also had specialists to represent them on technical issues, such as projected water needs and anticipated development. What impact did the use of such representatives have on the process of negotiation?

5. Initially, there seemed to be little room for negotiation. A Denver Water Board spokesman rebuffed the first mediation attempt, saying,

> Mediation and compromise are, of course, useful in many situations, but we do not believe that anything is to be gained, and much may be lost by attempting to pursue such a series of meetings.

What other considerations explained the board's posture?

MEDIATING LARGE DISPUTES 241

6. Early in the dispute, Alan Merson of the EPA publicly stated that the dam would never be built. So long as the dispute was cast in the simple terms of build or not build, is it understandable why there was no incentive to negotiate? How did the character of the dispute change in such a way so as to alter the incentives of the parties?

7. Early in the dispute, the Denver Water Board justified its refusal to negotiate by invoking the results of an election authorizing the sale of bonds that would be used to finance the construction of Foothills. Those who opposed the dam, they claimed, were trying to subvert the public will. Timothy Sullivan's proposal that referenda be used to resolve environmental disputes was introduced in chapter 7. Should the Denver bond election have closed the issue here? Does the fact that this dispute continued in spite of the election disprove Sullivan?

Mediation Study Questions

1. Representatives Schroeder and Wirth each tried to initiate mediation; the latter succeeded after the former had failed. To what extent does timing explain the difference in results? How did circumstances change from the time Schroeder made her proposal to the time Wirth made his? Did the previous false start help or hurt Wirth's efforts?

2. Another explanation for the different results lies in the way each representative tried to initiate mediation. Why do you suppose that Schroeder decided to use a press conference to float the proposal? The stratagem failed here. Might there be other instances in which such a move would be wise? How did Wirth's approach differ?

3. Schroeder nominated an experienced mediation team—Gerald Cormick's Office of Environmental Mediation from Seattle—but the disputants rebuffed the suggestion. Why do you think the OEM was accepted in the Snoqualmie case (discribed in chapter 6) but not in the Foothills dispute?

4. The mediating role that Wirth assumed was quite different from the one that David O'Connor played in the Brayton Point case (described in chapter 8). Was this simply a matter of personal style, or did the cases call for different treatment? Could a mediator operating the way that O'Connor did have resolved the Foothills dispute? Would a Tim Wirth have succeeded in Brayton Point?

5. When O'Connor mediated Brayton Point, very explicit ground rules were adopted at the outset. Would such rules have been desirable here?

6. At various stages of the dispute, the parties made wildly divergent assessments of the potential impact of the dam on population growth, water demand,

and wildlife. Were these differences based on contradictory information and models, or did they reflect a more fundamental conflict in personal and social values? Ultimately, Wirth enlisted the Corps of Engineers to resolve some of these differences. We have seen data mediation in Brown Paper (chapter 4) and Holston River (chapter 5). How was it different here? What was Wirth's goal in involving the Corps of Engineers? Who else might have served the same function? What effect did this data negotiation have on the larger negotiation process?

7. One of the central factual disputes in Foothills was whether construction of the dam would increase suburban sprawl and thus worsen air pollution. The private consultant retained by the Corps of Engineers concluded that attempts to control growth by limiting water service were futile. The EPA strongly disagreed with this analysis but ultimately did not challenge it in the *Foothills Newsletter*. Why did the agency surrender such an important point?

8. Wirth defied many common mediation precepts. In Brayton Point, the parties had agreed that everyone had to be present at mediation sessions. By contrast, Wirth began his efforts by meeting separately with some of the more powerful groups. Moreover, O'Connor was viewed as nonpartisan, but Wirth was known to be in favor of dam construction. Should Wirth's success be regarded as an exception to the rule, or does it call the rule into question?

9. The Foothills case is an example of what the late Lon Fuller termed a *polycentric problem*. There were numerous issues, each one of which was intricately laced to the others. Consider the way in which Wirth structured the mediation agenda: What issues did he choose to tackle first and why? He attempted to segment the problem, getting resolution on particular points before moving on to others. What are the advantages and risks of this approach?

10. The Foothills dispute had dragged on for years before Wirth intervened. What techniques did he employ to accelerate the mediation process? Are they applicable to other disputes?

11. To prod the parties into agreement, Wirth arranged an all-night negotiating session. What are the risks inherent in this approach?

12. The Bureau of Land Management (BLM)—excluded from Wirth's intense negotiation process—expressed concern that the conservation plan worked out in those sessions differed from the conditions it had imposed on the Denver Water Board in its permit. Wirth, however, was not worried about this discrepancy, which still exists. In a sense, it could be said that the agreement that the DWB made with some parties was inconsistent with the understanding it had with another. What might explain Wirth's lack of interest in resolving this discrepancy, particularly in view of the importance of the water conservation issue?

13. As with many environmental cases, there was considerable difficulty in binding the parties. At one point, the Denver Water Board brought suit against EPA officials, environmental groups, and individuals seeking large money damages against them. By putting pressure on many of the dam opponents, this tactic may have spurred settlement. It also expanded the number of litigants, thus potentially binding more parties to any judicial resolution. The board's move caused much antagonism, however, and easily could have polarized the disputants even more. Was there any less risky method the board could have used to the same end?

14. It is ironic, of course, that after all the parties were included in the suit and after they came to agreement, the federal district court judge sitting in the case was reluctant to ratify the agreement in a consent decree. Should judges be required to accept a settlement that is satisfactory to all the parties? What options are open to the parties if the judge refuses to issue a consent decree?

15. The material in the following chapter will explore the ethical duties of a mediator. One important question is the degree of the mediator's responsibility—if any—for the terms of settlement. Here, resolution was reached when project sponsors agreed to pay some of the attorneys fees for the environmentalists. Was it Wirth's function to determine whether in seeking payment for themselves, the environmentalists' lawyers had compromised their clients' interests?

Foothills Epilogue

This postscript on Foothills is adapted from the conclusion of the case study prepared by Heidi Burgess.

Some readers and Foothills opponents may question the quality and wisdom of the Foothills settlement. As mentioned earlier, the settlement appeared better than any alternatives that were possible, although an even better outcome might have been reached had the social and political arrangements been different. Given the overwhelming political power of the Denver Water Board and the extent of local support for the project, construction of Foothills was essentially impossible to stop. If the opponents (especially the EPA) had been better organized and had started negotiations earlier, they may have been able to delay the project further or force the DWB to accept the downstream dam. However, the support of local politicians and the media for Foothills, together with the growing list of people waiting for new water taps, practically eliminated the possibility of entirely blocking the project.

In view of this, the Foothills result was a good one. The DWB obtained its permits more quickly than it would have through litigation or further administrative action. The environmentalists received significant concessions, several of which almost certainly would not have been possible in a court judgment. Although they were not able to block the first stage of the Foothills project, the agreement increased their influence on future DWB projects and may have given them a better chance of prevailing in future water disputes.

The agreement also improved relations among many of the parties and set a precedent for open discussions and cooperation that will benefit both the parties and the public. Although disagreements of a similar nature are certain to recur, the pattern of cooperation and joint problem solving introduced by the Foothills negotiations may facilitate resolution of future disputes. Also, their success with negotiations will lessen the likelihood of stalemates.

On the other hand, the EPA and the environmentalists still believe that the project is unnecessary and that a much less expensive and less wasteful program of water conservation could accomplish the same ends. Unfortunately, a more extensive program for conservation was never given adequate study. Denver does not employ water meters in its older sections, and many residents still pay a flat monthly water rate. If a fraction of the money for Foothills had been invested in metering, inverted rate pricing, and other simple conservation stimuli, the environmental and economic savings might have been substantial. Conservation as an alternative to Foothills deserved serious consideration that it did not receive in the course of the dispute. After such consideration, the decision to construct Foothills with a program of impact mitigation and compensation might have been the wisest choice. But the decision was not a thoroughly informed one; and it is possible that the region's interests might have been better served by a stronger conservation plan.

The Role of the Press

The press, most notably the *Denver Post*, played several distinct roles in the Foothills dispute. First, it was a conduit for information. Because of local interest in water development, the dam proposal was well publicized from the start. The publicity, in turn, likely generated still greater interest. The emergence of environmental opposition made the story all the more newsworthy. As with many such issues, it is impossible to determine how much the press was reflecting public opinion and how much it was generating it.

Some of the other major cases we have encountered were largely unreported. The Brayton Point coal conversion case, for example, attracted far less media attention than did Foothills. In part, this contrast may be due to the differing nature of the disputes. A controversy over the proposed construction of a dam may appear clear-cut; the issues can easily be understood, albeit on a superficial level. By contrast, questions about the nonattainment status of the air shed and the potential effectiveness of precipitators can seem more obscure.

The difference in press attention can also be a function of media markets. Fall River, where Brayton Point is located, is near Boston and Providence, both of which have major television stations and newspapers. Yet the city is often overlooked by both. Had coal conversion been proposed for a generating plant in the immediate Boston vicinity, undoubtedly there would have been far more coverage. The character of an environmental debate thus is determined not merely by the technical issues at stake, but by the communications context in which they arise.

The press in Foothills was not only a conduit for information, but in the eyes of some disputants, it was a tool to be manipulated to further their interests. Representative Wirth used the press to keep the parties involved. His initial press releases, for example, gave the impression that the parties had agreed that the corps' study would be binding, although the parties actually felt that a firm agreement had not been made. (They agreed to abide by the corps' findings only if the study was fair and equitable.) Nevertheless, if either side had rejected the corps' conclusions without strong reasons, the public would have regarded it as backing out of an agreement. Had Wirth not used the press to publicize the corps' role, the EPA and the DWB could have repudiated the results of the engineering study more easily.

The Denver Water Board and the EPA both tried to use the press to obtain more coverage for their points of view. The EPA hoped to open up the corps' review to the scrutiny of the press and the public to legitimate their anti-Foothills stance. This strategy backfired, however, because the corps' findings were widely reported whereas the EPA's rebuttals were discussed less often. This gave the public a one-sided view of the final corps/EPA negotiated findings, further eroding public support for the EPA.

The press in Foothills also played a third role—that of a party of sorts. The *Denver Post* was not a mere neutral reporter of facts but a strong advocate of dam construction. Some environmentalists felt that the paper's position was reflected not only in strongly worded editorials but by one-sided general reporting. From their viewpoint, the *Post* was an important force with which they had to reckon.

It is difficult, nonetheless, to regard the press as simply another party. It is true that newspapers and television stations need not be mere bystanders in environmental disputes. They can certainly wield power by delivering support and controlling information. Yet, when the press is at the bargaining table, it is there ostensibly as an observer, not a participant. When the disputants deal among themselves, they may bargain over acceptable levels of particulate emissions or the amount to be spent for mitigation, but when they individually approach the press to try to influence its coverage and to win its support, the currency of exchange is often obscure. There was a period in American history in which developers would offer bribes to reporters and editors, and government officials and politicians might counter with special favors. Though this kind of yellow journalism is largely gone, other kinds of power brokering still occur. Just as public officials must nurture press support, newspapers and television outlets always remember that they are businesses that cannot afford to alienate too many advertisers and readers. Similarly, reporters sometimes subconsciously tilt their stories in favor of sources who seem likely to provide good news material in the future. In a sense, there may be two parallel negotiations, one in which the parties bargain over the environmental problem, and another in which they court the press. If agreement ultimately is reached, the press will not be a signatory; yet, the degree of its enthusiasm for any settlement may be key to implementation.

In short, the press plays a role that is unique in important respects. Distinct from being a mere bystander, it can wield substantial power to further its own priorities. Yet, unlike the other parties, it seldom has a formal place in the proceedings.

In some instances, the role of the press may be defined by the other parties. In the Brayton Point case, for example, the parties to mediation agreed at the outset that they would tell the press only that they were meeting but would not reveal the content of the negotiation sessions. Where there is high mutual trust, a self-imposed gag rule may work, but if one party grows dissatisfied with the process, there may be a strong temptation to leak information to the press in hopes either of winning public support or perhaps even sabotaging the negotiation. In some instances, open-meeting or "sunshine" laws may compel agencies to negotiate in public. Most of these statutes, however, contain exceptions that cover sensitive discussions of lawsuit settlement. Even when an agency is completely within its legal rights in closing the door to the public, an aggressive newspaper or television station can wave the banner of the open meeting law and make it seem as if the officials are trying to hide from the public.

Most students of negotiation believe that serious discussions of controversial issues usually can take place only in private. This certainly has been the case in collective bargaining of labor disputes. Tentative concessions that invite reciprocal concessions might not be made in the glare of the public eye. Negotiators do not want to be accused by their constituents of selling out their interests or of bargaining poorly. Not everyone, however, agrees with the proposition that publicity inhibits negotiation. Lawrence Susskind describes a case that he mediated in which there was substantial public involvement and press coverage (see Susskind, *The Negotiated Investment Strategy in Columbus*). He believes that the presence of the press can be salutary overall. The press can monitor and confirm information; it can give bystanders more confidence that their interests have been represented and accommodated; and, as in Foothills, the press can be used to commit parties to agreements that otherwise would be difficult to enforce. Susskind has particular hopes for the use of local cable television in environmental disputes. "Narrowcasting" would allow extensive information to be provided to communities that are too small to receive the attention of conventional broadcasters. Interactive systems could, in Susskind's view, be used to involve a greater segment of the public in the debate.

Thus far, the experience of cable television has not fulfilled this vision, but as vast areas of the country are now being wired, the technology will soon exist for improvising a new sort of political process, one based on negotiation among far more people than could fit around the traditional bargaining table.

It may well be that communications technology has advanced beyond our understanding of how best to use it. Social and political systems may have to be adapted before cable-based negotiation can take place. Still, there is a larger lesson in Susskind's musings: Disputants should not necessarily regard public participation and press scrutiny as negative. Parties may sometimes find it in their mutual interest to enhance the press's role rather than to diminish it.

10

MEDIATION ETHICS

INTRODUCTION

The last two chapters introduced mediation techniques and the special problems that arise in the mediation of large disputes. Regardless of the scope of a dispute or the specific methods that are used to resolve it, the practice of mediation can raise difficult ethical issues. In this chapter, we will explore the ethical dimensions of mediation and, at least implicitly, of dispute resolution in general.

We shall consider first the question of accountability: Is the mediator merely obliged to lead the parties to agreement, or does he or she have a broader duty to society to see that the terms of that agreement are efficient and just? Whatever may be the mediator's responsibility for end results, should there be ethical strictures that limit the means that are employed? Moreover, we shall analyze the mediator's own motives and incentives, particularly those that may conflict with those of his or her clients. Finally, we shall consider the status of the "mediator with clout": Should such a person be regarded not as a mediator but either as another party or as an adjudicator?

Unlike most chapters in the book, this one does not include a major case study. These ethical issues should be considered in the context of the Brayton Point and Foothills cases.

THE CONCEPT OF ACCOUNTABILITY

In collective bargaining of labor disputes and in some other fields, the mediator is not generally expected to be concerned about the quality and impact of negotiated agreements, so long as the parties to it are satisfied. If management and the union can live with a dollar-an-hour wage increase, then the mediator need not worry about inflationary effects on the general consumer.

Some observers believe that whether or not this notion has currency elsewhere, it should not apply to the mediation of environmental disputes, particularly when social impacts may be felt for years—even generations. The excerpts below represent opposing views on this broader view of mediator responsibility. To what extent were David O'Connor and Tim Wirth accountable in their mediation of the Brayton Point and Foothills disputes, respectively? Should their responsibility for the terms of settlement have been more explicit? Should the same standards of accountability apply to a private citizen and an elected official?

Susskind, Lawrence. "Environmental Mediation and the Accountability Problem," 6 Vt. L. Rev. 1, 4–8, 1981; citations omitted.

> One of the questions raised is to whom and how will environmental mediators be held accountable: More specifically, how can those affected by the actions of mediators effectively chastise, sue, or fire them? Labor mediators must abide by the rules established by the Federal Mediation and Conciliation Service or the American Arbitration Association. Mediators' efforts are policed by these associations to ensure conformance to their codes. Failure to comply can lead to disaccreditation. Labor mediators can be sued if they violate statutes or judicial decisions regarding proper mediation procedure. They can also be discharged by the parties to a dispute, thereby making it harder for incompetent mediators to find work in the future.
>
> There are no comparable statutes or judicial decisions that currently apply to environmental mediators. Most environmental mediation efforts have been undertaken by ad hoc mediation centers that are not bound by the codes of existing professional associations. Now, many environmental mediation efforts are undertaken by "one-time only" intervenors, so that attempts to discharge them will have little effect on their future mediation careers. In short, the moral, legal, and economic pressures that ensure the accountability of mediators in other fields do not apply to environmental mediators. This gap is of some concern.
>
> Even if it were clear how environmental mediators could be held accountable, debate about what their responsibilities ought to be would continue. The success of most mediation efforts tends to be measured in rather narrow terms. If the parties to a labor dispute are pleased with the agreement they have reached voluntarily, and the bargain holds, the mediator is presumed to have done a good job. In the environmental field, there are reasons that a broader definition of success is needed—one that is more attentive to the interests of all segments of society.
>
> If the parties involved in environmental mediation reach an agreement, but fail to maximize the joint gains possible, environmental quality and natural resources will actually be lost. If the key parties involved in an environmental dispute reach an agreement with which they are pleased, but

fail to take account of all impacts on those interests not represented directly in the negotiation, the public health and safety could be seriously jeopardized. If the key parties to a dispute reach an agreement, but selfishly ignore the interests of future generations, short-term agreements could set off environmental time bombs that cannot be defused. Although the key stakeholders in an environmental dispute may pay only a small price for failing to reach an agreement, their failure could impose substantial costs on many groups, who may be affected indefinitely. Finally, the parties to environmental disputes must be sensitive to the ways in which their agreements set precedents; even informal settlements have a way of becoming binding on others who find themselves in similar situations.

Stulberg, Joseph B. "The Theory and Practice of Mediation: A Reply to Professor Susskind," 6 Vt. L. Rev. 85, 110–114, 1981; citations omitted.

Why and how else, then, should the mediator pay special attention to those not represented, as Susskind is proposing? Presumably, the response would be along the following lines. In environmental disputes, decisions are being made that will irrevocably affect the future development and life-style that can occur. The widest possible consensus among those who will be affected by the development is therefore necessary, whether or not they possess the power as individuals or groups to block implementation of the agreed-upon plan. Susskind appears to suggest that it is the mediator's job to assure that all those interests are represented in the decision making process.

This procedure is hardly tenable, either conceptually or practically. Mediation is a dispute settlement process that requires the active participation of individuals or groups of people. They identify their concerns. They must find a way to work with each other once the solutions are identified and agreed upon. The nature of the solutions is not only context-dependent, but participant-dependent. The individuals participating in the process, both as advocates and mediators, are an important factor in what solutions are ultimately accepted. The mediator's role, as traditionally discharged, is to help those persons reach a resolution and then withdraw. He cannot deal with absent parties. He does not know who they are, what they would have said if they had been present, nor what their priorities are. . . .

Second, Susskind suggests that "[e]nvironmental mediators ought to be concerned . . . about . . . the possibility that joint net gains have not been maximized [and about] . . . the long-term or spillover effects of the settlements they help to reach." Susskind proposed that it is the mediator's responsibility as an objective observer to insure that the final solution secures the greatest overall net benefits for each party, without leaving any party worse off than it was in its original configuration (the Pareto-optimal principle). He further suggests that the solution agreed upon should have the least possible adverse impact on other aspects of present or future community life. Simply stating the proposed responsibility for the mediator in this way reveals how awesome the task is that Susskind is proposing for the environmental mediator. To insure the Pareto principle is met, the environmental mediator must be able to generate, or at least guarantee, consideration of every possible technical solution to the environmental problem. He must secure demographic information on all persons affected by the dispute and factor their interests, desires, aspirations, preferences and values into the solution. He must project alternative development plans for jobs, tax bases, population trends, aesthetic values, school development and recreational needs for each possible solution. He must calculate the advantages and disadvantages of each solution against retaining the status quo, including the costs involved in using alternative dispute settlement procedures. And the list goes on.

Although these tasks might constitute a city planner's dream, they involve a host of analytical problems concerning logical theories of probability, measurement, interpersonal comparisons, and contrary-to-fact conditionals. These problems catapult the mediator's task into an intellectual warzone which raises the serious possibility that Pareto-optimal outcomes in the context of an environmental dispute are not, in principle, possible. As such, Susskind's proposal that the mediator ought to insure such an outcome must, charitably speaking, be held in abeyance.

A more troublesome question arises, however, regarding the justification for a mediator to block an agreement that fails to meet the requirements of the Pareto principle. Who authorized the mediator to design or insure the attainment of the "optimal" outcome as so conceived? . . .

Susskind apparently contends that such a responsibility emanates from the nature of environmental disputes, particularly because the spillover effect of particular agreements could irrevocably preclude certain options from again being entertained. . . .

It is not unique to environmental disputes that decisions made today foreclose certain options for tomorrow. Life is replete with such instances. In the labor–management sector, for example, agreement on a particular wage settlement might retard the development of mass transportation and thereby irrevocably increase the level of air pollution resulting from the use of automobiles. . . .

If we were to accept the obligations of office that Susskind ascribes to environmental mediators with regard to insuring Pareto-optimal outcomes, then the environmental mediator is simply a person who uses his entry into the dispute to become a social conscience, environmental policeman, or social critic and who carries no other obligations to the process or the participants beyond assuring Pareto-optimality. It is, in its most benign form, an invitation to permit philosopher-kings to participate in the affairs of the citizenry.

McCrory, John P. "Environmental Mediation—Another Piece for the Puzzle," 6 Vt. L. Rev. 49, 77–79, 1981; citations omitted.

The criteria for judging the fairness of a mediation process and the quality of the mediation agreement proposed by Professor Susskind are the following:

1. "[T]he outcome is better if it is consistent with shared notions of equity and justice;"
2. The resolution should "well reconcile the interests of the parties;"
3. The resolution should be "consistent with principles reflecting pre-existing practice;"
4. The "agreement should set 'a good precedent for the parties involved as well as for other parties'";
5. The "agreement should be 'reached quickly at low cost'";
6. "The process of decision should be one that tends to improve rather than exacerbate the relationships among the parties;" and
7. The agreement should be readily acceptable to the parties to ensure acceptability and compliance with its terms.

For purposes of discussion the focus will be on the fairness and practicality of using these criteria as the basis for determining if a mediator should be liable for damages because the mediated settlement did not maximize joint net gains.

As a preliminary observation, Professor Susskind states: "Principled negotiation rests on the assumption 'that the proper standards for judging a process of conflict resolution are not those that may produce a particular result in a particular case, but rather those standards that will tend to produce desired results in an indefinite series of cases.'" Judging a process and judging an individual mediation effort are two different things, especially if the purpose of evaluation is to determine if the mediator will be liable for the payment of damages. When the focus is on a process, the concern is not an individual effort, which may or may not fit a pattern. When the focus is on judging the quality of an individual effort, the pattern is secondary. Thus, it is unclear how reliable those process-oriented criteria are for judging a single mediation effort. In addition, it is not stated whether, or how, the criteria would apply if the mediator's responsibilities were set forth in a contract establishing a different basis for judging his or her performance.

The explanatory comment which accompanies the first criterion states that emphasis should be on the results of the dispute-resolution effort. The explanation for the second criterion adds that a neutral observer must be convinced that joint net gains have been maximized. What emerges is a format under which liability would be determined in a judicial proceeding in which a judge, or a jury, must examine a settlement to determine if the best result were achieved. This evaluation would not be fair for at least two reasons. First, it must be remembered that the settlement is that of the

MEDIATION ETHICS 253

parties to dispute, not the mediator. Second, in context of the statutory scheme of environmental regulations, it would be inappropriate for a court to determine which result or decision would be best. It is likely that the court would have to deal with matters which have been delegated to an administrative agency by Congress or a state legislature. Under such circumstances, it would be the agency's job to seek the best solutions for the problem at hand. When a court engages in judicial review of the agency's determination, it applies a test of "reasonableness" not "rightness."

The fourth criterion relates to the precedential value of a settlement. Professor Susskind offers the following explanation: "It may well be, that the way to ascertain whether this criterion has been met is to see whether a precedent has been set that helps to achieve the first three criteria over time." This explanation suggests that the quality of a mediation settlement cannot be judged on the basis of information available to the mediator and to the parties at the time that it is made. Hindsight is always better than foresight. It would be unfair to determine the mediator's liability on the basis of information which was not available when agreement was reached.

With respect to the fifth criterion, Professor Susskind states: "When costs and benefits to the community-at-large are weighed, however, it is often difficult to prove that a particular outcome is efficient." Implicit in this observation is recognition of the highly subjective nature of environmental decision making. Reasonable minds may differ as to the solution to a particular problem. It would be unwise, under such circumstance, to have a judge [or jury] determine liability, thereby imposing his or her [its] notion of the best solution. It has been said that mediation is more reliable than a judicial proceeding for finding solutions for environmental problems.

In summary, it would not be fair, practical, or desirable to apply the criteria suggested in a judicial proceeding for the purpose of determining a mediator's liability. Recall that the agreement reached is that of the parties to the dispute. Professor Susskind says nothing regarding their responsibility or liability. In many cases, an administrative agency which has a regulatory function will be directly or collaterally involved in a mediation effort. Nothing is said in the explanation of the criteria regarding the mediator's liability when such an agency concurs in or adopts the mediated settlement.

Problems of Implementing Accountability

If you are persuaded in principle of the need for mediator accountability, you must next grapple with the problem of implementation. In the first excerpt in this section, Susskind outlines several different approaches. Which do you think would be most beneficial. Which would be the most costly to put in place? In the second excerpt, McCrory expresses strong disagreement with most of Susskind's recommendations. Which of his criticisms do you regard as most substantial? Can you imagine other solutions that might satisfy Susskind's concerns without raising the difficulties McCrory identifies?

Lawrence Susskind. "Environmental Mediation and the Accountability Problem," 6 *Vt. L. Rev.* 1, 40–47, 1981; citations omitted.

> Whether environmental mediation becomes a matter of course in the United States, or occurs only rarely when an impasse has been reached, guidelines are needed to ensure that mediation efforts are structured properly. Enforcement of such guidelines needs to be institutionalized, and there are at least three procedural approaches to holding environmental mediators accountable. The first involves licensing, certification, or registration. This procedure might be done at the state or federal level. Licensed mediators would be expected to subscribe to an explicit statement of their responsibilities. A second approach would involve creating environmental mediation offices attached to regulatory agencies or to the attorneys' general offices at the federal or state levels. This second approach could be augmented by having administrative law judges require mediation before accepting challenges to the decisions or actions of regulatory agencies. A third approach would involve shaping the public awareness of the risks and opportunities associated with environmental mediation—a well informed public can demand accountability.
>
> If environmental mediation proceeds under the auspices of an accredited organization such as the American Arbitration Association, their code of ethics could bind the mediators. For the foreseeable future, environmental mediation will probably not involve accredited mediators. Recent instances of environmental mediation have tended not to stress the formal credentials or the interventionists' skills of the mediators involved. Instead, mediators have had only to convince the parties involved that they might be able to provide assistance. Success to date has depended primarily on the mediators' capacity to maintain the trust of the participants. This capacity, and their ability to keep negotiations moving toward the preparation of written agreements, have been all that the participants have asked.
>
> ### A. CREDENTIALS
>
> The Office of Environmental Mediation in Seattle has suggested that the Code(s) of Behavior published by such agencies as the Federal Mediation and Conciliation Service and the Association of Labor Management Agencies could provide a basis for establishing standards for environmental mediators. Environmental mediators, however, require different standards which are tailored to their own situations. It is unlikely, therefore, that existing organizations which credential mediators would be helpful. Moreover, many environmental disputes are likely to be mediated by individuals called upon to intervene because of their positions and not their credentials as

mediators. Their success may well depend on their capacity to operate free from the constraints that mediators typically feel. Congressman Wirth and others like him will not be successful if they are bound by procedures promulgated by associations responsible for credentialing professional mediators.

Environmental mediators, to the extent that they adopt the broader view of their responsibilities suggested in this article, will probably need to possess substantive knowledge about the environmental and regulatory issues at stake. Effective environmental mediation may require teams composed of some individuals with technical backgrounds, some specialized in problem-solving or group dynamics and some with political clout. It would be difficult and probably inappropriate to credential such teams.

B. LINKS TO REGULATORY AGENCIES AND THE COURTS

Credentialing or licensing would probably be unimportant if environmental mediation were undertaken by independent agencies of government or by court appointed mediators. This agency involvement would also solve the financial problems that continue to plague ad hoc environmental mediation centers. Legislation could be enacted describing the circumstances and conditions under which mediation would take place. Those regulated and those doing the regulation would jointly select mediators. The mediators' responsibilities would be spelled out in the legislation. Potential and volunteer mediators could be required to disclose their views concerning the need to protect underrepresented groups, strategies for maximizing joint net gains, and the importance of considering the precedent-setting nature of mediated agreements. As mediation experience accumulates, mediation guidelines could be refined. Mediators could be paid as staff to the attorneys' general office.

Courts with responsibility for reviewing the administrative decisions of public agencies might rely more heavily on mediation. Indeed, judges could insist that mediation (or at least joint fact-finding) precede formal challenges to administrative actions. At the very least, this process would help narrow and clarify the issues requiring court review. Court appointed mediators would, of course, be accountable to the judges who appointed them as well as to the parties. The mediators would be paid in the same manner as judges.

C. CREATING AN INFORMED PUBLIC

The most effective way to hold environmental mediators accountable would be to increase the public's capacity to demand fair and effective behavior on the part of mediators. This strategy is long-term and could begin with the provision of government funds to ensure the representation of disadvantaged groups with a stake in a mediated dispute. To the extent that certain groups feel they are not competent to participate in technical aspects of negotiation, funds should be provided to appoint qualified agents to represent them. There are numerous public interest groups that could provide such assistance. Representatives of all stakeholding interests ought to be given funds to caucus with the people they represent. These are some of the costs associated with creating an informed public.

Several elected officials should probably participate in every environmental mediation effort. The larger the number of elected officials involved as interested parties or observers and not as mediators, the more likely that interests, indirectly or adversely affected, can find someone to hold accountable.

The community-at-large must keep abreast of the negotiations' direction, or underrepresented groups will be unable to assert their concerns more forcefully. Although mediation efforts in the labor–management field are conducted in private, this practice should not be the case with environmental mediation. At least some scheduled sessions should be open to the public and to the news media. Closed meetings are more acceptable if the agreements negotiated are subject to public scrutiny during parallel regulatory hearings, as in the Brayton Point case.

In summary, credentialing should not be insisted upon in the environmental field. The institutionalization of environmental mediation through formal links to regulatory agencies or the courts is much more likely to produce situations within which the responsibilities of environmental mediators can be defined and monitored appropriately. The process of building an informed public should begin, but the task promises to be very long-term.

A note on the payment of environmental mediators is probably in order. The most desirable situation is one in which all parties contribute equally to the cost of compensating a mediator. In many environmental disputes, this cost-sharing will not be possible. Thus far, environmental mediators have volunteered their services, but this practice will continue only as long as foundation support holds out. Some observers have suggested the need for a superfund through which corporate and government contributions can be channeled to provide ongoing support for environmental mediation centers throughout the country. The financing problem stems from the assumption that environmental mediation ought to be handled by ad hoc centers that do not need to bill their clients. It would not be impossible to have the federal or state governments (or the courts) contract with these private centers, but it would be easier to situate independent mediators in government itself. Given the relatively small number of cases likely to go to mediation and the special circumstances surrounding the selection of mediators appropriate to each case, it is unlikely that many individuals will be able to build an entire livelihood around environmental mediation cases.

Accountability requires that the parties to a dispute as well as the members of the community-at-large be able to hold environmental mediators to their responsibilities. Assuming that those responsibilities are clearly articulated and agreed upon in advance, accountability can only be achieved if the mediators can be effectively chastised, fired, or sued by the parties directly or indirectly involved. If a contract exists, accountability is easier to ensure. In the absence of a contract, the institutional framework within which the mediator is employed and rewarded is crucial. If mediators are licensed, accountability is presumably ensured by the licensing agency. If mediators were attached to courts or other government bodies that could effectively regulate their future employment, accountability would also be ensured. If most environmental disputes are mediated by individuals involved on a one-time basis, however, contracts detailing their responsibilities will be needed. Enforcement of such contracts would be much easier if mediators were bonded, but such insurance might be difficult to acquire. The means of institutionalizing the accountability of mediators will only be as effective as the parties to a dispute demand. If the parties fail to see the need for a broad definition of an environmental mediator's responsibilities, it will be difficult to ensure accountability on such a basis.

D. CONCLUSION

Environmental disputes will continue to erupt over the allocation of fixed resources, the setting of public priorities, and the setting and enforcement of environmental standards. All three types of disputes are susceptible to mediation although not in every case. Mediation will depend on the availability of a mediator or mediation team acceptable to all parties.

Environmental mediators ought to be concerned about (1) the impacts of negotiated agreements on underrepresented or unrepresentable groups in the community; (2) the possibility that joint net gains have not been maximized; (3) the long-term or spill-over effects of the settlements they help to reach, and (4) the precedents that they set and the precedents upon which agreements are based. To be effective an environmental mediator will need to be knowledgeable about the substance of disputes and the intricacies of the regulatory context within which decisions are embedded. An environmental mediator should be committed to procedural fairness—all parties should have an opportunity to be represented by individuals with the technical sophistication to bargain effectively on their behalf. Environmental mediators should also be concerned that the agreements they help to reach are just and stable. To fulfill these responsibilities, environmental mediators will have to intervene more often and more forcefully than their counterparts in the labor management field. Although such intervention may make it difficult to retain the appearance of neutrality and the trust of the active parties, environmental mediators cannot fulfill their responsibilities to the community-at-large if they remain passive.

The institutionalization of guidelines and procedures for holding environmental mediators accountable would be best handled by linking environmental mediation directly to the prosecutorial and judicial branches of federal and state governments. Efforts to license or credential environmental mediators are not likely to be effective. The long-term task of building the capacity of the public to participate in environmental mediation should begin in earnest.

John P. McCrory, "Environmental Mediation—Another Piece of the Puzzle," 6 Vt. L. Rev. 49, 64–68, 70–71, 73–99, 1981; citations omitted.

Professor Susskind states that environmental mediators are not subject to the moral, legal, and economic pressures of accountability which apply to mediators in other fields. To fill this perceived void, he proposes that guidelines be established to ensure that mediation efforts are structured properly and that these guidelines be institutionalized so that they are enforceable.

A. CREDENTIALING

Credentialing of environmental mediators, whether by government or by private agencies, would provide an effective means for accomplishing an objective of the Susskind proposal. It could be used as a mechanism to institutionalize standards for environmental mediators. Although there would be danger of destroying the flexibility of the process if the standards were made too specific, a code of ethics for environmental mediation would be helpful. Such a code has been adopted by agencies which provide collective bargaining mediation services. A similar document might be devised for environmental mediators.

Credentialing agencies could also be used to ensure that there there is a readily available source of qualified mediators with adequate "substantive knowledge about the environmental and regulatory issues at stake" in a dispute. Agencies could catalog panelists according to their special knowledge, experience, and areas of interest to facilitate locating those individuals best able to handle particular types of disputes. Although it would be unwise to require that all environmental mediation be done by credentialed persons, disputants who need a neutral would have established sources and procedures for selecting a mutually acceptable person.

Credentialing agencies can also sponsor publications and programs to improve the quality of mediation services. Newsletters, printed summaries of documented mediation efforts, and similar publications containing information relating to current events and important developments could be distributed to mediators and other interested persons. Seminars and workshops at which mediators could share their experiences, discuss common problems, and focus on emerging issues and problems would foster an informed and studied development of the process of environmental mediation.

Educational programs could also be sponsored for potential consumers of mediation services. . . .

B. LINKS TO REGULATORY AGENCIES AND THE COURTS

In his discussion of linkages to courts and agencies, Professor Susskind suggests that potential mediators should be required to disclose their views in writing on certain ethical and procedural issues. These issues include the need to protect any underrepresented groups, strategies for maximizing joint net gains and the importance of taking into account the precedent-setting nature of mediated agreements. This suggestion is impractical. The precise strategy for identifying interests to be protected, to the extent that they are identifiable, will vary from case to case. A mediator will not be able to devise strategies for achieving the optimal resolution of a dispute until he or she has become involved in the mediation effort. An attempt to make such decisions in advance, and premature disclosure, could destroy the procedural flexibility which a mediator must have. It might also limit his or her ability to interject innovative substantive proposals for resolution and to deal with unanticipated issues and situations which may arise after mediation is underway.

If mediation is to achieve acceptability as a technique for resolving environmental disputes, governmental linkages and support will be required. This aspect of the Susskind proposal is valid. Governmental support could satisfy two important needs: providing systems to deliver mediation. Due to the nature of environmental disputes and the parties involved, it is not realistic to expect that the parties can or will share the expenses of mediation. Most environmental mediation to date has been done by individuals or private agencies which were able to donate their services because they operate with funding from private sources or government grants. The use of mediation for environmental disputes will remain limited unless a broader base of support is established. . . .

A governmental mediation agency, whether at the state or federal level, could also serve important administrative functions. It could screen disputes according to established criteria to determine if free or partially subsidized services should be provided, or if referral should be made to a private agency. The agency could also resolve initial questions regarding identification of parties for mediation and could insulate mediators from making decisions during the course of mediation. For example, an interest group may seek to intervene after mediation is underway. Having to rule on the entry of a new party may jeopardize the credibility and perceived impartiality of the mediator. A decision permitting or denying intervention may be seen as favoring one faction or another. The need to cope with the issue could also be a significant distraction to the mediation effort. If the agency could make the intervention determination, it would preserve the mediator's credibility and insulate the process from time-consuming and distracting influences which might diminish the quality of the mediation effort.

C. AN INFORMED PUBLIC

There can be no quarrel with the notion that there is a vital public interest which should be considered in environmental decisionmaking. The extent to which the mediation process and individual mediators can and should be responsible for protecting that interest is, however, another matter.

Professor Susskind is critical of environmental mediation because so-called "public interest representatives" have not been "duly appointed" by the interests they represent. If there is no homogeneous identifiable public interest, there are obvious procedural problems associated with appointing its representatives. Moreover, making a mediator legally accountable for protecting such vague and amorphous interests is difficult to justify. Such difficulty should not imply, however, that a mediator should ignore identifiable underrepresented interests or need not be sensitive to a broad sense of the public interest. These considerations are important and the parties to a dispute should be reminded of them at appropriate times.

D. SUING THE MEDIATOR

The most provocative aspect of the Susskind proposal is contained in the following statement: "Accountability requires that the parties to a dispute as well as members of the community-at-large be able to hold environmental mediators to their responsibilities. Assuming that those responsibilities are clearly articulated and agreed upon in advance, accountability can only be achieved if the mediators can be effectively chastised, fired, or sued by the parties directly or indirectly involved."

Mediators may always be chastised or fired. Making them suable adds a new dimension to the process of mediation. To explore that dimension, four questions must be answered: (1) suable by whom? (2) suable for what? (3) What remedies would be available? and (4) What standards would be used to determine if a mediator has breached his or her responsibilities so that a remedy could be imposed?

The foregoing quotation provides the answer for the first question. A mediator is accountable to the parties to the mediation effort and to anyone else in the community at large who is directly or indirectly affected by the settlement agreement. The pool of potential plaintiffs would be large and, very likely, undefined during the course of a mediation effort. Governmental agencies would probably be among parties which would be eligible to sue.

The duty which environmental mediators would have, which could be breached and become the basis for suit, relates to "the fairness of the processes in which they engage as well as the quality of the agreements they helped to reach." The emphasis "should be on the results of the dispute resolution effort and not just on the fairness of the negotiations process." Thus, the answer to the second question is that a mediator could be sued on both procedural and substantive grounds.

Procedural grounds might include the ground rules established by the mediator, the nature and substance of his or her communications with the parties in joint or private meetings, the mediator's decisions regarding intervention of an interest group and the level of participation accorded to parties and nonparties during a mediation effort. Substantively, the agreements reached would be expected to maximize the joint net gains of various interests, including the interests not represented in the mediation effort. In other words, the mediator must help the parties find the best balance for all of the competing interests. The broad sweep of the Susskind proposal would suggest that the remedies available in the suit against a mediator would include an order enjoining or requiring certain mediation procedures, an order requiring a mediator to allow intervention, an order setting aside a mediated settlement, and recovery of money damages resulting from the implementation of a settlement agreement that did not maximize joint gains.

The most complex inquiry relates to the standards which would be applied in determining when a plaintiff would prevail in a law suit against a mediator. Professor Susskind states that a mediator's responsibility could be spelled out in legislation or in a contract between the parties to a dispute and the mediator. He also proposes specific criteria for judging the fairness of the mediation process and the quality of settlement agreements. The following discussion examines the feasibility of establishing mediators' responsibilities by statute or contract and the utility of Professor Susskind's proposed criteria.

It is unlikely that an enforceable statement of a mediator's responsibilities could be set forth in statute. If mediation is to maintain its character as a flexible process, rather than a structured procedure, a rigid statutory framework will do more harm than good. Due to the nature of environmental regulation, substantive standards by which to judge the quality of the mediated agreement and the potential liability of a mediator would be even more difficult to establish by statute.

The suggestion that a mediator's responsibilities can be defined in an enforceable contract is also troublesome. It would indeed be optimistic to assume that parties who cannot resolve their dispute could agree upon a contract that would spell out, with enforceable precision, mediation procedures to be followed or standards for judging the quality of the mediated agreement. Even if such an agreement could be drafted, there remains the matter of protecting the interests of the underrepresented. Must the mediator represent those interests when the contract is drafted? If so, he or she would be placed in an awkward position for two reasons. First, the interests which need protection may not be identifiable, if at all, until after the mediation is underway. Second, the mediator's status as a neutral, actual or perceived, would be jeopardized. He or she would have to represent interests which might be adverse to those of disputants who would be parties to the contract. . . .

Even if the foregoing problems could be overcome, the idea of specifying a mediator's responsibilities in a contract is inconsistent with the nature of the mediation process. A mediator needs the flexibility to adjust procedures to meet specific needs which may arise from time to time, including the intervention of new parties and the interjection of new issues. Flexibility is also needed with respect to the substantive outcome of the mediation effort. An important part of the mediator's job is to reorient the thinking of the parties so that they will be more receptive to new ideas and compromise.

He or she must have the latitude to move the parties toward solutions which they could not anticipate.

Study Questions

1. What economic and social forces are likely to make one form of institutionalizing accountability more likely than others?

2. Will accountability stifle mediation? Some observers believe that potential Good Samaritans are somedays deterred from rescuing people in peril by the fear of a lawsuit, should their efforts fail. In response, some states have enacted laws to insulate Good Samaritans from suits in cases involving anything short of wanton recklessness on their part. In seeking greater mediator accountability, is Susskind moving in the opposite direction? Is it possible to impose mediator liability without discouraging intervention? What solutions to this dilemma can you fashion?

3. Is the analogy to the Good Samaritan an apt one: Should mediators, instead, be likened to lawyers, who are retained by clients to resolve disputes? Lawyers, after all, are supposed to be accountable for their errors. (To give this question full attention, you should consider the note below on mediators' incentives.)

4. Lawyers who are accused of malpractice or unethical conduct may defend themselves against such charges, even if doing so means revealing confidential client statements. What defenses should a mediator have? What if the mediator can justify his or her action, but only by revealing the confidences not of the complainant but of the other parties to the agreement?

5. Statutes of limitation apply in most areas of law, so that a wrongdoer is liable only for a stipulated number of years. When the impacts of an important environmental decision may be felt for decades, should the mediator continue to be responsible?

6. What remedy should be available in the case of mediator malpractice? Making the mediator personally liable may be of little use where his or her assets are modest and the damage enormous. Should the injured party be able to repudiate an agreement that was poorly mediated?

Related Ethical Issues

Incentives to Mediate

Much negotiation literature postulates that mediators must be strictly nonpartisan, but mediators clearly have their own goals and priorities, some of which may conflict with those of their clients. The nature and impact of these incentives must be identified if the mediation process is to be understood.

What factors may induce a third party to intervene in a dispute? Some conflicts, after all, are so destructive that no one wants to get caught in the cross fire. Even where there is some hope for reconciliation, the mediator must determine whether the prospects are worth the effort. Essentially, the mediator must weigh the costs and benefits of involvement. In some instances, the mediator must share the costs even though others will reap most of the important benefits.

One inducement is financial. In the Brayton Point case, the parties agreed to underwrite a $20,000 mediation fund, a portion of which was for the mediator. If the dispute requires substantial time and energy, the mediator simply may not be able to forego compensation. No matter how payment is arranged, however, some conflict with the interests of the parties will exist. If, for example, he or she is paid by the hour, the mediator profits from a protracted dispute. (It is small comfort to note that lawyers frequently bill on the same basis, for they have long been accused of fomenting and prolonging conflict for their own benefit.) A lump-sum payment is not necessarily better for the parties, however, for then the mediator has no monetary inducement to spend more time on a problem even when the parties' interests require it. A third possibility—the provision of mediators by an independently funded servie—mitigates this problem but does not eliminate it. Salaried mediators still must measure the extent of their involvement in any one case against the value to them of participating in other disputes.

Dedicated mediators often say that it is their duty to put the interests of their clients ahead of their own. To note that a mediator's professional well-being does not always coincide with the interests of his or her clients is not to say that personal well-being will dictate conduct. Nevertheless, it does underscore the need for other incentives that will encourage the mediator to enter disputes and facilitate settlement even when doing so may not be personally advantageous.

For some people, the nonmonetary rewards for mediating may be significant. Successful resolution of seemingly intractable disputes may bring the mediator prestige—a commodity that may be valued for itself or that may be exploited for advancement in other fields, notably law and politics. Yet here, too, the interests of the parties and the mediator are potentially at odds. One of the tenets of mediation practice is that the mediator should keep a low profile. Timothy Sullivan has observed that parties in mediation must trust the mediator to honor their confidential disclosures. As a result, "strong professional taboos exist discouraging the mediator from revealing either the course of negotiations, the communications between the negotiators, or even the actions which he took to promote a settlement" (see Sullivan, *Resolving Development Disputes through Negotiations*, New York: Plenum Press, 1984). Thus, mediators who fashion an ingenious settlement may be in a catch-22 situation: If they seek to take credit for their success, they risk spoiling their reputation by appearing indiscrete.

Conflict between the mediator and his or her clients may be even more pronounced in other circumstances. Sullivan notes that it is sometimes necessary for negotiators to "scapegoat" the mediator so that they can preserve their reputations with one another or with their respective constituencies.

> In the etiquette of bargaining, yielding to an adversary's demands can convey weakness, but yielding to a mediator's request shows statesmanship and reasonableness. . . . Negotiators [can] return to their constituents and claim that circumstances required the concessions. (Sullivan, p. 102)

This can enable them to gracefully move away from their previous positions. If settlement can be reached only if the mediator agrees to take the blame for forcing the parties to accept it, he may be committing an act of professional harikari.

Mediators sometimes intervene, not for personal gain, but out of a sense of social responsibility. Even without arguing the metaphysics of altruism, it is important to distinguish the mediator who has some interest, albeit indirect, in promoting settlement from the mediator who is truly disinterested. A public official who intervenes in hope of saving jobs or lessening pollution may be considering personal political gain. In a sense, such a person becomes a party in interest. None of us is purely neutral. We all feel the consequences of unemployment, higher energy costs, and dirty air, even if only slightly and indirectly. Of course, there are many cases in which the mediator is only minimally affected by the terms of a particular settlement. As a general matter, a party in interest will more likely be concerned with the substance of settlement, whereas a mediator's prime interest is in the process by which those ends are reached.

The weighing of these incentives to intervene, tangible and otherwise, clearly is an individual matter. The calculation of costs and benefits of involvement necessarily depend on a person's opportunity costs: Is there a better use of his or her time? Then, too, mediators may have different utility functions; prestige may be more important to one than to another. Finally, as a mediator's decision to enter a dispute may rest heavily on his or her estimate of the chances of success, the judgment necessarily is subjective.

Thus, some mediators might walk away from disputes that others would attempt to settle. It is also conceivable that there are conflicts that everyone would agree were ripe for intervention. In this later circumstance, parties theoretically could bargain with potential mediators to get the best service at the most favorable terms. In practice, however, market imperfections make this kind of shopping unlikely.

Mediators themselves may not agree on what constitutes successful intervention. Parties who fail to settle their principal conflict may still be able to narrow their differences and create the basis for future negotiation. If it is hard to define successful mediation, then it is at least as difficult to judge a mediator's performance. Clearly, a mediator's record cannot be reduced to a simple settlement/no-settlement percentage because some disputes are much more amenable to resolution than are others. Then, too, not all settlements are equally just or efficient. A person who subscribes to Susskind's notions of "accountability" might avoid a dispute that appears likely to produce a socially undesirable outcome, whereas other mediators who feel no responsibility for the terms of settlement would not hesitate to intervene.

The same considerations that may draw a mediator into a dispute may affect his or her conduct as negotiation transpires. Assessment of particular costs and benefits may change, however, as negotiation proceeds. As the mediator learns more about the issues in dispute, he may revise the prospects for settlement. Even if he concludes that his participation will prove fruitless, however, contractual obligations and matters of reputation may make it difficult for him to withdraw.

In sum, any complete analysis of mediated disputes should take into account the fact that the mediator, even if scrupulously *neutral* in the commonly understood sense of that term, will have his or her own goals and priorities. To say that a mediator operates merely out of loyalty to the process avoids this truth only by falsely distinguishing between means and ends. To be loyal to the process of mediation, one must either believe that in the long run it tends to produce better results or that as a process it has value in itself. Often these beliefs will, in the abstract, be compatible with the interests of disputing parties, but the means by which the beliefs are put into practice carry costs that someone must bear. Time and money are scarce resources. How they can be applied in any one dispute will depend on competing private and public demands. To state that intervenors have their own priorities and must work within certain limitations is to clarify the mediation process, not to disparage it. Just as negotiation can be encouraged by enhancing disputants' incentives to bargain and diminishing obstacles, mediation can be promoted broadly or, in particular cases, by manipulating the incentives to intervene.

Ethical Standards for Mediators

In the second section, we encountered the notion of *mediator accountability*. There, the focus was on the mediator's responsibility for the settlements he or she facilitates. In the materials that follow, the emphasis is more on the ways in which mediation is carried out, though, as we shall see, the distinction between substance and procedure is not always easy to respect. Moreover, examination of ethics requires consideration both of the moral implications of various mediation techniques and of their effectiveness.

Much of the environmental mediation that has taken place has involved free-lance mediators, who are not guided or constrained by any professional standards. Although there is increasing sentiment in support of a formal ethical code, there is little agreement on what it should require. The problem lies with the elusive nature of the mediation process. Means must be judged, not ends. The outcome in any case may be idiosyncratic; it is not necessarily the mediator's fault if the parties cannot come to agreement. Medical malpractice may offer a parallel. The quality of a doctor's performance is determined not by whether the patient lived or died, rather, by whether the attending physican followed established medical practice. The ethical standard thus might be good mediation practice. As mediation is even more an art than the practice of medicine, however, it is even more difficult to establish professional consensus on what constitutes proper procedure.

Gerald Cormick, founder of the Insitute for Environmental Mediation in Seattle, Washington, poses some general principles in the next excerpt. They rest on the premise that there are three central concerns to the mediator: achieving a settlement, the justice of that settlement, and its stability.

Cormick prefaces his list of principles with the observation that the extent of the mediator's ethical responsibilities may vary from dispute to dispute.

>   1. The more naive the parties, the greater the ethical responsibility of the mediator.
>   2. The greater the impact of the issues in dispute on parties not at the table, the more critical the responsibility of the mediator.
>   3. The less proportional the relative power equation between the parties, the greater the ethical burden on the mediator.

Thus, in collective bargaining, where the parties are experienced and have a continuing relationship, the mediator need only be concerned with achieving settlement, particularly if impacts on others are minimal. By contrast, environmental negotiations often introduce one or more of the variables that Cormick believes require greater ethical responsibility.

To what extent are the criteria Cormick enunciates moral precepts, to be valued for their own sake? Are they simply procedural rules for efficient action? To the extent they are of the latter, do you agree that they are effective? Finally, are his criteria consistent with one another? Other questions follow the excerpt.

Cormick, Gerald. "The Ethics of Mediation: Some Unexplored Territory," Unpublished paper presented to the Society of Professionals in Dispute Resolution, 1977.

> [Cormick notes that the first five criteria relate to the "justice" of possible settlements and the next four to "stability." The tenth relates to both justice and stability.]

1. *The mediator should be explicit as to the basic elements of the mediation process.*

Mediation should be demythologized. To call mediation an art or to suggest that what the mediator does cannot really be described because it is so highly personal in nature is to make the parties less able to control the negotiation process when a mediator intervenes. On the other hand, perhaps the most important single control over the activities of the mediator would be to make the parties themselves more aware of what mediation can and cannot accomplish and what types of actions are likely to result in what ends.

Any mediation tactic or strategy which relies on the naivete or ignorance of the parties for its effectiveness must be rejected.

2. *The mediator should foster and protect the proportional relative power relationship between the parties in decisions regarding entry, strategy and tactics, and the shaping of agreement.*

A basic social principle is that proportional power provides individuals and interest groups a basis upon which to pursue their own best interest. Mediators, in their decisions relating to their entry into, behavior within, and exit from a conflict situation should be conscious of the need to protect this proportional power relationship.

Where the proportional power relationship is sufficiently unequal that a mutually acceptable agreement is unlikely to emerge, the mediator should not enter the dispute. To do so might only result in lending credibility to a unilateral solution imposed by one party on another.

3. *The tactical decisions of a mediator should be based on an explicit, conscious rationale capable of later explanation and evaluation.*

"Seat-of-the-pants" mediation is just not good enough. Perhaps the worst enemy of parties involved in a conflict is a benign "do-gooder" who fails to carefully assess the impacts of his or her intervention on the parties, their relationship or the perceptions of the broader public. It is only by perceiving mediation as an intervention process that the mediator may properly consider the cumulative effects of his or her actions or the way in which a person's action-choices may preclude later opinions.

4. *The mediator must be concerned with enhancing the ability of the parties to jointly administer any agreement which is reached.*

This criterion is of particular importance where, as in many community and environmental disputes, there is no pre-existing relationship between the parties to the dispute. In such a case, the mediator may find it necessary to prolong the negotiation–mediation process in order to give the parties an opportunity to develop a working relationship. In some instances, the mediator may even find it necessary to assist the parties in finding third-party sponsorship and/or funding to underwrite and support joint implementation of an agreement over the longer run.

5. *The mediator should not permit him or herself to be a party to any agreement which violates the basic principles of freedom, justice and proportional empowerment.*

Simply put, the mediator should not permit him or herself to continue as intervenor in a situation where the goal of the parties is to reach an agreement which abrogates these basic principles either for one of the parties in the dispute or for some party not at the table. The mediator should withdraw.

6. *The primary responsibility of the mediator is to enhance the collective bargaining or other relationship existing between the parties.*

This criterion goes to the very nature of the mediator's role as an intervenor. That is, his or her insertion into a relationship can change that relationship for better or for worse. A mediator is merely an extension of the negotiating relationship. The ultimate aim of any intervention into that relationship should be to enable it to emerge better able to proceed without further intervention by the mediator.

7. *The mediator should promote the ability of the parties to negotiate joint agreements.*

This criterion may require a conscious effort by the mediator to train one or more of the parties in such basic negotiation skills as organizing a negotiating team, phrasing demands, or listening to the other party(ies). It may also involve providing one or more of the parties with access to enter into good-faith agreements, or otherwise better equipping the parties to operate on a good-faith, knowledgeable level in their dealings with one another.

8. *The mediator must familiarize him or herself with the specific dynamics of the dispute situation in which he or she is intervening.*

In the labor–management sector there are important differences in the attitudes of the parties toward one another and toward the broader public interest. Such differences occur both between the public and private sectors and between blue-collar employees and professional or semi-professional employees in the public sector itself. In disputes in which race is a factor, there are certain dynamics with which the mediator must be familiar if he or she is to recognize certain basic concerns of a minority population in the larger society. Mediators can worsen situations by actions as simple as using specific "trigger words" which can set off a chain of actions and reactions. The mediator in non-labor management disputes should be familiar with the complexities inherent in developing viable relationships and agreements in relatively unstructured situations and where decision-making authority is unclear or diffused.

9. *The mediator must have a concern with the viability of any agreement reached by the parties in his or her presence.*

The viability factors in an agreement are fourfold: (1) technical feasibility, (2) legal feasibility, (3) political feasibility and (4) financial feasibility. Clearly, these are of varying importance and arise in different combinations in different dispute arenas. Settlement or accommodation per se is not the only responsibility of the mediator. Where an agreement is reached which is not viable it may discredit the negotiation–mediation process. It may also serve to cripple any party whose power is based on its ability to confront an established institution and who finds the apparent legitimacy of the process has decreased public sympathy, making it more difficult to mobilize supporters.

10. *The mediator should keep before the parties a consideration of the realities of the broader public interest.*

This criterion is related to the concern with the viability of agreements discussed above. It is not suggested that the mediator attempt to impose the interest on the parties, but, rather that he or she has an ethical responsibility to ensure that the parties consciously consider the public interest and ways in which it may affect any arrangement which they might reach.

Study Questions: Incentives and Ethics

1. Using Cormick's list as a guide, identify the ethical decisions David O'Connor encountered in mediating the Brayton Point dispute, described in chapter 8. Did O'Connor honor all of Cormick's principles? Should he be faulted for not insisting on greater participation by consumers of electricity and neighbors affected by its generation?

2. In that dispute, the four principal parties agreed to make equal contributions to the $20,000 mediation fund. There certainly will be cases, however, in which parties with a large stake in the dispute simply cannot afford to pay for mediation. In such instances, it might seem equitable to allocate the expense according to ability to pay. What problems would this raise? How might they be solved?

3. There was no similar financial pool in the Foothills case, described in chapter 9. What were the costs of mediation in that instance and who bore them?

4. Representative Shroeder tried to get an environmental mediator involved in Foothills; indeed, Gerald Cormick was one of the nominees. The Denver Water Board was opposed to mediation at that point. Suppose, however, that Cormick had been invited to intervene. In what ways would his ethical principles have caused him to follow a different course than that taken later by Tim Wirth? Who properly might have carried the costs of such a mediation?

5. Should we regard Cormick's list as simply a private statement of principle or are his criteria precise enough to be the foundation of a code for environmental mediators generally? One can imagine several ways in which such a code might be applied: A dissatisfied disputant might claim he or she was injured by a specific violation. A mediators' association might wish to expel an unethical member. Or, the mediator's employer might wish to use a code to evaluate his or her performance. To which context could Cormick's criteria be most easily adapted?

6. Cormick suggests in his second criterion that a mediator must not disturb the power relationship between the parties. Yet, where inequality in power stems from historical injustice, how is it ethical to perpetuate the imbalance? To remain "neutral" in such a situation is really to side with the more powerful interests, is it not? Cormick declares that if "the proportional power relationship is sufficiently unequal that a mutually acceptable agreement is unlikely to emerge, the mediator should not enter the dispute." Does this solve the problem?

7. The basic concerns of the mediator may be in conflict. What, for example, is the mediator's responsibility if he or she realizes that a party has miscalculated the impact of the terms it is about to accept. A company may have vastly underestimated the cost of an antipollution device; a regulatory agency may have overlooked an incidental by-product of the proposed solution. Should the mediator raise the mistake when doing so will require a recalculation that may jeopardize the settlement? Is it relevant that the negotiators are experienced or naive, or that they enjoy substantial technical resources or none? Does the mediator's duty depend on whether the miscalculation will merely diminish profits or bankrupt the company?

8. What, if any, leverage does a mediator have to avoid being placed in a compromising position? For example, if the mediator feels that one side is stonewalling or trying to subvert consensus, what countermeasures may he take? If the mediator feels that the agreement the parties are about to ratify is irresponsible, is it enough for him or her to walk away quietly? Are there any instances in which it would be proper for a mediator publicly to repudiate a proposed settlement?

The Mediator with Clout

Introduction

Professional mediators often claim complete neutrality in the disputes they handle. Although the mediator may have no stake in the outcome, however, it is hard to imagine a case where he or she is neutral in every respect. At the least, the mediator's intervention usually reveals a bias in favor of settlement: even mediators who profess absolutely neutrality tend to measure their success by whether they have enabled the parties to reach closure. Peace, of course, does not always come without a price. Moreover, as we have seen, mediators who are utterly indifferent as to outcomes invariably must have their own agendas about process: Other cases compete for their attention. Time can be costly. Their reputations may be enhanced or diminished, depending on the outcome of the mediation.

Thus, even in the purest of cases, the term *neutral* does not describe the mediator's role with complete accuracy. As a matter of common usage in the field, the term stands for the proposition that the mediator has little or no interest in the substance of the dispute. The word *neutral* also reenforces the image of the mediator as a person skilled in the negotiation process, but powerless to impose a solution on the disputing parties.

In this section, we consider instances in which the mediator's "neutrality"—as commonly understood—is somehow compromised. This may occur when a person who has an identifiable interest in the outcome of a dispute tries to serve as a mediator or where the mediator has the power to order an outcome if the parties cannot find resolution on their own.

Some students of mediation might insist that whatever such parties should be called, they should not be regarded as mediators. In the realm of practice, however, where the roles of the parties may shift in the course of a negotiation, semantic distinctions often break down. From a functional analysis, it is clear that stakeholding negotiators do sometimes engage in a process that looks very much like mediation, no matter what we choose to call it. In the same vein, a judge or bureaucrat who has the ultimate power to settle a dispute may nonetheless try to prod the parties into reaching agreement on their own. Our focus in this section will be on the special issues and problems that arise when a mediator also has another significant role in the dispute.

Some experienced observers believe that good negotiators often function as quasi mediators. This can be particularly true when a member of a bargaining team must mediate differences among his or her own colleagues. In a case of collective bargaining, for example, some union representatives may wish to settle whereas others are intent on holding out. The lead representative may have to shuttle from constituent to constituent to fashion a unified coalition. Even when he or she deals with the other side, he or she may be both a negotiator and a mediator. In public, such a figure may stake out strong positions, but in private meetings with the other side, he or she will work jointly to try to facilitate agreement. Just as a negotiator may be required to use the skills of a mediator, so does a mediator depend on negotiating techniques. The successful bargainer often prevails by getting the other side to doubt its position. For example, a plaintiff strengthens his hand by demonstrating to the defendant the risks to him of going to court. Similarly, the artful mediator gets parties to move from impasse by raising doubts about the consequences of sticking to their current positions. A review of the principal cases in the preceding chapters should demonstrate the common elements of negotiation and mediation.

Mediators also serve dual roles when they have the ultimate capacity to impose a resolution on the disputing parties. In some fields, there has been growing interest in the practice of *med-arb*, a hybrid process in which the intervenor who is selected is given the power to settle the dispute if mediation fails to yield agreement. Judges traditionally have been reluctant to try to mediate cases that they may ultimately be required to decide, though some members of the bench are now taking a more active role in trying to help the parties reach agreement.

What are the advantages of having one person serve as both mediator and arbitrator? Are there any drawbacks? How might it affect the strategy and behavior of the negotiating parties? Think back to the Brayton Point case in chapter 8: Would the mediation process have been different if David O'Connor had held the power to impose a resolution? Are there particular kinds of environmental disputes that are appropriate for med-arb and others that are not? Does it matter whether the parties have chosen to undertake med-arb or have had it imposed by law?

Comparison Case: Gospel-Hump

The following case is adapted from Timothy Sullivan's *Resolving Development Disputes through Negotiations*, New York: Plenum Press, 1984. It involves an Idaho land use dispute in which Frank Church, then a United States senator, intervened. Sullivan terms Church a "powerful mediator"; Susskind similarly might call him "a mediator with clout." Read the case and decide whether Church should be thought of as a mediator, stakeholder, or adjudicator. How was his role different from that played by Congressman Timothy Wirth in Foothills, described in chapter 9.

> Gospel-Hump is a roadless area of national forests located in Idaho. After an initial lawsuit over the use of this land in 1972, the Forest Service agreed to prepare an environmental impact statement before taking any action in this region that could reduce the area's wilderness potential. For the purposes of study and management and to comply with their agreement, the Forest Service divided the region into eight planning zones, and prepared land management plans for these zones. Idaho conservationists challenged two of the land management plans. This challenge halted the Forest Service decisions affecting any part of this whole region while the appeals of the two management plans were in process. Sawmills in the region were running out of harvestable trees, and without some Forest Service action, they would have to shut down.
>
> In a decision on March 8, 1977, the Chief of the Forest Service reviewed the original study plan, disallowed the Forest Service's piecemeal approach, and required the development of a comprehensive plan for the whole region. No action was possible before completion of this study. This new comprehensive study would take at least four years to finish, and could also face appeals and possible litigation. The length of this process threatened to shut down the sawmills in this region since the supply of harvestable

trees would remain frozen during the preparation of these new studies. The lumber industry is the major employer in this remote region; closing the lumber mills would bring substantial economic hardship to the region.

Members of the local Chamber of Commerce asked Senator Frank Church to try to help them resolve the dispute. One member asked if the Senator could help them by incorporating the core of the area into the national wilderness system while freeing peripheral lands. Senator Church sat on the Committee on Energy and Natural Resources which originates legislation which can determine the use of federal lands, and he had the power to help them. Church decided to convene a meeting between a committee of local members of the Chamber of Commerce who represented the wood products industry and conservationists to see if they could work out an agreement. At the first meeting, both sides found that they had much in common. They all enjoyed the outdoors and shared love for the wilderness. They immediately decided that 45,000 acres of the most productive timberland should be excluded from wilderness designation. They agreed to try to negotiate a full solution.

Senator Church left members of his staff to mediate the dispute between the two parties. Negotiation focused on topological maps, with each side drawing lines to include and exclude areas from the wilderness region. In the course of negotiation, the dispute was reformulated, changing from an ideological confrontation over land use to a cooperative planning endeavor aimed at maximizing joint benefit. Topological maps served as a natural negotiating text.

After a long period of negotiation, 220,000 acres, consisting mostly of high alpine country, were classified as wilderness. This land had little timber value. The most heavily forested portions, about 123,000 acres, were allocated for timber harvest and development; 45,000 acres of forest were made immediately available to wood products industries. In addition, a seven member citizens' advisory committee was formed to work with the Forest Service to plan development for this region.

This compromise plan underwent revision during the legislative process. The original groups did not represent mineral interests; this apparently was a simple oversight. Senator McClure, Idaho's other senator, added an amendment addressing this problem. The Forest Service requested a change in the boundary of the area under consideration, and the senate committee deleted 14,000 acres for the provisions of this act. Finally, other revisions enabled the Secretary of Agriculture to allow snowmobile use in certain regions of the wilderness area. The legislation was passed by Congress and it solved a problem that had been trapped in agency reviews for several years.

The Gospel-Hump Wilderness Dispute illustrates how a mediator with power can change the nature of a conflict, and make new alternatives available to the disputants. Senator Church had great prestige with the conservationists groups in his state because of his sponsorship of the original wilderness legislation. The Chamber of Commerce request for help offered the Senator the possibility of expanding his base of support in the Idaho business community. His senate committee position gave him the power to resolve the conflict. Without his legislative intervention, the dispute would have twisted slowly through agency studies and judicial reviews. Thus, his involvement made possible a settlement that the conflicting groups could not achieve on their own.

The argument, in the following excerpt, states that there are costs as well as benefits when an elected official serves as mediator. Which of these costs do you regard as most substantial? Are they more likely to be felt in certain kinds of cases than in others?

Stulberg, Joseph B. "The Theory and Practice of Mediation: A Reply to Professor Susskind," 6 *Vt. L. Rev.* 85, 107–109, 1981.

> Susskind explicitly states that the Congressman who served as the mediator in the Foothills Water Treatment Project discussions was publicly committed to a particular position advanced by one of the parties prior to his entry as a mediator. . . .
> Susskind is very candid on this point: "Wirth's political position gave him clout, even when he chose not to use it. This clout was apparently respected by all the participants." The message is obvious: if the parties [did] not cooperate, the Congressman could hurt them in Washington on this or some other matter. That is, the Congressman who intervened had the power by virtue of his position to prevent any party from obtaining its goals. Hence, they agreed to negotiate with him in order to secure a desired objective. Certainly, there is nothing wrong with proceeding in such a manner, and the account of the resolutions suggests that his intervention was successful. One must question, however, how the parties thought that Wirth's intervention was most helpful. They might have been uncomfortable sharing confidential information with him regarding the lack of group consensus on a particular proposal. A party may not have informed him of all possibly acceptable alternatives, knowing that his legal or political posture might be compromised. The need of the parties to talk and work together would be substantially reduced if the simple persuasion of the other parties by the mediator to [his] one viewpoint would be sufficient to gain resolution. If the Congressman's political power is an ingredient necessary to the implementation of the agreement, then the agreement is only as stable as the Congressman's continued political success. The parties can legitimately wonder how the Congressman's political needs affect his efforts to prod the parties into an agreement. And the parties surely cannot be criticized if they view the Congressman's intervention as an opportunity to secure a variety of political objectives, thereby making trade-offs on the environmental matters in dispute in order to gain benefits on other matters of personal interest.
> Clearly, this type of intervention is quite different from that of a medi-

ator who derives his power from his very commitment to neutrality. This difference is more than a terminological quibble about what constitutes "mediating." At issue is an understanding of, and respect for, what the parties to the mediation sessions are entitled to expect from the intervenor. Will confidences be honored? Who sets the agenda in terms of issues to be discussed? Will the order in which the issues are discussed be skewed so as to insure the mediator's desired outcome? Will meeting times be scheduled for the convenience of the parties or might they be arranged by the intervenor in order to make it difficult for some (i.e., "obstreperous") parties to attend and voice objections to the intervenor's preferred position? Will the mediator refuse to schedule meetings if the one party whose position the mediator supports demands that future meetings be conditional upon the other parties having made particular concessions?

One can certainly offer answers to these various problems, but Susskind's burden is more substantial. First, he should demonstrate how a mediator committed to neutrality cannot render effective service in an environmental dispute. Second, he should explain the obligations of office that the "mediator with clout" assumes when he renders his services. Is it appropriate, for example, for the "mediator with clout" to threaten a recalcitrant party with political retaliation? If not, why not? Third, Susskind should illustrate the value derived, if any, by labeling such intervention "mediation," since it differs in so many striking ways from mediation in labor–management collective bargaining, community dispute negotiations, court-diversion programs, and countless other private dispute settlement systems. Clarification is necessary to insure a degree of consistency in program posture and purpose among those encouraged to experiment with "mediation" programs as an alternative dispute settlement procedure.

Conclusion

The following fable was written—probably in triplicate—by Peter Lovi, town planner of Ithaca, New York. It nicely poses the broad social and philosophical issues raised by mediation.

To: Solomon, Mediator
From: The Enemy, Religion
Re: Your Mediation Efforts

I believe that our group has a fundamental disagreement with your method. How can you expect us to fairly sit down and discuss these important issues with people who are wrong? It is as if we do not speak the same language. Answer me one question and I'll reconsider mediation: where do we begin; what first principles should we establish which will not hopelessly compromise our efforts from the outset?

To: Solomon
From: The Enemy, Ethics
Re: Your Mediation Efforts

    I really don't know if I should write this letter. After all, what right have I to speak for the group of which I am a member? In making the effort to speak for the present members, I also speak for future members, for they will have to live with the history of our actions. Some people may not join us as a result of the actions we take; others will. Not only do I presume to speak for these people, I have the weight of our group's history to consider. Should I be true to the ideals of our founders or should I break with tradition, just as they once did in response to a new and ever changing world?

    Anyway, we in the groups have a problem with your method, even though we believe that you would be willing to sacrifice goodness upon the altar of fairness. We are looking for an outcome which is fair, not only for the present members, but for those yet to come and those who have passed on. Who speaks for these people? Who has the right to judge our actions ethically legitimate? Answer these questions and we will return to negotiations.

To: Solomon
From: The Enemy, Politics
Re: Your Mediation Efforts

    Let me begin by stating that, regardless of my feelings towards any other people in this bitter difficulty, I respect and admire you a great deal and trust your judgment completely. However, in discussions with others in the negotiations and members of the interested news media I detect a restlessness, an uneasy air, a doubt that your process is moving us along toward any substantive agreement. Let me be frank—are we making any progress? I am a practical man, unused to treading the rarefied theoretical heights you so nimbly scale. Like a tyro I feel myself lagging behind, panting and sweating. I fear that my group may decide to drop out of this intellectual steeplechase; not because the scenery isn't lovely, but because we are getting the familiar sense that the course is leading us nowhere. We're not engaged in some idle game for the amusement of the public and the exercise of their commentators. Again, and I write this note as a friend and in the strictest confidence, shouldn't we consider pressing the opposition into giving ground? Beware! The mandate is slipping from your grasp. Already we hear talk in the shops and markets, denouncing these proceedings as a charade and a gross waste of the taxpayers' money.

    Answer me one question and I'll be willing to continue bargaining. When will we get to vote on something?

To: Solomon
From: The Enemy, Aesthetics
Re: Your Mediation Efforts

I know you mean well, and your efforts to mediate this quite intractable problem are laudable and unique. However, I regretfully inform you that my group and I must withdraw from the negotiations. As you are evidently a man of good faith and great wisdom I will state our complaint simply—what is the purpose of our efforts if the solution, though workable, lacks beauty? Should we all labor these many days and nights only to bring forth a camel; a graceless misshapen beast obviously designed by committee? If the solution we must bring forth is to be something for everyone, it is apt to do little for any.

I, sir, believe that fairness is not enough. The solution should have elegance, that wedding of function to form which ennobles the human spirit and separates the artist from the cipher. We will not allow the consideration of beauty to be sullied by the rough hand of the economist, the sharp tongue of the lawyer, or the arched brow of the philosopher.

Answer me but one question and I will return to the table. Where does your principle leave the pursuit of beauty?

To: All the Enemies
From: Solomon
Re: Our Mediation Prospects

Recently, several of you have written to me individually, expressing personal reservations about the state of our mediation efforts. I am taking the opportunity to respond to your criticisms collectively; not only will this save time and effort on my part, but it reduces the possibility for misunderstandings between us. Reducing misunderstanding is the name of the game here—if nothing else, I want every one of you to be able to walk away from these negotiations with a greater and more subtle appreciation of the issues involved in this dispute. In the end I hope that a fair solution will be proposed. However, experience has taught me not to hold my breath waiting for it. Agreement is the most evanescent of gifts; part of this procedure's aim is to allow each of you to demonstrate to each other that you are worthy recipients. When that time comes, we will move quickly. Until then we must talk, play, work, and think deeply about which aspects of our problems are most important to us. Equally important is the ability to listen carefully to what others have said in order to understand which aspects are most and least important to them.

One of you has asked, "Where do we begin . . . how do we establish first principles?" To this question I must first answer that we have already begun. Our first principles are intimately bound up with the acts of speech and the process of active listening. The active listener does not regard the words of others lightly. Rather, the listener assumes that the speaker has something new and important to say. If the words or phrasing are awkward, ask for clarification. If the meaning is indistinct, request greater precision. If no word is expressive of a concept, jointly invent one. Part of the mediation process is the creation of a language common to all the participants. This language will have its own vocabulary and syntax as befits the peculiar needs and requirement of the case.

Another of you question, "who are we to discuss these issues and what authority gives us the presumption to speak for those who may not be here?" To this person I ask, "Are you not a man like other men; do you not have joys and fears, anger and anguish? Do you not laugh and cry?" The decisions we make here are not written in stone; we are only men, not gods. It is sufficient merely that we be representative of those who might discuss the issue. This is why we try to get as many people involved as possible in our discussions. We will make errors in our solutions; we should accept our fallibility graciously but not shirk our responsibilities because of it.

A third approaches me in confidence and fears the process moves too slowly. This person wishes we could take sides, make proposals, platforms and agendas. The ballot box is this person's altar; he worships the fickle god Polis. To this believer I counsel patience; do not try to create your god through the artifice of majority rule. Rather, wait for HIM or HER to appear. Perhaps the appearance will be in a word or phrase which sets our negotiations in a new light, a light in which everyone's position is better illuminated.

The last of you asks me perhaps the most difficult question. How do we, as a group, select that which is not beautiful and elegant? In our pursuit of the fair do we abandon pursuit of the beautiful? The example of the camel is mentioned as what can happen to a project designed by committee.

Of course, I consider the design of the camel to be an extremely elegant adaption to the harsh necessities of its native environment. If our group proves as wise as the committee responsible for the camel, then we shall truly have accomplished something remarkable. I make this point to emphasize that beauty is not only in the eyes of the beholder but in the act of creation. If the sole criterion is the passive beauty of the finished product, then we may be judged well or badly on the merits of our work. But if our search for beauty extends to the act of mediation itself, to leave the mediation is shirking of one's aesthetic responsibilities. Rather than asking where our procedure leaves the pursuit of beauty, I would rather say that it is a beautiful pursuit of a fair compromise.

11

Negotiated Rulemaking

Introduction

In the cases we have studied so far, negotiation and mediation have been practiced in a variety of contexts. For example, the Holston River case (chapter 5) involved Tennessee Eastman's application for a National Pollution Discharge Elimination System permit to discharge liquid chemical waste. By contrast, the Brown Paper case (chapter 4) began when the EPA brought an enforcement action in response to the company's apparent violation of air pollution laws. The Jackson Hole case (chapter 7) pivoted on the EPA's power to make grants to improve water treatment facilities.

All of these disputes took place within the context of established laws and administrative regulations. At issue was their application. Significant environmental disputes can also arise over the laws themselves. When a new environmental law is introduced in Congress or state legislatures, interest groups who expect to be helped or hurt compete with one another to persuade or pressure the elected representatives. The stakes are high because the standards that are adopted in the new law establish the framework for subsequent bargaining over its application. A strict law strengthens the hands of the environmentalists; a lenient one, the position of industries and developers. The legislative process itself can be seen as the sum total of many small negotiations in which votes are secured or lost through amendments, support on other measures, and general influence trading. This aspect of the law-making process is familiar, and has been well described in Eric Redman's *The Dance of Legislation* (New York: Simon and Schuster, 1973) and Richard Harris's *Decision* (New York: Dutton, 1971).

Even after the legislature has acted, however, the law-making process often is not over, particularly when the subject matter has technical dimensions as complex as does protection of the environment. The legislature endorses a fairly broad policy, then delegates to an administrative agency the task of promulgating regulations that will carry it out. In theory, the agency merely implements the legislative will, but in practice that intent often is rather ambiguous. By manipulating the substantive and procedural rules it adopts, the agency can significantly strengthen or dilute the apparent impact of the law. As a consequence, the same interest groups that lobbied hard during the bill-drafting stage are very likely to remain involved in the administrative rulemaking process.

In some respects, the two law-making functions are similar; there often are, for example, many parties, each one with its own agenda, jostling one another for the decision-maker's favor. In the case of administrative rulemaking, however, this lobbying takes place in the formalized setting governed by statutory procedures. Disgruntled parties can seek judicial review of administrative actions. Moreover, the rulemaking agency is both a decision maker and an affected party with its own set of priorities.

The principal case in this chapter describes an instance of rulemaking by the EPA under current administrative procedures. As we shall see, negotiation did take place but within constraints that seemed to promote polarity rather than joint problem solving. The chapter also includes proposals for reforming the administrative process to allow for rulemaking that is more explicitly based on negotiation. The legal obstacles to such an approach are described—and the reader should be able to assimilate them. However, one's main concern in this chapter is to understand the administrative process from the perspective of negotiation. What are the advantages of reform? What incentives might lead parties to negotiate their differences? Are there disadvantages that must also be tallied?

Although the thrust of this chapter is rulemaking, the reader should be alert to the fact that the principal case is rich in examples of multiparty bargaining, particularly the behavior of coalitions. Quick reference to chapter 6 may renew your familiarity with analytic devices that facilitate understanding of multiparty bargaining.

CASE STUDY: WATER TREATMENT RULEMAKING

This account is based on a more extensive study prepared by Heidi Burgess, Diane Hoffman, and Mary Lucci; it has been substantially condensed.

Introduction

Most cities in the United States process their municipal wastewater (including raw sewage from residential, commercial, and industrial sources) in one or more publicly owned wastewater treatment works (known in the field as POTW's). These plants continuously treat incoming water to remove suspended particles, floatable materials, and organic matter. The treated product has two components: a solid waste material (disposed of by land spreading, landfill, or incineration) and a treated liquid effluent that usually is discharged into a natural body of water.

Municipalities that discharge waste into public waters are subject to federal regulations, just as are private corporations. Over the last several decades these antipollution requirements have become increasingly stringent. Some communities contend that the costs of required treatment can exceed the benefits and have vigorously opposed federal imposition of extensive water treatment requirements.

At present, two major stages of wastewater treatment are widely utilized: primary and secondary treatments. Primary treatment involves three steps. First, the wastewater is passed through a screen to remove large floating objects. Next, the sewage passes into a "grit chamber," where sand, grit, and stones are allowed to sink to the bottom. Finally, the sewage passes into a sedimentation tank. When the velocity of flow through the tank is reduced, the suspended solids sink to the bottom. Sedimentation tanks can remove up to 50% of suspended solids and 30% of organic matter in sewage. If chemical coagulants are added to the sedimentation tank, these values increase to 75% and 50%, respectively,. Almost all cities in the United States now treat their wastewater with at least this primary stage.

Secondary water treatment uses organic processes to purify wastewater further. It adds considerably to costs. The process begins when the primary treatment water enters an "aeration tank," where it is mixed with air and bacteria-containing sludge. The sewage remains in this tank for several hours, during which time the bacteria break down the organic matter in the sewage. The bacteria then can be removed to a sedimentation tank similar to that used in primary treatment. Some of the settled sludge is recycled through the aeration tank, which causes the bacteria to consume even more of the remaining organic matter. The excess sludge is then discharged along with the other solid waste produced by the primary treatment. (This description applies to an activated sludge plant; other processes can also be used to achieve secondary treatment.)

The efficiency of wastewater treatment is usually measured by the level of suspended solids and the biochemical oxygen demand characterizing the discharged liquid. Good treatment leaves the wastewater with few suspended solids and low levels of oxygen demand. If sewage discharge contains relatively high levels of organic matter, the bacteria in both the sewage and receiving water demand a high quality of dissolved oxygen to consume the organic waste, leaving less dissolved oxygen in the receiving water for fish and plants.

The extent of waste treatment is but one factor that determines area water quality. Another important consideration is the nature of the water into which the discharge is flowing. Deep water, with strong tidal flushing, may be able to tolerate greater discharges than more stagnant areas. The type of discharge also can be significant: An industrial city may have far more toxins in its waste than a neighboring residential community of the same size. Likewise, the wastes of an isolated community may have far less impact than one that is surrounded by other public and private dischargers.

It is difficult to design and enforce a water quality standard that takes into account all of these variables. A uniform standard that applies to all dischargers, whatever their size or location, is far easier to administer than one that requires case-by-case analysis. A uniform standard is also a bulwark against the unraveling of other regulations. During the 1970s, the EPA argued hard for a blanket requirement that all publicly owned sewage plants employ secondary treatment, and Congress adopted such a policy in 1972. Five years later, however, Congress amended the law to allow waivers of the secondary treatment requirement in some cases. This study describes the rulemaking process in which the EPA defined just which municipalities would qualify for the coveted waiver.

Federal Water Pollution Control

Although Congress authorized grants and technical assistance for the construction of municipal treatment plants in the 1940s and 1950s, the 1965 Water Quality Act was the first federal attempt to require the states to establish strict quality standards for interstate waters and their tributaries. The standards—subject to federal agency approval—were to include criteria for water quality and a plan for their implementation and enforcement. Typically, the plans included limitations on effluent for all major pollutors.

The 1965 act gave the states 2 years to develop water quality standards. The Federal Water Pollution Control Administration took 18 months to develop regulations regarding standards, however, leaving the states little time to submit adequate standards. Developing water quality standards presented significant technical problems. The interaction between pollutants and the receiving water is extremely complex; simply collecting the data on both concentrations of pollutants and their effects is a difficult task. The problem is compounded when many polluters discharge into the same water.

Enforcement of water quality standards also proved troublesome. As delineated in the Water Quality Act, enforcement involved a lengthy, three-stage process consisting of an enforcement conference, a public hearing, and court action. The conference stage alone could easily take years to complete. If the government's case ultimately ended up in court, its success depended on adequate substantiation of the violation through scientific data regarding levels and effects of the pollution. This was often difficult.

By the early 1970s, Congress and a growing number of environmentalists had lost faith in the water quality standards approach and began to consider new ways to deal with the growing need for water pollution control Some observers supported the states' efforts and blamed most of the problems on mismanagement by the federal government. By contrast, most members of Congress and environmentalists believed that the states' implementation of water quality standards had failed and that the ambient water quality in the nation was declining. They sought a new, stronger water pollution control program on a federal level.

Congress began revising the legislation in 1971. By then the environmental movement was gaining momentum, and several major environmental bills had been enacted. The National Environmental Policy Act—passed in 1969 and signed January 1, 1970—was heralded as a cornerstone of federal environmental legislation, even though the full import of the act was not known. New and much stricter air pollution legislation also had been adopted in 1970, and the Environmental Protection Agency had been established to administer and coordinate most of the federal government's pollution control efforts. Thus, the stage was set for an expanded role for the federal government in water pollution control.

The 1972 Federal Water Pollution Control Act debate inherited much of its tone and substance from the dispute associated with the 1970 Clean Air Act amendments (CAA). The CAAs were much stricter than earlier air pollution control legislation, and they drew fire from industry, the states, and even the executive branch of the government. However, the senators who initiated this legislation enjoyed the support of environmental groups—a growing constituency—and the law was passed with considerable favorable public reaction.

As a result, the 1972 FWPCA also was very tough and, like the CAAs, it generated much debate. Most controversial were the two goals set forth at the beginning of the act:

> That discharge of [all] pollutants into the navigable waters be eliminated by 1985 . . . wherever attainable, an interim goal of water quality which provides for the protection and propagation of fish, shellfish, and wildlife and provides for recreation in and on the water be achieved by July 1983.

In order to achieve these goals, the FWPCA switched the emphasis from *water quality* standards to uniform technology-based effluent standards. Unlike water quality standards, which first set criteria for acceptable pollution levels in receiving waters and then set effluent restrictions on a case-by-case basis, uniform effluent standards limit the amount of pollutants that can be discharged by each type of polluter—regardless of the character of the water into which the pollutants are released. Under the new law, the focus was to be on any pollutants put *into* the water rather than on their incremental effect on water quality.

The 1972 law represented a fundamental change in the philosophy of water pollution control. According to one authority, the earlier approach had been based on the view that

> waste disposal is an acceptable use to make of a body of water, so long as it

does not interfere with other desired uses. In contrast, effluent standards are generally based on the view that all pollution is undesirable and should be reduced to the maximum extent that technology will permit.

Supporters of the new effluent standards approach claimed that the old water quality standards program created by the 1965 Clean Water Act had failed because it was not feasible to administer discharge restrictions according to the nature of the receiving water. Effluent limitations based upon technological feasibility would be much easier to implement than the water quality standards, they believed, because effluent limitations could be set for entire categories of polluters without assessment of individual cases.

The effluent limitations established by the 1972 FWPCA were assigned by industry and were based on the pollution control technologies that were available to the industry. For example, the act required that private industry utilize the "best practical control technology currently available" by 1977 and the "best available technology economically achievable" by 1983. (The restrictions for each industry were to be set by specific regulations.)

Thus, in the case of municipal dischargers, secondary treatment of discharge material was part of the 1977 uniform technology-based standard. By 1983, municipal dischargers were required to utilize the "best practicable waste treatment technology over the life of the works."

Public Reaction to the 1972 FWPCA

Both the goal of zero discharge and the uniform technology-based effluent limitations set forth in the new law drew immediate criticism. Predictably, some of the opposition came from communities that claimed that local conditions made secondary treatment a costly yet unproductive burden.

Opposition also came from more impartial quarters. The National Water Commission (NWC)—created by Congress in 1968 to review water resource problems—attacked the zero-discharge goal as being conceptually unsound, financially extravagant, and potentially damaging to the overall efforts toward water pollution control. The NWC questioned the advisability of uniform effluent standards, stating that more flexible and cost-effective approaches to pollution control were necessary in order to attain maximum benefits.

Although the NWC praised congressional efforts to provide a more effective pollution control program, it suggested several significant changes to the 1972 act:

> The shift away from reliance on water quality standards and economic practicability as bases for regulation should be reversed. The new Act's establishment of a no discharge goal to be achieved through application of the best available waste treatment technology is unsound in theory and will

prove unworkable in practice. The Congress should revise this misconceived goal now and reaffirm its commitment to the water quality standards approach and economically practical minimum treatment requirements. . . . The Administrator should be authorized to encourage those local expenditures which will produce the greatest improvement in water quality and constitute the most effective use of limited funds. The uniform requirement for secondary treatment could cause clean water moneys which have been squeezed out of a tight budget to be expended for facilities with minimal impact upon the receiving waters while leaving raw sewage outlets without interception. (*Legislative History of the Clean Water Act of 1977*, vol. 3, October 1978, Serial No. 9514, p. 320)

A second entity, the National Commission on Water Quality (NCWQ)—created by the 1972 FWPCA to assess the impact of that legislation—concurred, particularly in respect to the uniform requirement of secondary treatment:

> The Commission addressed the problems of a relatively small pot of money having to be stretched a long way. In California alone . . . the cost of achieving secondary treatment will be well over $1 billion.

Therefore, the NCWQ suggested that

> Congress authorize waiving. deferral or modifications of the 1977 requirements on a category-by-category basis for near shore ocean discharges of POTW's. (National Water Commission, *Water Policies for the Future*. Washington, D.C.: Government Printing Office [1973]: 88–90)

In response to these contentions, the EPA administrator established a task force in March 1974 to investigate the need for secondary treatment of municipal ocean discharges. Its members advised against eliminating the uniform secondary treatment requirement on the grounds that the EPA knew too little about the effects of pollutants on the ocean environment to administer a case-by-case approach. Earlier case-by-case approaches had not worked, and it was felt that there was no reason to believe this type of approach would work now.

Although the task force agreed with critics that biochemical oxygen demand was not a major concern with ocean water, members noted that other pollutants, especially toxic organic materials and metals, still posed a serious hazard to ocean ecosystems. Although secondary treatment originally was developed to remove biochemical oxygen demand and suspended solids, the task force concluded that it was also the best and most cost-effective technology for removing the toxic pollutants of greater concern. The task force proposed that the secondary treatment requirement be maintained but that alternative technologies be explored. In order to accomplish this, it recommended that Congress amend the 1972 FWPCA to allow the EPA to extend the 1977 deadline for secondary treatment cases where its cost-effectiveness was questionable. Such a bill was introduced in Congress, but it did not pass.

Early in 1977, the 95th Congress began to consider the secondary treatment issue and other alleged problems associated with the 1972 FWPCA. The ultimate result was the 1977 Clean Water Act, including section 301(h).

In the legislative hearings, environmentalists and the EPA felt compelled to defend the secondary treatment requirement as it stood. They shared, in varying degrees, three concerns. One, of course, was continued protection of water quality through better treatment of municipal waste. There was also concern over a possible "domino effect," that is, the granting of waivers for cost-benefit reasons in some cases would result in granting of waivers for many other reasons and eventually would paralyze the entire program for water pollution control. There also was serious concern about the burdens of administering a waiver system. The EPA expected that many municipalities would seek waivers if only to delay compliance with secondary treatment requirements.

In short, as one agency official stated,

> because of our limited understanding of the effects of effluents on oceans, decisions would necessarily be highly judgmental and would, of course, be appealable in the courts. The net effect of this process would be a return to the requirements of proof of harm to receiving waters before controls can be required.

The 1977 Clean Water Act

Congress passed the 1977 Clean Water Act in December of that year. For the most part, the act reflected a broad congressional consensus that the 1972 legislation was sound and that, with fine tuning, it should be continued without interruption. Funding was extended for many programs, and most of the amendments were relatively minor.

Notwithstanding the testimony of the EPA and environmental groups, however, Congress did make a concession to municipalities under pressure to install secondary treatment by including a waiver provision—section 301(h).

The new law did not call for a full-scale cost-benefit analysis before requiring secondary treatment. Instead, the wavier provision was intended, according to the legislative history, "to provide a very narrow opportunity for certain municipal dischargers, if they can meet a specific burden of proof, to qualify for a modification of the secondary treatment requirement." Thus, the burden of proof was placed on the municipality—not on the EPA.

To ensure that the burden of proof lay with the municipalities instead of the EPA, Congress established a set of criteria that applicants had to meet in order to be eligible to apply for 301(h) waivers. According to that section, the EPA administrator was empowered to issue a secondary treatment waiver for "any pollutant in an existing discharge from a publicly owned treatment works in marine waters," if the applicant can demonstrate that:

1. There is a water quality standard applicable to the pollutant for which the waiver is requested.
2. The resulting discharge will not interfere with the protection of public water supplies or with the protection and propagation of a balanced indigenous population of shellfish, fish, and wildlife and allows recreational activities in and on the water.
3. The applicant has established a system for monitoring the impact of the discharge on the aquatic biota.
4. The modified requirements will not result in additional requirements for other polluters.
5. All pretreatment requirements applicable to sources that introduce waste into such treatment works are enforced.
6. To the extent that is practicable, the applicant has developed a program to eliminate the entry of toxic pollutants from nonindustrial sources into the treatment works.
7. There will be no new or substantially increased discharges from the polluter greater than the levels specified in the permit.
8. Any Title II funds available to the owner of the treatment works will be used to achieve the effluent reductions required by this section or other sections of this act.

The law also gave municipalities that desired waivers less than a year to submit their applications. This was enough time for cities that had gathered data on their discharges to make a presentation to the EPA, but it gave little leeway to municipalities that had not already collected the technical information. The level of treatment required of POTWs applying for waivers would be determined by the effect of the effluent discharge on ambient water quality and on the ecology of the receiving water. Thus, all decisions would be made on a case-by-case basis. Although the burden of proof was supposed to be placed on the municipalities, the case-by-case approach ultimately gave the EPA much more administrative responsibility than the agency had under the uniform effluent standards.

The EPA saw this situation as the equivalent of a return to the pre-1972 regulatory scheme and a major step backward in their pollution control efforts. Many environmental groups, who also were distressed with the implications of the amendments, shared the EPA's views and concern that the case-by-case approach to pollution control was still unworkable and presented a major threat to the strong regulatory stance of the 1972 technology-based effluent standards.

By contrast, most municipalities were very pleased with the 301(h) amendments. The legislation provided an opportunity for them to bypass the high costs of secondary treatment. Their optimism was tempered, however, by the fact that the impact of the new provision ultimately would depend on the regulations promulgated by the EPA. In light of the agency's opposition to waivers, there was concern about the way in which it would interpret and implement the new statute.

The Rulemaking Process for Section 301(h)

The EPA saw both the language of section 301(h) and the guidelines as open to wide interpretation. As Lisa Friedman, of the office of EPA's general counsel, noted, there was not "a great deal of legislative history to hang your hat on." The record that was available charted a contradictory course.

If the EPA could have operated completely autonomously, it would have made the waiver criteria as strict as possible. Only the few municipalities (such as Seattle, Los Angeles, Honolulu) that had lobbied for the waiver would have been eligible to apply. Yet, though the agency had the responsibility of preparing the regulations, it was subject to both political and legal constraints. It could not blatantly disregard the congressional mandate. Moreover, administrative laws open the rulemaking process to public comment and review (e.g., by industries and interest groups). Although the informal rulemaking procedures limit the amount of bargaining and negotiation that can take place, they nearly guarantee some compromises.

> The notice and comment rulemaking procedure is explicitly designed to solicit information and views from interested environmental groups and other concerned state and federal government agencies . . . the results cannot help but represent a compromise between the interests of competing groups. (Resources for the Future and Urban Institute. "Environmental Regulation in Theory and Practice: EPA's Process of Setting Best Practicable Control Technology Standards," Report No. NBS-GCR-ETIP79-63 [1978]: 3–21)

The two ultimate questions were: Who would be allowed to apply for waivers, and who would be awarded them? The answers, however, would be defined in highly technical terms. Preparing the regulations required a detailed evaluation of effluents and their impacts (including impacts on biota and relations to other discharges), issues that were technically complex and matters of scientific debate. Both the EPA and the NCWQ had warned about the relative lack of available data concerning the impact of pollution in the marine environment. Nevertheless, the agency now was confronted with that task. It also faced a potentially enormous number of applicants.

One means for the EPA to limit its administrative burden was to construe the statutory language narrowly. Although authorizing waivers, Congress had limited them to communities whose "existing discharge" met eight water quality criteria. Thus, the EPA's definition of existing discharge became crucial, as did their evaluation of the technical proof demonstrating the applicant's compliance with the eight limiting factors.

If the EPA interpreted *existing discharge* to mean only those discharges that met the eight criteria *at the time the statute was enacted*, then few municipalities would be eligible. By contrast, if the EPA chose to interpret the phrase also to include proposed outfalls or improved discharges, then the number of eligible applicants would be greatly increased. A city that could meet the criteria but had yet to do so would nonetheless be ineligible to apply for a waiver—let alone get one. In short, a narrow definition of *existing discharge* was one way that the EPA could shut the door on potential applicants. The EPA also had some discretion in defining the precise technical conditions that would justify a waiver. Necessary ocean depth and geographic criteria were open to debate.

Primary responsibility for developing the 301(h) regulations was delegated to the Office of Water Program Operations at EPA headquarters in Washington. An external task force was formed to initiate the rulemaking process and to assemble the diverse individuals needed to write the regulations.

Tom Jorling, then assistant administrator for water and hazardous materials, selected the members of the task force. Before he joined the EPA in 1976, Jorling had served for 5 years (1968–1977) on the House Public Works Committee. Consequently, he was very familiar with the actors and issues involved in the dispute over the secondary treatment of marine discharges. Other key members of the task force were Lisa Friedman, a lawyer with the EPA's Office of General Counsel; Thomas O'Farrell, an engineer with the municipal technology branch of EPA's Office of Water Planning and Standards; and Donald Baumgartner, a physical oceanographer with EPA's Corvallis Laboratory in Oregon. This team comprised the nucleus of EPA's technical and legal expertise in the 301(h) rulemaking effort.

Other individuals who advised the task force on marine science issues included Rick Schwartz, also from the Corvallis laboratory; Jan Praeger, from EPA's Rhode Island laboratory; and Frank Herbard and Ed Myers, from the National Oceanic and Atmospheric Association in Boulder, Colorado.

Constraints on the Task Force

The task force members were constrained by law and circumstance. All the members held other jobs, and the task force had limited time to do its work. The new law required that all applicants for waivers be filed 270 days from the date of enactment, which meant that the EPA regulations for filing and processing applications had to be written as quickly as possible. Moreover, because the EPA could not bring enforcement actions against violators until the waiver decisions had been made, enforcement actions against POTW's discharging into marine waters were likely to be delayed until the final regulations were determined. Thus, a prolonged rulemaking process would constrain the EPA's enforcement activities.

The conduct of the task force was circumscribed also by the EPA's statutory authority and by the Administrative Procedure Act (APA). Because the Clean Water Act did not specify that the 301(h) rulemaking process be formal, the EPA followed guidelines in section 553 of title 5 of the APA governing informal rulemaking.

Although less rigid than formal rulemaking (where there is an adjudicatory hearing before an administrative law judge, testimony is recorded, and cross-examination of witness is permitted), informal rulemaking is still highly regularized. The agency must notify interested parties of its rulemaking activities and offer them an opportunity to participate through documented information and arguments, with or without the option for oral presentation.

Moreover, a recent federal court decision, *Home Box Office v. FCC* (567 F.2d 9, 1977) had extended the bar against *ex parte* communications in formal rulemaking to informal processes as well. The APA defines an *ex parte communication* as an "oral or written communication not on the public record with respect to which reasonable prior notice to all parties is not given."

Before the *Home Box Office* case, such off-the-record contacts between agencies and parties had often been an important part of agency administrative procedure. The practice was criticized, however, for inviting "inaccuracy in agency decision makers, and improper political influence." ("Note: Due Process and Ex Parte Contacts in Informal Rulemaking," 89 *Yale Law Journal* 199, 1979.) The court specifically provided that once notice of proposed rulemaking has been issued, any agency official or employee who might reasonably be expected to be involved in the process should

> refus[e] to discuss matters relating to the disposition of a [rulemaking proceeding] with any interested private party, or an attorney or agent for such a party, prior to the agency's decision. (*Home Box Office*)

Should an *ex parte* contact nonetheless occur, the court further provided that the document—or a summary of any conversation—must immediately be placed in the public file established for each rulemaking docket so that interested parties may comment on it.

The rationale for the court's holding is "the inconsistency of secrecy with fundamental notions of fairness implicit in due process and with the ideal of reasoned decisionmaking on the merits." (*Home Box Office*) Indeed, by requiring that contacts take place through formal channels and be officially recorded, the decision greatly decreased the amount of bargaining that had traditionally occurred, at least that bargaining that had taken place behind closed doors. (As is explained in a note that follows this case, another court ruling, handed down after the 301(h) rulemaking, has narrowed the prohibition against ex parte communication and thus has opened up the rulemaking process.)

The Preliminary Public Meeting

The first action of the task force took place on February 8, 1977, with the announcement in the *Federal Register* that the EPA would hold a public meeting 2 weeks later in San Francisco to receive comments on the implementation of section 301(h).

According to the EPA notice,

> the purpose of this meeting is to receive the public's views on how EPA should interpret and apply the statutory criteria which an applicant must meet in order to obtain a modification of the secondary treatment requirement.

The notice went on to describe the criteria and suggested a number of specific interpretation problems that needed review. Issues included the use of surrogate parameters for the state water quality criteria, definition of balanced indigenous population, and the requirements for the treatment and discharge of toxic waste.

At the meeting, 37 individuals representing municipalities, special interest groups, and citizen constituencies gave their comments. Although many of the participants responded to the issues of interest to the task force, almost all of them also had an agenda of their own. Each municipality seemed to have sent its representative in order to argue for its own waiver eligibility, and each urged the EPA to write the regulations accordingly.

Many cities also were concerned about the definition of existing discharge. For instance, a San Francisco spokesman had heard that the EPA would grant waivers only to dischargers whose existing outfalls met the stated criteria. He felt that the exclusion of cities and towns that had planned modifications to their outfalls was unfair.

A speaker from Massachusetts argued against the EPA's limiting the depth of the receiving water for dischargers who otherwise would be eligible for a waiver because, on the East Coast, there are no deep waters in which to discharge. Depth, he contended, was not as important a factor as circulation and tidal flushing—conditions that can prevail in eastern waters.

Individuals from Guam, American Samoa, and the Mariana Islands claimed that their respective territories were special cases and should be eligible for waivers. A representative from Guam explained that his island was only a few miles from the 35,000-ft-deep Marianas Trench and that the absence of modern conveniences and significant industry meant few toxic wastes in that area. He pointed out that implementation of secondary treatment would be very costly to the inhabitants of these islands, many of whom exist at subsistence level.

Each municipality had its own concerns, and each sought to widen the EPA's potentially narrow interpretation of the eligibility criteria—at least in a way that would qualify its own POTW. The environmental groups took the opposite stance and urged the EPA to minimize eligibility by tightening the regulations. A speaker from the National Wildlife Federation argued for a tight geographic limit that would only allow certain communities on the Pacific coast to apply for waivers. In short, this intitial meeting produced no consensus among the interested parties and the EPA.

The Preliminary Concept Paper

After the public hearings, the task force set out to write a preliminary concept paper (i.e., a draft proposal of the regulations). The effort required 7-day workweeks with sessions often starting at 7 A.M.

Tom Jorling met with the group about once a week. His mandate to the group was to develop as stringent a set of criteria as possible. He believed that the waiver provision set a bad precedent and should provide, at most, a narrow opportunity for a very few dischargers to obtain a secondary treatment variance. Thus, he wanted to write the regulations so as to minimize the number of waiver applicants and to leave little flexibility in the eligibility criteria. Others in the EPA shared his concerns.

At the same time, Jorling also wanted to ensure that the waiver procedure did not become a confrontation between municipalities and the EPA. Given scientists' uncertainty about pollution's effects on the marine environment, Jorling wanted the criteria to be as objective as possible. This was almost an impossible task. Don Baumgartner remembers the internal pressure that accompanied the attempt to develop objective scientific criteria. He says that Jorling sought a simple set of chemical and biological tests that would determine the effects of the discharge on the receiving water. However, Baumgartner argued that such tests were not valid because different waters have different degrees of sensitivity and resilience. Because the EPA knew that they probably would have to defend their technical criteria in court, Baumgartner wanted to be certain the regulations were technically and legally defensible.

The task force considered various means of limiting the number of applicants. One was to set a minimum depth of the receiving water because both the statute and the legislative history referred to discharges in deep ocean waters. However, Baumgartner and other scientists at the public hearings argued that depth was not necessarily a factor of concern. Other characteristics of water (e.g., ocean currents, tidal flushing rates), they stated, were more important than depth.

Another possible limiting factor was geography. The task force would have liked to limit applicants to Pacific Coast communities, but this appeared to contradict congressional intent. In the legislative history, Senator Mike Gravel (D-Alaska). who introduced the secondary treatment waiver provision, had stated:

> I did not intend to limit the application of the provision to Anchorage, Seward, and a few other cities that can meet the geographical requirements to come forward and attempt to prove their case. (That might even include communities currently under a schedule to provide traditional secondary treatment.) That is my understanding of the conference's intention as well. I can understand the EPA's concerns about the administrative burden this provision might place on the agency that we should not legislate unreasonably so as to accommodate an agency. (*Legislative History*, 1978. p. 535)

On the basis of this and other testimony, the EPA abandoned its attempt to establish a geographical limitation.

Having excluded the possibilities of depth and location as constraints on waiver applicants, the task force sought to develop a narrow definition of *existing discharge*. The group defined *existing discharge* as that discharge actually flowing into marine waters on or before December 27, 1977 (the date the 1977 Clean Water Act was enacted). This meant that all discharge generated as of that date would have already had met the eight criteria listed in the statute. Under this definition, many municipalities in the process of modifying treatment facilities would not qualify for a waiver.

In mid-March the task force issued a copy of the preliminary concept paper to all individuals who had attended the public hearings in San Francisco. In addition to proposing a very strict definition of existing discharge, the paper also outlined technical standards that would be used to determine whether that discharge was having a negative impact on the receiving water. Surrogate parameters for biochemical oxygen demand and suspended solids were established. The paper also defined balanced indigenous population as

> that ecological community of marine organisms that: (1) might reasonably be expected to repopulate the edge of the zone of initial dilution from nearby polluted water if the outfall were removed; and (2) exhibits characteristics of natural nearby unpolluted, healthy communities existing under comparable environmental conditions."

The draft also set out extensive requirements for chemical and biological testing of toxic wastes. POTWs would be expected to meet the same criteria regardless of the composition of its discharge.

Public Response to the Preliminary Concept Paper

Public response to the preliminary concept paper was almost entirely negative. There was loud opposition from many cities. The chief engineer of Boston's

Metropolitan District Commission was afraid that the EPA's reference to the 301(h) amendments as a "narrow opportunity for certain dischargers" to obtain a waiver, would prohibit East Coast communities from qualifying. The city manager of Seward, Alaska, strongly recommended that the definition of existing discharge be expanded to include proposed improvements to existing treatment and discharge facilities.

This view was echoed by a Seattle official. A narrow definition of existing discharge, he warned, could actually hamper improved water quality.

> Our concern is that we will be precluded from consideration for a waiver in relocation of a major discharge off Alki Point in West Seattle. In this case the existing outfall is inadequate and a new larger outfall serving a larger service area is being considered in our facility planning. Many environmental benefits have been identified with this alternative.

The same spokesman also questioned the paper's chemical and water quality assessment criteria. The EPA had proposed that all applicants demonstrate that their discharge meets marine water quality criteria for 65 toxic chemical substances and compounds. "The chemical and water quality assessment is not currently workable," he said, "because marine water quality criteria are not available and almost no information exists on test procedures."

Environmental groups were equally quick to voice their objections to the EPA draft. Typical was a statement on behalf of Citizens for a Better Environment (CBE):

> CBE is concerned that Section (301)h opens the door for environmental harm. We feel that the exemption policies represent an ill-advised retreat from the federal water pollution control program and an unnecessary threat to the achievement of our national water pollution abatement goals. The transformation of eastern coastal waters near New York into a "dead sea" attests to the fact that even the ocean can be grossly fouled by pollution. As an environmental group located in San Francisco, we are anxious to prevent the same degradation of our coastal waters.
>
> CBE takes heart only in the fact that Congress recognized the environmental risks tied to secondary treatment exemptions. It therefore outlined in the law eight specific criteria which must be met by operators of treatment plants before an exemption can be issued. CBE believes Congress intended these criteria to limit the exemptions to the few special cases where environmental repercussions will be at an absolute minimum.

Harsh criticism also came from another quarter. Representative Glenn Anderson, a member of the House Committee on Public Works and Transportation, informed EPA administrator Douglas Costle that

> To be perfectly frank, this paper contains statements which are personally

quite disturbing. To that end I am requesting that our Full Committee Chairman, Harold T. Johnson, authorize our committee staff to make these proposed regulations its number one priority.

The preliminary concept paper had drawn fire from all sides.

The Proposed Regulations

The proposed regulations were written in approximately two weeks and were published in the *Federal Register*. Because many respondents had criticized the preliminary concept paper and suggested changes, the task force attached several questions to the proposed regulations.

In spite of the earlier opposition, the proposed regulations did not differ substantially from the version in the preliminary concept paper. The preamble warned:

> It cannot be emphasized too strongly that the modification provided under Section 301(h) represents a highly restrictive departure from the technology-based pollution control philosophy of the 1972 Act. It is a limited exception, a narrow opportunity for certain municipal dischargers to qualify for a modification of the secondary treatment requirement. Applicants are reminded that they bear the burden of demonstrating to EPA's satisfaction that they qualify for the modification requested.

Thus, in order to carry out this strict policy, the definition of existing discharge remained very demanding as did the other criteria for application.

The proposed regulations were circulated much more broadly than had been the preliminary paper. Within the EPA, both the Office of Planning and Management and the Office of Enforcement offered comments. The former was concerned most with the cost of application for small communities in Alaska and the Pacific island territories. The quantity and sophistication of the data required to substantiate the application would involve a very expensive undertaking. The Office of Planning and Management recommended that either the amount of data required be reduced for all applicants or that applicants belonging to this group be considered according to a separate policy.

The Office of Enforcement was concerned that the requirements and terminology of the regulations be consistent with those in the NPDES permit application (because the waivers would consist of modified NPDES permits). They also felt that the regulations should spell out the procedures for processing applications and issuing permits. In all other areas the two offices concurred.

The task force solicited comments on the definition of existing discharge, specifically:

A. Whether section 301(h) should be construed to allow EPA to issue

waivers based on modifications made after the deadline for preliminary application;
B. If so, the types of modifications that should be permitted;
C. How the EPA could make an accurate predictive judgment as to whether future discharges would meet the stringent water quality, physical, chemical, and biological criteria;
D. The deadline the EPA should impose on applicants for completion of any modifications necessary to the discharge;
E. What action the EPA should take if the modifications are not completed on time or if the actual impacts of the discharge are more severe than the applicant had predicted.

Interested parties had 45 days to comment on the proposed regulations. During this time the EPA held two sets of public hearings—one in Washington and the other in Seattle.

Public Response to the Proposed Regulations

Although the environmental groups active in the area of water quality generally supported the proposed regulations, they continued to assert that anything less than a hard-line stance by the EPA would threaten the environment. A CBE spokesman stated:

> Although we see no reason for the EPA to exempt municipal treatment plants from complying with secondary pollution control standards, we are pleased that the proposed EPA regulations recognize the adverse environmental harm which would be the consequence of the indiscriminate granting of exemptions.

Stephen Schroeder, a representative of the Natural Resources Defense Council, went further, saying that the proposed regulations were not stringent enough. The provisions were "deficient," he contended, in that they omitted certain eligibility criteria required under the section's legislative history. Moreover, Schroeder added:

> The effectiveness of the proposed regulations cannot be properly judged without a number of documents that do not yet exist. These include marine water quality criteria for toxic pollutants, 403 ocean discharge regulations, and a support document explaining the basis for the technical aspects of the proposed regulations. We urge EPA to modify the proposed regulations in conformance with these comments, and to publish the missing material as soon as possible.

If Schroeder was excessively critical, it was because he knew that municipalities would argue forcefully that the regulations should be less stringent; he attempted to counterbalance their arguments and hoped that the EPA would not then be forced to loosen their stance.

The municipalities as a whole were indeed hostile to the proposed regulations, especially to the definition of existing discharge because the proposed definition rendered the majority of them ineligible. Their comments pressed for expanding the definition to include planned modification in discharges. Many argued that they had been planning improvements for several years but that the EPA, the grants administrator that many municipalities relied on for construction of treatment facilities, had slowed their efforts.

The West Coast municipalities based their argument on legislative history that, they contended, clearly indicated their inclusion in the waiver provision. A spokesman from San Diego accused the EPA of having drafted regulations that were "biased in favor of secondary treatment at the exclusion of any other possibility and in violation of congressional intent." He claimed that the legislative record revealed not merely specific criteria, but that it defined eligible municipalities, namely those on the West Coast.

The small dischargers had an agenda of their own: Not only were they opposed to the definition of existing discharge, they also argued that they should be subject to less rigorous data requirements. Otherwise, they pointed out, the application cost of small municipalities would be prohibitive. They also argued that the smaller volumes of their effluents were of much less concern than those of major dischargers.

East Coast communities were still trying to persuade the EPA that they deserved equal consideration for waivers. The city manager of Portland, Maine, expressed fear that the EPA would give preference to West Coast municipalities that had been active earlier. At the Washington, D.C., public hearing, he stated:

> We do not feel that the so-called legislative history has any bearing whatsoever on the amount or nature of the evidence that must be provided by applicants nor would it affect the attitude of the EPA staff responsible for technical review. There are many reasons why East Coast communities such as Portland did not participate in early EPA hearings to the extent that the West Coast cities did [among them, the cost of travel and technical expertise].
>
> In conclusion, a decision to approve or disapprove an application for the secondary treatment waiver should be made solely on the basis of its technical merits, not on the basis of which areas of the country were the most vocal opponents of proposed regulations.

At those same public hearings, concern was expressed about the EPA's restrictive definition of existing discharge and its implied requirement of deep receiving waters. Although the proposed regulations did not specify a minimum depth as such, the preamble did state that "initial dilution must be of the order achieved by accepted designs of multiport ocean outfalls at depths of approximately 200 feet or greater." East Coast municipalities, fearful that this statement might become a strict criterion in the final regulations, argued that the term *marine waters* was defined by the statute as either deep waters or estuarine waters with strong tidal movement. The proposed regulations had eliminated depth requirements and geographic limitations. East Coast communities did not want to see them slip back into the final version.

The native Alaskan villages, territorial possessions, and Puerto Rico also opposed the definition of existing discharge in the proposed regulations. They felt that the issue of secondary treatment did not apply to them at all. Many of the villages had only two hundred to three hundred inhabitants and did not have even primary treatment facilities. Thus, they thought the EPA should make a special provision for them in order to bypass the waiver requirements altogether.

The arguments of individual municipalities were echoed by organizations like the Association of Metropolitan Sewerage Agencies (AMSA) and the Pacific Legal Foundation (PLF). Both groups argued strongly for changes in the EPA's definition of existing discharge. At the Washington, D.C., hearings, an AMSA spokesman chided the agency for trying to implement a policy that would bar many communities from even applying for a waiver:

> AMSA believes that a far more sensible approach to implementing Section 301(h) is to adopt a policy that will encourage municipal systems to discuss directly and candidly with EPA staff whether the particular circumstances that prevail in their areas make it worthy of exploring, whether an application can be prepared and submitted for consideration that has a fair chance of approval. EPA's staff should be instructed to work with communities constructively to achieve a modified permit if the physical properties make it possible.

The EPA had anticipated strong criticism of the proposed regulations. The strict set of criteria would qualify only a very few municipalities to apply for waivers, and it would disqualify many communities with discharges of minimal environmental impact. The municipalities that did apply would face an exacting, complex review process. Extensive data were requested, and the evaluation was based on subjective criteria such as a balanced indigenous population.

In taking a strong stance in favor of secondary treatment, the task force hoped to make potential applicants provide hard data that would prove why either the definition of existing discharge or the eight qualifying criteria should be changed. The EPA then could use these data to rewrite the proposed regulations rather than have to generate all the data themselves. The EPA's underlying goal was to limit the number of potential applicants so that they could spend more time reviewing each application. They hoped that this would enable them to make better decisions than they could if they were overloaded with large numbers of applications requiring rapid review.

This strategy, however, carried a price. The EPA was now being attacked from two sides: the environmental groups and the municipalities. The environmental organizations were fewer in number and had scant resources to devote to the battle. Municipalities possessed limited resources, but they were numerous and enjoyed legislative leverage.

Congressional pressure on the task force began immediately after the comment period, particularly from a few members of the House of Representatives (Bizz Johnson and Glenn Anderson from California and Ray Roberts from Texas). In late May, the House Committee on Public Works and the Transportation Subcommittee on Water Resources held hearings on the scientific and technological considerations of modifying the secondary treatment requirement for municipalities that discharge into marine waters. The tone of the hearings was totally critical of the EPA and its handling of the issue. No one who in any way supported the EPA was invited to the hearings.

One marine specialist contended that

> it is a major problem of our society that the regulatory process has been increasingly used to frustrate, to bypass, or to expand upon the legislative intent of Congress and thereby create an entirely new domain of executive legislation.

The witness chided the EPA for perpetuating "well-meaning but counterproductive environmental dogma" and called for corrective Congressional action. "And as a people who care about our environment," he concluded, "we cannot continue to deny the coastal waters the abundance of protective biological nutrients, nor can we continue to assault them with nonproductive waste waters."

Senator Gravel of Alaska also was similarly critical of the EPA. He was particularly concerned about the agency's narrow definition of existing discharge. He had sponsored the waiver provision, he testified, in order to allow communities that were still in the midst of improving their treatment and discharge of waste water to be exempted from having to install unnecessary secondary treatment facilities.

The task force subsequently met with several legislators and the staff members of others to discuss concerns raised at the congressional hearing. Within the group, there was fear that if the EPA did not abandon its restrictive stances, Congress would rewrite the law to make it easier for all dischargers to apply for waivers.

The EPA realized that they would have to yield some ground in response to these attacks. Yet Jorling wanted to concede as few points as possible. Challengers questioned how the EPA could accurately judge the effect of dischargers, the types of modifications the agency should consider, and the agency's course if a given discharge proved to be more harmful than the applicant had predicted. Changing the definition of existing discharge could mean rewriting many regulations to allow for the effect of planned outfalls. Allowing more municipalities to apply for waivers would slow decision making and enforcement efforts. Administrators in the enforcement office warned that they lacked the resources necessary to manage a broader waiver program. One official noted that if waivers for proposed outfalls were to be considered, the review process would be much more difficult

> because both the agency and an applicant would be put in the predictive mode with neither in possession of firm evidence to support conclusions.

The task force concluded that the program would not be legally defensible unless the application process included POTWs that had been part of the history of the legislation. Many large West Coast municipalities (including San Francisco) that had been heavily involved in the congressional lobbying for section 301(h) and were planning wastewater treatment improvements were ineligible, according to the EPA's definition of existing discharge. The EPA decided to broaden the final definition of existing discharge to include in the application process at least those POTWs that had been involved in the initial lobbying.

Once the task force had made its decision, its members conducted work load analyses in order to estimate the number of applicants likely to apply for waivers on the basis of the new definition. They were very concerned about applicants who would qualify under various interpretations but who would not have adequate biological and water quality data on which the EPA could assess the impact of their effluents. Because the EPA did not have the resources to perform its own assessments, the prospect of receiving "qualified" applicants without data to support their position was viewed as a threat to the agency's enforcement efforts.

Final Regulations

The EPA had planned to publish the final regulations before the September 24, 1978, deadline for waiver applications, but the decision to redefine existing discharge required extensive changes. Therefore, the EPA failed to meet this goal. The delay was prolonged by the reassignment of two task force members who had done the bulk of the drafting of the document. As a consequence, the EPA had to announce in early September that applicants would be required to submit only preliminary requests for secondary treatment modification. These forms included only name and address, a copy of a current NPDES permit, and a brief description of the outfall and its location. The agency received approximately two hundred such requests.

The final regulations were published on June 15, 1979. In response to municipal and public pressures, the final definition of existing discharge allowed application on the basis of the existing discharge or a proposed improvement in outfall or treatment (e.g., upgrading of treatment from primary to advanced primary, or relocation of an outfall that had been thoroughly studied and planned by the applicant). Applicants would have to demonstrate that their modified discharges would not interfere with the water quality needed to protect both water supplies and propogation of a balanced, indigenous population of shellfish, fish, and wildlife; the water also had to be safe enough to allow recreational activities. Thus, water quality and the impact of the discharge effluent on the receiving water became critical in granting waivers.

Because the impacts of pollutants are so difficult to predict, the EPA made it clear that applicants seeking modifications based on future improvements would bear the additional burden of demonstrating that their proposed discharge would meet not only the requirements of section 301(h) but also the criterion of maintaining a balanced indigenous population.

The final regulations withdrew the requirement that applicants demonstrate conformance of their discharges either to the EPA's criterion for marine water quality or to secondary treatment levels for each of the 65 toxic wastes. Instead, applicants were required to meet state water quality standards, to determine which of the 65 toxic wastes exist in their effluent, and to monitor the effluent for the presence of any of these wastes.

Several factors precipitated this change. If the effluent met secondary treatment standards, the dischargers would not need a waiver. In addition, the proposed regulations had surpassed the EPA's legal and technical capabilities. The requirement that applicants meet water quality criteria for all 65 toxic wastes was not legally defensible. The list of wastes had originated in section 304(a) of the Clean Water Act due to a requirement that the EPA publish a list of toxic wastes and water quality standards for interested parties. The list was to be solely for informational purposes and not a standard that states must adopt or implement. Therefore, there was no rationale for including them as requirements in section 301(h). Requiring such standards would present the EPA with great technical difficulties because water quality criteria did not exist for all 65 toxic wastes. Also, technical capabilities for monitoring so large a number of wastes were either nonexistent or subject to enormous error.

Other major changes were incorporated in the final regulations. A separate policy for native Alaskan villages, the territorial possessions, and Puerto Rico was added. This provision effectively eliminated regulations concerning the 65 toxic wastes from their waiver process and left the matter open for later resolution.

The new regulations required primary wastewater treatment of all applicants. The legislative history "provided clear evidence that Section 301(h) was intended to allow municipal marine dischargers to provide less than secondary treatment." Testimony given during congressional hearings supported this position.

Finally, the task force added two other provisions to limit their applicant pool. First, POTWs already applying secondary treatment to their wastewater could not be considered for a waiver. This was implied in the proposed regulations but was spelled out clearly in the final version. The EPA argued that this was necessary to prevent backsliding. Second, only those dischargers who had submitted preliminary requests would be considered. This way the agency knew it would not have to review more than two hundred applications. Although it cost the EPA a large concession, the agency succeeded in limiting waiver applications.

Outcomes: Views of the Interested Parties

Although the EPA had had to make some major concessions, it did succeed in reducing the final number of applications for waivers to 70. The EPA hired an outside contractor (Tetra-Tech) to do the initial reviews. Many applicants submitted applications that were several volumes in length, making the review process extremely time-consuming. After Tetra-Tech examined the applications, EPA's scientific advisors and enforcement officers reviewed them and prepared a recommendation that was referred to the administrator, who made the final decision. The waiver decision was incorporated in a draft NPDES permit and subject to public notice, comment, and administrative appeal. On September 8, 1981, the agency announced decisions for an initial group of eight applicants: Six 301(h) waivers were approved.

The EPA has faced two lawsuits: one from the Pacific Legal Foundation (PLF), which wanted more concessions from the EPA; the other was from the Natural Resources Defense Council (NRDC), which preferred that the agency had made no concessions. The outcome of the latter case was announced in May 1981. The court upheld the regulations with three exceptions. It invalidated the prohibition against waivers for the discharge of less-than-primary treated effluent, the prohibition against waivers for the discharge of sludge, and the prohibition against waivers for communities already achieving secondary treatment. (NRDC v. EPA, Cir. No. 79-1639, D.C. Cir., May 7, 1981.)

The West Coast fared well in the short term. Although the EPA did not make it easy for all West Coast municipalities to apply for waivers, it expanded the definition of existing discharge. Of the 70 applications submitted for review, there were 12 from California, 19 from Washington State, 11 from Alaska, and 5 from Hawaii. If these applicants receive waivers, the objective of the large West Coast municipalities will have been met in full.

The East Coast dischargers received all that they had requested (i.e., elimination of the restriction on depth and location of the receiving water and redefinition of existing discharge). The EPA has received 12 applications from East Coast municipalities.

The small dischargers were not as successful as the larger municipalities. They are still required to perform expensive chemical analyses to determine which of the 65 toxic wastes are present in their effluent and collect all of the information required of large dischargers. In response to the complaints by small dischargers, the EPA claimed that section 301(h) did not authorize EPA to categorically exempt on the basis of size or volume. Congress thus recognized that volume of discharge is not in and of itself an indicator of its toxicity or its effect on the marine environmental. As a matter of practice, however, the agency has been reviewing the applications of larger dischargers first, so that the data and experienced gained from these reviews may be of value to smaller communities. Nevertheless, these dischargers will likely have to hire consultants to prepare their waiver applications.

The native Alaskan villages, territorial possessions, and Puerto Rico enjoyed a successful outcome. Their requests for special consideration with regard to secondary treatment was granted.

Case Study Questions

1. Those drafting the rules were concerned with two distinct thresholds: the definition of those municipalities that could apply for waivers and the establishment of standards to determine which applicants would actually get them. Under the initial proposal the application threshold was very high. Would it not have been far more politic for the EPA to draft rules that ostensibly would give many municipalities the chance to apply, even if the substantive standards remained very strict? Why did not the EPA follow this approach? How did the EPA's decision affect the bargaining strength of the affected municipalities?

2. The municipalities which were particularly active in the rulemaking process were those who had yet to implement secondary treatment and were seeking exemption from having to do so. What bearing might their position have on those communities which had already undertaken secondary treatment?

3. The initial public hearing to gather reactions to the preliminary concept paper was held in California in spite of the fact that the EPA is headquartered in Washington, D.C. One explanation for the choice of this forum might be the season (it was late winter, a gloomy time in the capital). What bargaining or political considerations might also explain the location?

4. In response to the proposed regulations, a representative from the city of San Diego argued that the legislative history showed a congressional intent to exempt West Coast municipalities from the secondary treatment requirement. Obviously, such an argument was intended to be persuasive in its own right. What other function might it have been meant to serve?

5. Citizens for a Better Environment submitted a statement to the EPA

attacking the basic wisdom of granting any exemptions; yet Congress had already mandated waivers over similar objections that groups like CBE had voiced during the legislative debate? Was CBE beating a dead horse?

6. As one of the EPA's own attorneys noted, the statutory language was loose and nonspecific. Thus, the agency had latitutde in drafting the regulations. Given this discretion, why then did the EPA ultimately promulgate rules that would allow far more applications for waiver than it apparently preferred?

Competition or Cooperation?

Recall the multiperson prisoner's dilemma illustrated by the problem in chapter 2 involving the neighbors who were all polluting their small lake. The 301(h) rulemaking case was another multiparty conflict. Was it also another example of the prisoner's dilemma? What further information would you need to know in order to be fully confident of your answer?

According to the authors of the case study, there were at least three groups of municipalities actively concerned with insuring their right at least to apply for a waiver or, at best, to get one: the West Coast municipalities, the East Coast municipalities, and the communities in Alaska, Puerto Rico, and various territories. Yet, though they shared the common goal of a liberal waiver policy, there is no indication that they coordinated their activities or pooled their political clout. On the contrary, the arguments that some of these groups advanced seemed to bolster their own positions at the expense of other communities. The West Coast cities claimed, for example, that legislative history showed that the waiver provision had been specifically intended for them. The city manager of Portland, Maine, consequently was forced to take issue with the arguments advanced by those who might have been expected to be his allies.

What was there about the situation that created competition instead of cooperation? To be sure, the various classes of communities advanced their own justifications for special consideration. The West Coast cities claimed the deep water discharge negated the need for secondary treatment, whereas those on the East Coast contended that greater tidal fluctuations more than made up for the shallower discharge area. These do not, however, appear to be mutually exclusive propositions. It is true that the communities could be classified into other groups: large versus small dischargers, dischargers with industrial waste and those with none, and communities with existing discharge and those where it was proposed. Are the interests of these groups so opposed that coordinated effort is impossible?

Study Questions on Negotiation and the Rulemaking Process

1. Consider the procedural constraints under which rulemaking occurred, specifically the hearings that were required under the Administrative Procedures Act. Under the APA, the agency must invite commentary from any interested parties. Who is likely to respond to such an invitation? Are there important parties or interests that may not be represented?

2. Much of the commentary that was solicited appeared to be quite hostile to the EPA's preliminary concept paper and, later, to its proposed regulations. What is there about the process that encourages such polarity of opinion? If the rulemaking had been *formal* instead of *informal*, would this tendency have been aggravated or tempered?

3. Under *Home Box Office v. FCC*, 567 F.2d 9 (1977), an agency is prevented from communication with interested parties individually, once the comment period has run. What effect does this rule have on attempts to reach a consensual decision? Is it possible to loosen restrictions on communication without opening the possibility of unequal access to (and influence on) the agency?

4. Evaluate the following proposal, drafted by the authors of the original case study and presented in the first portion of the next section. The comments of the case study authors have been altered somewhat to incorporate subsequent revisions in a cited article by Richard Stewart, "Regulation, Innovation, and Administrative Law: A Conceptual Framework," 69 *Cal. L. Rev.* 1256, 1981. Stewart's section on negotiated standard setting (at pp. 1341–1353) is particularly pertinent to this chapter. His thesis is illuminated in the citations in the proposal to be discussed later, and it is developed more fully in another section.

Reform

Negotiation in Rulemaking

The 301(h) rulemaking case illustrates how the rulemaking process incorporates negotiation in a restricted form. Some bargaining occurs tacitly between parties as they reformulate their initial positions in light of reactions. Jorling's initial position on 301(h) was quite firm: Very few communities could apply for a waiver. The final position, although not a return to case-by-case decision making, represented a substantial concession on his part. How did this shift occur?

Regulators recognize that the rulemaking process entails bargaining, but they often do not embrace it enthusiastically. Richard Stewart has noted that agencies' hesitation over negotiation also reflects their frequent reluctance to 'lose control' of the rulemaking process. This reluctance reflects agency staffs' ideological premise that the agency represents the public interest and that it would accordingly be an abdication of its responsibilities for the agency to turn standard setting over to a private negotiating process. (Stewart, "Regulation, Innovation, and Administrative Law: A Conceptual Framework," 69 Cal. L. Rev. 1256, 1346, 1981)

Stewart has also observed that this attitude may be manifested by the frequency with which an agency adopts "a firm, even extreme, position at the outset, expecting to be whittled down during the rulemaking process." If a compromise position is offered at the outset, it may be further diluted in favor of the regulatees. Moreover, as Stewart observes, agency "administrators wish to maintain flexibility to deal with shifting political pressures and bureaucratic exigencies" (Id.).

The EPA's initial position on 301(h) reflected its policy to avoid case-by-case decision making and maintain maximum protection for the environment, according to the vague standard articulated by Congress. This initial position was made more tenable by the relatively strong support it received from the environmental groups. Negotiation occurred—albeit tacit negotiation—when this position was exposed to public and congressional scrutiny. In effect, the EPA tested the waters to learn whether enough support existed for its interpretation of *existing discharge* to sustain a subsequent challenge. The public-hearing process, the comment period, and the contacts with Congress gave Jorling and the task force an opportunity to refine their positions and settle some issues before litigation.

But to say that tacit negotiation occurred is not to say that the negotiation worked well. After all, both the municipalities and the environmental groups challenged EPA's final regulations in court. The tacit negotiations that did occur were constrained by EPA's interpretation of the *Home Box Office* case. As long as the agency believed that it could not talk informally to the interested parties, it was impossible to assemble an interpretation of "existing discharge" that would be acceptable to the EPA, the municipalities, and the environmental groups.

Although it is not certain that better communication would have resulted in a mutually acceptable agreement, the prospects for such an agreement are usually improved when the parties are free to sound out positions informally—out of the public eye. Unfortunately, the public character of the tacit negotiation process described by Stewart often encourages posturing. Each group publicly commits itself to an extreme initial position from which compromise is difficult. In contrast, nonpublic negotiation gives the parties considerably more room to explore alternatives and find acceptable outcomes. As Lawrence Bacow has written,

> There is good reason why the public is always excluded from serious

bargaining sessions: The give-and-take that is the essence of successful bargaining is inhibited if the parties must be concerned with how their constituencies will interpret intermediate positions taken on specific issues. Since it is not possible to win on every issue, negotiators prefer to present the fruits of their labor as a package instead of piecemeal. Secrecy permits this. (Bacow, *Bargaining for Job Safety and Health.* M.I.T. Press, 1980, p. 128)

Fortunately, future EPA officials will have considerably greater latitude to engage in informal discussions during the rulemaking process. In a recent case interpreting the Clean Air Act, the Court of Appeals has narrowed the *Home Box Office* decision, ruling that EPA is free to engage in informal discussions during rulemaking. If the agency relies upon these discussions in formulating a rule, a record of the discussions must be included in the formal agency record. (See, *Sierra Club v. Costle*, No. 79-1565, D.C. Circuit, April 29, 1981.)

Unfortunately, it is difficult to predict how the rulemaking process will change with fewer restrictions on ex parte communications.

An explicitly negotiation-oriented rulemaking process would require a number of additional legal and institutional changes to be effective. Nonetheless, it offers a number of potential advantages, some of them identified by Richard Stewart: extension of the agency's informational and analytical capabilities, fostering agency understanding of compliance burdens, and hence promotion of technically sound and efficient regulations. Stewart contends that this approach could serve the interests of the regulated municipalities and industries and, by reducing conflict and court challenge, ease the regulator's burden as well:

Negotiated rulemaking could also reduce decisional costs and delays, which would encourage innovation. Informal discussion and negotiation could accelerate identification of the key issues and of the data and analysis required for their resolution. The present system often requires several rounds of formal comments or judicial remand for agency reconsideration of an issue that has been inadequately addressed. If a consensus process promotes agreement by all interested parties in the outcome, formal comment procedures could be substantially shortened and judicial review avoided altogether. If participants to the informal process could not agree, the agency could determine the proposed standard. Even in the latter situation, the informal process could help the agency frame a more workable and acceptable standard, diminishing the scope and complexity of the rulemaking proceedings and the likelihood of judicial review. (Stewart, pp. 1344–1345)

Negotiated rulemaking thus provides a potent alternative to the adversary approach, one which offers incentives for all interested parties. One major incentive would be relaxing of judicial attitudes. Negotiated rulemaking would supplant the current "hard look" approach with a more lenient process of judicial review. Stewart contemplates a system under which courts could accept less detailed agency explanations for decisions, including rebuttal of outside parties' criticisms; decline in successive "rounds" of comment in response to new data or issues; relax requirements that agencies provide a comprehensive 'record' of the data and analysis justifying the decision; and forego a detailed examination of the consistency of the agency's decision with such record. On the other hand, if the process does not yield consensus, courts should apply the "hard look" approach and associated procedural formalities to ensure effective review of agency decisions. Finally, courts should heavily discount claims that could have been raised during the negotiation stage. (Stewart, p. 1348)

In sum, there is scholarly support for optimism about negotiation-based reform of the rulemaking process. Stewart's article suggests in far greater detail how it might be accomplished. (He suggests, for example, that agencies could compensate advocacy groups for possible loss of formal legal leverage by funding their participation in the informal rulemaking process.) It is thus possible that the administrative process could be streamlined; yet, in a way that would insure greater environmental protection. These gains are speculative, of course, because the process has not yet been attempted. But the problems of the existing approach suggest the wisdom of considering such an alternative.

Regulatory Reform

Stewart's analysis (1981) is fully developed in his previously cited article. In it, he notes that there are precedents for negotiation-based standard setting. For example, manufacturers have voluntarily banded together to promulgate product and fire safety codes through a process of consensus building. Such activities demonstrate the way in which industry expertise can be efficiently tapped to create useful standards.

Stewart cautions, however, that the voluntary standard-setting model is not appropriate for environmental regulation:

> Industry-wide standards present the danger that the consensus process will be dominated by regulated industries seeking to reduce regulatory standards to the lowest common denominator acceptable to all firms in the industry. (*Id.*, pp. 1342–43)

This has not been a major problem in product safety, he contends, where firms have a substantial economic interest to adhere to voluntary standards. Such an incentive is usually absent in environmental problems. He adds that,

> the adoption of uniform environmental standards for industrial pollution could have drastically different effects on firms within an industry. The disparity could create serious obstacles to consensus.

Stewart believes that negotiated rulemaking, a process in which the government agency would continue to play a major rule, holds more promise. It will be most likely to succeed, he suggests,

> with regulatory decisions that are neither so narrowly focused that they afford little opportunity for horsetrading and compromise, nor so open ended that they present an unmanageable number of issues and parties. (*Id.*, p. 1345)

In his view, effluent limitations, new source performance standards for particular industries, and automobile safety and emission standards are appropriate candidates for such an approach.

Stewart's observation that agencies may be reluctant to enter negotiation out of fear of losing control of rulemaking has already been noted in the proposal in the previous section. He also observes that private parties may likewise be hesitant:

> Private parties may also be reluctant to engage in a process of open discussion and negotiation. If they can be harmed by delay or have limited resources, they may fear informal negotiation is a device to postpone decisions and wear them down by multiplying the number of proceedings in which they must participate. Also good faith negotiation necessarily involves some disclosure of the parties' true positions and priorities. Such disclosure may compromise later assertions of more intransigent legal positions seeking to overturn the agency's decision on judicial review." (*Id.*, p. 1346)

Stewart observes that agency and private party reluctance can be overcome only if the incentives to engage in negotiation are made sufficiently large to outweigh the disincentives:

> Three steps must be taken to strengthen the incentives for all parties to participate in good faith in a negotiation/consensus process. First, the opportunities for delay in the present system of agency procedures and judicial review must be substantially reduced as a quid pro quo for participation in a successful process of negotiation and consensus. Second, the responsible agency must be willing to run the risks involved in giving up a measure of control over the rulemaking process and invite an active role by outside parties in the earlier, more fluid stages of policy formation. Agencies may be willing to do so if existing procedural formalities and standards of judicial review are correspondingly relaxed. Third, to compensate advocacy groups for the relaxation of procedural formalities and to equip them to participate effectively in informal processes, funding for such participation should be provided. (*Id.*, p. 1347)

Stewart's views on implementing these steps have been cited in the earlier section. "Broad-scale agency use of negotiated standard setting cannot be mandated," he concludes.

It must grow out the agencies' perception that the process will advance their self-interest by reducing decisional costs, delays, and court challenges and by promoting policies that are more easily implemented because they are more acceptable to the parties involved. But congressional legislation would help remove inhibitions on negotiated standard setting by clarifying uncertainties in existing law and by signaling legislative encouragement. (*Id.*, p. 1353)

Legal Considerations

Introduction

Much of the foregoing material emphasizes the shortcomings of the traditional rulemaking process. Referring to procedural requirements as *constraints* on agency action may give them an unduly negative character. These requirements, after all, have been implemented to serve important legal, and ultimately, social policies, chief among them being protection of individuals' due process rights. Is it possible to improve the substance of environmental rulemaking without riding roughshod over other important goals?

The following excerpts from "Rethinking Regulation: Negotiation as an Alternative to Traditional Rulemaking," (94 *Harv. L. Rev.* 1871, 1981; citations omitted) describe the legal underpinnings of traditional administrative procedure. Some arise from statutory law, like the APA; others find their source in cases like *Home Box Office*. Statutes and court decisions can be modified with relative ease. More fundamental constitutional precepts, like due process and separation of powers, are less amenable to change.

> Traditional rulemaking lies toward the adversary end of a spectrum that ranges from purely adversary dispute resolution techniques, like litigation, to methods relying solely on bargaining, like legislation. Although APA procedures for informal rulemaking are flexible, the statute assumes parties will participate in rulemaking through the characteristically adversarial techniques of formal argument and proof. . . .
>
> Although rulemaking by negotiation might take many forms, this Note suggests two models for purposes of analysis. . . .
>
> 1. *Agency Oversight Model*. Under the agency oversight model, an agency would initiate informal rulemaking by publishing in the Federal Register not only a description of the topic, but also a general invitation to participate in negotiations. It would specifically invite affected groups and offer to assist participation by unorganized interests. From those responding, it would select a manageable number, while seeking representation for all interests with distinct viewpoints. The agency would then invite the representatives to a closed bargaining session. Agency officials would not be present at this session. After the group reached agreement, standard APA informal rulemaking procedures would begin. The agency would publish the agreement as a proposed rule along with a statement of basis and purpose

composed by the negotiators. Though more abbreviated than the explanation that currently accompanies proposed rules, the statement would summarize negotiators' arguments for the rule that emerged, opposing arguments, and the reasons the negotiators rejected them. The agency would then receive and respond to comments on the rules, as it does in current rulemaking. Although it would accord the negotiated agreement considerable weight, the agency would examine anew, and in light of the governing statute and its policies, the data, comments, and statement of basis and purpose; it would then reach an independent conclusion on the final rule.

2. *Agency Participation Model.* Under the agency participation model, the process would begin as in the agency oversight model, but the agency itself would participate in the negotiation. It would present to the negotiators its policies and its interpretation of the statute, and would respond to their suggestions. As one of the negotiators, the agency would have to agree to all bargains before they could be promulgated as rules. If the parties could not agree, notice-and-comment would begin as it does under the current system. If all agreed, however, the agency would publish the bargain as a proposed rule and then accept public comment. If the comments indicated that the session had omitted a distinct interest or ignored a possible solution, the agency would remedy the flaw and reconvene the negotiation. The agency would repeat the cycle until a rule emerged that drew no significant, novel comments. . . .

Negotiation would yield better rules than current informal rulemaking for several reasons. First, rulemaking involves polycentric problems—conflicts in which the resolution of any part of a dispute affects that of all other parts, leaving a complex fabric that adversary proceedings cannot unravel. A process that brings interested parties together to consider all parts of a dispute at once can better accommodate such an interaction of concerns. Second, while the adversary system encourages "exaggerated, inflexible posturing," negotiation yields a pragmatic search for intermediate solutions. Because negotiators learn other parties' economic and political constraints, they may realize the impracticability of their own bargaining positions and discover more common ground than the would as adversaries. . . .

[Moreover] parties to negotiation identify with and defend the resulting agreement and are less likely to resist its enforcement or to challenge it in court, especially if the resulting rules are substantive improvements over those the adversary process would have generated.

The oversight model is less likely to improve post hoc acceptability than is the participation model. This is so because oversight model negotiators must guess whether the agency will approve their agreement and because the agency may hesitate to approve solely on the recommendation of interested parties an agreement in which it played no part. In the participation model, parties may discuss proposals with agency representatives; as a result, the process is more likely to generate a rule acceptable to society. . . .

The claimed advantages of regulatory negotiation assume the presence of a number of favorable conditions. If these conditions are not present, negotiation will simply add a useless layer to rulemaking.

1. Adequate Yet Manageable Representation. Although complex issues inevitably affect many groups, negotiators must be few enough to keep the negotiation manageable. On some issues, however, the number of distinct policy positions or interests may be unacceptably large, even though some groups may be willing to economize by joining forces. To limit participants, the agency should require groups with a common viewpoint to choose a single representative. . . .

2. Inducing Good Faith Negotiation. Groups who benefit from the status quo or who believe notice-and-comment would treat them better than negotiation would rather obstruct than bargain. Agencies must thus devise incentives for good faith negotiation. In the agency participation model, the agency negotiator and reviewing courts could look suspiciously at comment and challenges by parties showing bad faith. If the negotiators failed to agree in the agency oversight model, they could send the agency the rule that drew widest support along with dissenters' reasons for opposition. The agency could ignore bad faith dissents. Such a process would make good faith negotiation the only road to regulatory influence and would persuade obstructionists to make concessions of their own so that they might extract concessions from others. Finally parties are likely to cooperate when they must maintain a long-term relationship.

In addition to practical considerations, legal principles must guide the design of a regulatory negotiation system. The major legal limits on negotiation are those the nondelegation doctrine imposes on private assumption of public authority and the requirements of judicial review under the APA, including the judicial prohibition of ex parte communications. . . .

Because society is complex and the process of legislative compromise difficult, Congress can legislate only in general, leaving agencies to resolve particulars. But this delegation of authority has constitutional limits. Under the "contractarian" theory of democracy, laws derive their legitimacy from the consent of the governed. Since members of Congress are elected, the governed can be said implicitly to approve the laws Congress passes. The actions of agency officials, by contrast, do not rest on public approval, but gain legitimacy only through congressional enactments. To ensure the legitimacy of administrative action, courts have demanded that Congress pass guidelines that provide agencies with meaningful standards.

Judicial scrutiny of congressional delegation intensifies when private groups replace presumably neutral agency officials and gain power themselves. For one thing, courts suspect that private representatives favor their supporters and thereby violate the due process rights of unrepresented individuals. More importantly, courts fear that delegation to private individuals may further attenuate voter control of government; private representatives owe allegiance only to their supporters, while administrators must account to the elected officials who appointed them. . . . Despite the nondelegation doctrine's ebb since the high water mark of [cases like *Carter v. Carter Coal Co.* and *Schecter Poultry Corp. v. United States*, citations omitted], the doctrine itself, and the court's hostility to private exercise of public authority survive to this day.

When courts believe that private groups play only an advisory role—when, for example, the groups propose rules for a neutral agency's approval—they turn back delegation challenges. . . .

Under the literal requirements of this doctrine, negotiation would have to stop short of granting de jure rulemaking authority to private groups. This limitation poses no problem for the participation model, for agency assent is a prerequisite to the model's agreements. The oversight model, though, is caught in a scissors—agency oversight must be sufficiently strict to calm nondelegation worries, yet sufficiently relaxed to make the negotiation meaningful. In practice, agency supervision in the oversight model would probably satisfy courts. The agency would review all data de novo and would not defer to the negotiated rule if it conflicted substantially with the public interest. . . .

Even if it involved a significant delegation, negotiation might nonetheless avoid nondelegation problems if all interests were effectively represented. By replicating the process of pluralistic decision at the agency level, adequate representation would calm the fear that agencies will evade popular control and would thus satisfy the underlying concern of nondelegation cases, if not their precise holdings. . . .

Challenges to negotiated rules would come either from unhappy negotiators or from parties excluded from the process. Both groups would face obstacles to their challenges. Courts might look suspiciously at suits by dissenting negotiators and require some special explanation for their inability to influence the negotiation. If absent groups declined an opportunity to participate, courts would not receive their challenges kindly. . . .

1. *Notice-and-Comment Rulemaking Requirements.* Despite the apparent simplicity of the APA vision of informal rulemaking, courts have added procedures that have made rulemaking significantly more formal. One important requirement is that the agency construct a record containing all the facts on which the agency based its decision. The agency's decision must result *only* from material in the record; the courts have required that ex parte communications be placed in the record and have reacted hostilely to agency use of nonrecord material. In addition, the agency must make the record complete early enough in the proceeding to allow interested parties to comment on, and thus test the strength of, relevant facts. These requirements allow parties to comment fairly on all data, and provide the basis for intelligent review by the courts.

In addition, the agency must explain its rule in a concise general statement of basis and purpose. . . .

2. *The Record Requirement's Application to Negotiation.* The requirement of an adequate record may threaten the oversight model. If negotia-

tions are private, a crucial part of the model's rulemaking will be unrecorded—namely, the data employed in negotiations, on which the agreement will be based. Although technically the oversight model does not meet the mandate for a complete public record, it might still satisfy the purposes behind the record requirement—guaranteeing that the court know enough about the issues to judge whether the agency acted arbitrarily and allowing public examination of data. It might do so by requiring negotiators to release all data that would not damage the privacy of the negotiations, along with a summary of the discussions. Nevertheless, the data package might lack vital information, since the most important data could easily be the most sensitive. Courts would thus lack sufficient information to judge agency decisions.

A preferable solution would have courts examine the record in camera. . . .

3. *The Statement of Basis and Purpose.* Both models fail to satisfy the literal requirements courts have established for the statement of basis and purpose. The presentation of negotiators' reasoning process is impossible; the give-and-take of a negotiation yields agreements based as much on horse-trading and bargaining skill as on expert analysis. To impute reasoned logic to a negotiated settlement is to rewrite history.

Negotiation will thus have to comply with the purposes of the statement. One of the purposes is to ensure that the agency gave fair consideration to all interests. . . .

To satisfy this concern for balanced participation, negotiators should compose a statement of basis and purpose summarizing the arguments and facts supporting the negotiated rule. Like a legislative history and preamble, the outline would trace the rule's development and the arguments for and against it. These efforts might not satisfy reviewing courts, which lack the agency's expertise and may be unsure of the rule's implications. This uncertainty would prevent them from determining whether the rule is consistent with other rules and the authorizing statute. Looking for the logic that genuinely motivated the choice, a court might dismiss the compromise statement as merely a post hoc rationalization. Yet a properly drawn statement could meet the concerns that representation be balanced and that all views be adequately considered.

The agency might also accomplish the goals of a statement of basis and purpose by holding an abbreviated notice-and-comment proceeding, specifying before the negotiation a spectrum of acceptable rules and justifying this range in a statement of basis and purpose. In the oversight model, the negotiating agency would announce the range beforehand and not accept an agreement that exceeded it; in the participation model, the agency would employ its veto power to keep the agreement within the range. If the range were sufficiently narrow to be within the agency's nonarbitrary discretion, courts would view it as the equivalent of a rule; the agency would simply be announcing the options it finds acceptable before choosing the best. Yet the spectrum would have to be broad enough to leave room for flexible negotiation. In addition, the setting of acceptable guidelines might be costly and time consuming for the agency, because it would require a brief notice-and-comment period before negotiations began. These disadvantages might undercut support for, and dissuade agencies from experimenting with, negotiation.

   4. *The Ban on Ex Parte Communication.* Courts have limited private contacts between agency officials and affected groups. . . . The rule against ex parte communication poses substantial problems for the participation model; if negotiations are secret, agency participation arguably involves ex parte communications. The model could survive the rule, however, in either of two ways. First, courts could eliminate the ban. Because the doctrine is still unsettled, this is a possibility, though not a strong one. Second, courts might accept a procedural analog that satisfies the function of an ex parte ban. . . . The following process will ensure full representation: Upon promulgation of a rule, an absent party, by examining the statement of basis and purposes, would decide if its interest had been adequately represented. If it decided in the negative, the party would petition to be represented at a reconvened session. If the agency refused, a reviewing court would scrutinize the statement to determine whether the party had made a colorable showing of lack of representation. If it had, the court would inspect a transcript of the session in camera or would require that a summary be made available to the party. The court would determine from this information whether the party had a spokesman at the bargaining table. If it did. the challenge would be dismissed. If it did not, the court would order the party admitted to the reconvened session. In this way the purposes of the ex parte ban would be met, while publicity would be kept to a minimum and the selection of negotiators would be open to judicial scrutiny.

   The oversight model would fare better under the rule against ex parte communications for two reasons. First, although the ban forbids agency officials to *receive* private communications, it appears to allow them to *speak* to the parties on an ex parte basis. Thus, the agency could stimulate bargaining by notifying the parties of the issue and rules the agency is considering, summoning them to a session, and suggesting areas of compromise. The agency's expertise would permit it to offer wise suggestions that might prod negotiators to agree. Second, *Home Box Office* prevents private communications with officials "involved in the decisional process." Agency mediators could therefore participate fully in negotiations if a "Chinese wall" divided them from the rulemakers. The Chinese wall would prevent them from communicating what they had learned in these negotiations to those involved in the decision.

   The oversight proposal might tread on the ex parte prohibition if courts viewed the agreement itself as an ex parte communication. Although the agreement would become public, its significance to the agency might exceed its public significance; in other words, the agency would accept the agreement not on its merits but simply because all affected groups had agreed. To block this back door influence, the agency could publicize the special status of the agreement, allowing other parties to criticize it, for example, as the product of an unbalanced negotiation. . . .

Regulatory negotiation faces major legal problems. Although it would probably survive nondelegation challenges, the procedural strictures that reviewing courts have imposed may strangle negotiation. Three possible solutions exist. First, negotiation might be made public. This would satisfy reviewing courts, because the record and the reasons for the decision would be open to public scrutiny. Although the glare of publicity might wilt negotiations, open negotiations might succeed on technical and noncontroversial issues. Second, standards could limit negotiators' discretion. By means of a brief, informal rulemaking process, the agency could define a range of acceptable rules, supported by a record and statement of basis and purpose. Negotiators would then settle on a rule within the range. If the range were no broader than the spectrum of rules a reviewing court would find to be within the agency's nonarbitrary discretion, the procedure would survive. Of course, the initial rulemaking and the narrowed scope of negotiation would limit the value of negotiation.

As a preferable solution, courts could devise a new set of procedural safeguards for negotiation. Because the current safeguards arose in an environment of adversary rulemaking, they may be inappropriate for regulatory negotiation. In designing safeguards, the courts would balance negotiators' need for privacy against the fear that representatives might co-opt the agency at the expense of unrepresented groups. Such safeguards might include scrutiny of the choice of negotiators to ensure balance and effective representation of constitutents. Courts could demand that the agency review the agreement and justify its approval with a statement of basis and purpose.

If . . . courts are willing to relax judicially imposed procedural requirements, regulatory negotiation may offer an opportunity to improve our slow, expensive, and ineffective system of regulation.

STUDY QUESTIONS

Under the various proposals discussed so far, judicial review of regulation would still occur, though, it is hoped, less frequently. Review could be sought under either the agency oversight or the agency participation models. Without it, administrative power would be largely unbridled. Review of an administrative rule might be sought where there had been no consensus or where an agreement was challenged by a party that was not at the bargaining table. Is it possible to create judicially manageable rules to resolve the problems that are likely to arise?

1. The author of the excerpt in the previous section suggests that if there are many interested parties, "the agency should require groups with a common

NEGOTIATED RULEMAKING   317

viewpoint to choose a single representative." Who is to say that groups have a common view point? Could that have been said, for example, of the municipalities in the 301(h) case? Given the economies of concerted action, should we assume from the fact that certain parties have not banded together that they have somewhat different interests? Even if some parties have a common agenda, what should the agency do if they cannot agree on a representative?

2. The author further suggests that courts "look suspiciously at suits by dissenting negotiators and require some special explanation for their inability to influence the negotiation." What if the dissenter's principal complaint is that he or she was poorly represented at the bargaining table? Is it enough to say that success in the courtroom or in the traditional rulemaking process also often turns on the competence of legal counsel? Where there is a formal record and established procedures, a poorly represented party may be able to demonstrate that the lawyer made an egregious mistake and bring a malpractice suit. Can a lawyer ever be guilty of malpractice in negotiation? Even in less extreme cases, judges sometimes admit that they try to compensate for disparities in legal representation. Should an agency assume this responsibility in negotiated rulemaking?

3. To deter obstructive behavior, the author recommends that reviewing courts (and agencies, in the case of the oversight model) "look suspiciously at comment and challenges by parties showing bad faith." How can bad faith be proved? Should an unwillingless to revise an initial bargaining position raise a presumption of bad faith? If so, will not parties be careful to make high initial demands, just so that they have room to make token concessions? What other tests of bad faith might be adopted? Given the difficulty of establishing bad faith, should sanctions against it be stern or lenient?

4. In the case of the agency oversight model, it has been suggested that an "abbreviated notice-and-comment period" could generate "a spectrum of acceptable rules" for the negotiators to choose among. To the extent that one of the claimed virtues of negotiated rulemaking is the tapping of the expertise of affected parties, can the agency be expected to draft the most efficient and just rule *before* negotiation has taken place? Will the notice-and-comment procedures invite polarized statements that will taint the negotiation process?

5. Again, in respect to the agency oversight model, the author suggests that when consensus proves impossible, negotiators send the agency "the rule that drew widest support," along with a statement by dissenters. Should the agency give more weight to a rule that draws 90% support than to one that draws only 70%? What if many industries favor the proposal, but the party chosen to represent environmental interests strongly opposes it? Should the agency be

concerned not merely with the substance of support and opposition but with its intensity. How might it gauge this quality?

Legislative Proposals

Introduction

Interest in regulatory negotiation has also been expressed in Congress; several enabling bills have been filed, though none have yet been adopted. One of them, The Regulatory Negotiation Act of 1980 (S. 3126, introduced by Senator Levin), is summarized below. As you read the proposal, you should contrast its procedures with those followed in the drafting of the 301(h) regulations. Also, consider the following questions:

1. What incentives would the Levin bill provide to parties to participate on a Regulatory Negotiation Commission?
2. Is the voluntary nature of participation on these commissions likely to encourage or inhibit consensus?
3. Suppose you are chairman of the Administrative Conference. Under section 201(d) you are to approve applications for commission status only if there is "reasonable likelihood" of success? What evidence do you consider in making this assessment?
4. Recall the terminology in the Note "Rethinking Regulation" cited in the previous section. Would you say that this bill is based on the "agency participation" or the "agency oversight" model? Why do you suppose the particular model was chosen; will it encourage consensus?
5. If you were called upon to redraft the bill, what changes would you make?

The Regulatory Negotiation Act of 1980

>   Sec. 101. FINDINGS. The Congress finds and declares that—
>   (1) Government regulation of the economy has increased rapidly in recent years due to an increased awareness of the environmental, social, and health effects of a variety of economic practices;
>   (2) although such increased regulation has commendable purposes, it has frequently resulted in contradictory, inefficient, unjustifiably expensive, and often counterproductive regulatory requirements;
>   (3) unnecessary regulation has reached a level where it is having a significant adverse effect on the economy;
>   (4) ineffective regulation has prevented the attainment of important national goals in the areas of the environment, health, and safety;
>   (5) some of the problems in Government regulation are attributable to the adversarial process of setting regulatory policy, a process in which the best solutions to problems are often ignored by all parties to a dispute in order to maintain their bargaining positions; and

(6) an adversarial regulatory process frequently ignores the expertise and understanding of people working in the affected areas.

Sec. 102. PURPOSE. The purpose of this Act is to establish a pilot program to encourage the voluntary formation of regulatory negotiation commissions as an alternative to the adversarial process of establishing regulatory policy. Commissions receiving assistance under this Act shall be composed of a balanced representation of industry, public interest groups, labor, State and local officials, or other participants with a vital interest in the areas under consideration by the commission, and shall meet to negotiate recommendations on regulatory policy which represent a consensus of the viewpoints of the participants in the commission.

Sec. 103. DEFINITIONS. For the purposes of this act—

(1) The term "regulatory negotiation commission" means a group formed on a voluntary basis by private individuals and organizations to study one or more regulatory policies which (a) contains representatives of all or most of the major positions on the issues under consideration by the commission, and (b) attempts, through negotiation, to reach recommendations on regulatory policy which represent a consensus of the viewpoints of the participants in the commission;

(2) the term "Conference" means the Administrative Conference of the United States;

(3) the term "Chairman" means the Chairman of the Conference. . . .

Sec. 202. PILOT PROJECT AUTHORIZED. (a) In order to carry out the purposes of this Act, the Chairman of the Conference shall establish a pilot program to make grants to five regulatory negotiation commission projects during each of the fiscal years 1980 and 1981. Grants made under this Act shall be for the payment of administrative expenses of regulatory negotiation commissions, shall be in an amount not in excess of $250,000 for each commission, shall remain available without fiscal year limitation, and may be augmented by funds from non-Federal sources.

(b) (1) By April 1, 1981, the Chairman shall announce . . . the availability of grants under this act . . . The chairman shall only make grants for such projects for matters pertaining to regulatory policy in the areas of health, safety, and the environment, and for which—

(A) a major law has been enacted, but proposed rules and regulations have not been issues;

(B) final rules and regulations have been issues, but are likely to undergo major revision; or

(C) basic statutory changes are contemplated.

(2) The selection of areas of regulatory policy for regulatory negotiation commission projects shall not be subject to judicial review.

(c) Individuals and organizations with an identifiable interest in any regulatory area selected by the Chairman under subsection (b) may make an application to the Chairman for a grant under this Act. Each such application shall—

(1) be signed by all proposed members; . . .

(2) include a description of the regulatory area to be discussed, the need for a regulatory commission in such area, a proposed membership list for the commission and justification for that list, proposed rules for the operation of the commission, a statement of purpose for the commission, a proposed time period for the completion of the work of the commission, and an organization plan and an agenda for the commission.

(3) contain a written commitment signed by all proposed members of the commission to negotiate the issues under consideration in good faith, and to produce a report on the negotiations with a time period appropriate to the regulatory area under consideration. . . .

(d) (1) The Chairman shall only approve an application under this Act if the Chairman determines that there is a reasonable likelihood that the regulatory negotiation commission applying for a grant under this Act—

(A) is able to produce a report that will significantly expand the existing areas of consensus among major affected parties in a regulatory area;

(B) is able to significantly increase cooperation between such parties;

(C) will include a balanced representation of the major affected interests in an area, including business and public interest organizations, in accordance with subsection (e) (1); and

(D) can recommend policy alternatives that will provide significant improvement over existing policy. . . .

(e) (1) In determining whether a proposed commission meets the requirements of subsection (d) (1) (C), the Chairman shall consider whether the regulatory negotiation commission applying for a grant under this Act contains sufficient representation of the major positions of interest in the area of regulatory policy to be considered by the commission in order that each such interest is able to effectively express its views during the deliberations of the commission. The Chairman may not approve an application for a grant under this Act—

(A) in the case of a regulatory negotiation commission which will consider regulatory policy in the area of the environment, unless at least one-third of the members of the commission are representatives of business and at least one-third of such members are representatives of environmental organizations; . . .

(C) the Chairman is satisfied that major interests other than the interests specified in subparagraphs (A) and (3), including labor, consumer organizations, and State and local officials who have a significant contribution to make to the commission are provided with an adequate opportunity to make such a contribution. . . .

(f) Grants made under this Act may be used—

(1) to employ an administrative director for a regulatory commission, who—

(A) shall be responsible for the administrative operation of the commission and such other mediative or facilitative duties as the commission finds appropriate;

NEGOTIATED RULEMAKING 321

(B) shall not represent any member of the commission with respect to a particular viewpoint in a regulatory area;

(C) shall be compensated at a rate which is not in excess of $45,000 per year; and

(D) shall not be considered an employee of the Federal Government;

(2) to pay travel expenses for members of a commission and per diem expenses; . . .

(3) to pay other administrative expenses. . . .

(g) Each commission receiving a grant under this Act shall issue a report at the conclusion of its negotiations, outlining areas of consensus, areas of disagreement, and recommendations, and constraining any background material the commission may consider appropriate.

(h) Any regulatory negotiation commission receiving a grant under this Act may change its membership, rules, or agenda at its discretion [but subject to approval of the Chairman]; . . .

(i) Any meeting of a regulatory negotiation commission receiving a grant under this Act shall be open to the public unless a majority of the commission members vote to close the meeting. . . .

Sec. 201. GOVERNMENT PARTICIPATION IN COMMISSION NEGOTIATIONS. An agency shall send an observer to any regulatory negotiation commission requesting an observer. An agency observer shall report to the agency concerning commission activities, shall provide information to the commission, and may make suggestions to the commission. An agency observer may not negotiate regulatory policy positions on behalf of his agency, and the view of the agency observer shall not be considered to represent the formal position of his agency. An agency observer may not be present at a closed meeting of a regulatory negotiation commission.

Sec. 203. GOVERNMENT COMMENTS ON COMMISSION REPORTS. (a) An agency shall comment on the report of the regulatory negotiation commission receiving a grant under this section within 60 days after the receipt of such report. . . .

(c) All Federal agencies engaged in the areas of health, safety, and environmental regulation shall assure that the recommendations and reports of any regulatory negotiation commission which is voluntarily established by private parties and is not receiving a grant under this Act are seriously reviewed by the appropriate agencies.

Sec. 301. (a) The provisions of any law or rule relating to prosecution for ex parte communications shall not apply to any communications between an agency of the Federal Government and any regulatory negotiation commission.

Problem

Reconsider the 301(h) case in light of the preceding sections on regulatory reform. If Senator Levin's bill had been in effect in 1977, who might have attempted to put together a commission for the 301(h) regulations? Under the legislative standards, would this problem have been an appropriate one for a grant? Who would have been the likely commission members? What sort of person would have been an appropriate director? Would the fact that the Levin bill is based on the agency oversight model have encouraged or deterred negotiation?

Consider also the agency participation model that was described in the excerpt "Rethinking Regulation" in Section 3b. What incentives would the EPA have had to take part in such a process? What, if any, risks might it have incurred by doing so? What reasons are there for believing that the resulting regulations might have been somewhat different than those that were derived under the traditional process?

Congress has yet to enact the Levin bill or legislation like it, but in 1983 the EPA undertook a demonstration project in which it will attempt two modes of regulatory negotiation—one with the agency acting as a mediator, the other with an outside neutral acting in that capacity. Assuming that the EPA has the legal authority to engage in such a project, is anything lost when there is no explicit legislative authorization?

12

Institutionalizing Negotiation

Introduction

Environmental negotiation and mediation are still very much in their infancy. Each effort to resolve a dispute using these techniques involves educating the parties and convincing the skeptics. Often, procedural hurdles must also be overcome. Our system is set up to facilitate judicial review, and in doing so, it can discourage the parties from settling their differences out of court. Rules governing adjudicatory hearings, ex parte communications, access to public meetings, and standing tend to put parties into combative stances. That people nonetheless manage to overcome these obstacles to reach negotiated settlements is evidence of the appeal of alternative techniques.

Our system need not be set up to discourage consensual agreement in environmental disputes. Indeed, in other contexts (most notably labor relations), bargaining is the rule rather than the exception. There the fundamental goal is to facilitate face-to-face negotiation. Recently, as environmental negotiation and mediation have begun to gain credibility, proposals have surfaced to institutionalize these techniques in a similar manner. Generally, these proposals fall into four categories: (1) those that systematically attempt to remove procedural obstacles to consensual agreement; (2) those that seek to anticipate and avoid conflict; (3) those that mandate negotiation in specific situations; and (4) those that attempt to create an atmosphere that is conducive to consensual agreement through the use of incentives.

In a sense, this chapter synthesizes themes that have been introduced earlier. For example, any evaluation of proposals to promote consensual agreement requires an understanding of the incentives that can lead individuals or organizations to litigate or negotiate. Likewise, identification of obstacles to negotiation (such as the difficult task of binding a government agency to a long-term agreement) implicitly suggests possible remedies. (When particular obstacles or disincentives are removed, parties may recalculate the advantages and disadvantages to negotiating.) In the previous chapter, we discussed the barriers to negotiation in rulemaking and how they might be overcome. In this chapter, we shall analyze obstacles that arise in other contexts as well as specific remedies that have been attempted or proposed.

As you read the following materials, consider how you can intelligently judge the promise of the various proposals. Would you, as a party, be more or less inclined to reach an out-of-court agreement or avoid litigation entirely as a result of the suggestions that follow? The authors of the various excerpts catalog the expected benefits. Can you identify any costs?

Removing Procedural Obstacles

Introduction

The great majority of lawsuits instituted in the United States are ultimately resolved by the parties. Some 90% of all divorces are uncontested, for example. An even higher proportion of personal injury claims are settled out of court. Likewise, only a handful of felonies go to full trial: the rest are all plea bargained.

People settle cases because they think on balance that it is better to come to agreement than to bear the expense—and risk—of pursuing a lawsuit to the often bitter end. The agreements that people reach, of course, are shaped by decisions that are ordered in those few cases that actually go to trial. This practice of "bargaining in the shadow of the law" is not just advantageous to the parties, but it is probably essential to society at large. The judicial system labors hard simply to keep up with its current trial schedules. As some prosecutors have remarked in response to criticism of plea bargaining, if all claims had to be heard in a full trial, the court system might collapse under the strain.

Rules of court procedure typically empower courts to require that contesting parties go through some sort of conciliation process to see if it is possible to avoid a trial. The procedures vary from jurisdiction to jurisdiction and from court to court. Many states, for example, statutorily impose waiting periods on parties to divorce actions in the hope that the parties may reconcile their differences. Michigan requires plaintiffs and defendants in civil actions to submit their claims to nonbinding arbitration prior to going to trial. The procedure encourages settlement by providing each side with an objective assessment of the strength of their cases without the expense of a full-blown trial. Some judges also take an active role on their own in counseling the parties to encourage settlement, although this type of activity has received criticism from some commentators. The danger, of course, is that the parties may feel coerced to settle by a judge who has the power to threaten a litigious party with unfavorable rulings from the bench if the case should go to trial.

Although procedures vary, the parties usually are free to settle at any point during a lawsuit, whether at the outset, when the complaint has been just filed, or years later, when the jury is deliberating. When parties do settle, judges usually approve their terms with only cursory review. (This is not true in class action suits, where courts scruntize proposed settlements to ensure that the representative plaintiffs have not compromised the interests of the overall class in return for their own personal gain.)

In spite of this tradition of encouraging agreements and approving settlements, few courts have been imaginative in promoting out-of-court settlements of environmental controversies. In the Foothills case, we saw one instance where a judge refused to endorse the terms of a negotiated settlement. In their article "Toward a Theory of Environmental Dispute Resolution" (9 B. C. *Environmental Law Rev.* 311, 1981), Lawrence Susskind and Alan Weinstein argue that courts should assume a much more active role in encouraging consensual agreement in environmental disputes. Specifically, they suggest that judges could (1) routinely appoint mediators during the pretrial phase of lawsuits, (2) supervise the bargaining process to ensure its procedural fairness. and (3) act as gatekeepers to determine who is admitted to the bargaining process. As a judge, what risks would you see in the Susskind/Weinstein proposals? What criteria would you employ to determine who is admitted to the bargaining process? Are there other measures you might take to encourage privately reached settlements?

Susskind and Weinstein themselves acknowledge that government agencies also face powerful bureaucratic disincentives to negotiate. Can you think of ways to overcome the obstacles that they catalog in the next excerpt? If you were the head of a state environmental agency with enforcement responsibilities, what steps would you take to encourage out-of-court settlements of enforcement actions? How successful do you think you are likely to be?

Excerpt from Susskind and Weinstein (1981, 352–353):

> Because negotiations and mediation, at least at the outset, will be perceived as novel procedures, the agency that participates in a bargaining effort—even if court-supervised—may lay itself open to charges that it is exceeding its legitimate authority. Critics may claim that the agency is shirking its duty—particularly in enforcement actions—and, rather than attempting to make "deals" with those who violate environmental laws, should be seeking to enforce the law in the manner officially prescribed. Further, the agency may risk charges that a bargaining effort shows that it has been "captured" by the very interests it is supposed to be regulating.
>
> Agency officials may also be hesitant about offending powerful elected

officials who influence policy, control agency resources through the budget process, and suggest appointments. When an agency participates in a large-scale bargaining effort involving numerous parties, its activities may be perceived by some elected officials as an intrusion on their own political "turf": they may view such activities as just the sort of political "log-rolling" which they believe to be their private bailiwick.

Agency officials may also be reluctant to participate in bargaining because it involves a lessened role for themselves. Rather than being the central figure in a formal process, with attendant media coverage throwing a spotlight on agency personnel, the official finds himself engaged in "behind-the-scenes" discussions where discretion, not publicity, is the rule. Further, in a consensual process, the agency official neither sits in judgment nor enforces the law (both positions of power and prestige): instead, he becomes merely another actor in an often frustrating and tedious process with no guarantee of success.

To make matters worse, even though the agency is only one party to the bargaining process, because of its high "visibility" it risks being held solely responsible for an unpopular agreement or blamed if negotiations break down. An agency may also find it extremely difficult to exit from bargaining sessions, no matter how reasonable the action might be, without being accused of damaging the prospects for settlement.

Supplementary Note

In addition to the obstacles noted by Susskind and Weinstein, other problems may frustrate agency participation in bargaining, especially in enforcement actions. Agencies responsible for enforcing environmental laws tend to employ people who are fervently committed to the mission of the agency. Individuals who seek to settle enforcement actions may be seen as ideologically weak by their own peers. Personnel policies also may discourage settlement. Young lawyers seek out jobs in enforcement offices because such jobs offer the promise of valuable litigation experience. With trial experience in hand, it is often easier to land a lucrative job with a private law firm. Comparable employment opportunities may not await the attorneys who have spent the bulk of their government practice systematically trying to stay out of court. Moreover, the attorney who settles cases may not even advance as rapidly within government. Attorneys often are evaluated on the basis of their litigation record. A well-publicized court victory usually will do more for one's career than a carefully negotiated settlement.

Conflict Anticipation and Its Kin

This book has emphasized negotiation and mediation as processes to resolve environmental disputes. In the course of considering these approaches, we have necessarily encountered other processes like litigation and arbitration. Although these modes of dispute resolution presently are preeminent, there are other approaches. This brief note serves to introduce one of them—conflict anticipation.

Some observers use the terms *conflict anticipation, conflict avoidance,* and *conflict management;* others draw distinctions among them. In fact, there is no consensus on their precise definition. In some instances, the practice of conflict anticipation is not really different from the negotiation and mediation that we have seen in the cases throughout the book. People who claim to practice conflict anticipation (as opposed to mediation) simply are underscoring their belief in the importance of early intervention in disputes. Why, they ask, should intervenors wait until resources have been wasted, parties polarized, and opportunities lost before attempts are made to settle a dispute? Indeed, a retrospective look at most of the cases in this book suggests that oftentimes everyone would be better off if disputes could be settled early rather than late. Conflict can have significant costs. If these can be avoided or minimized, the resulting savings can be split among the contending parties.

No one argues with the desirability of the abstract goal of reducing the costs of conflict, but some environmental mediators are hostile to the concept of conflict anticipation both on practical and philosophical grounds. On the first count, they contend that mediation is often futile if it is attempted too soon. Until the issues are clearly defined, the disputants are at an impasse, and if some important deadline is imminent, intervention by third parties may complicate matters and actually prolong the controversy. Moreover, they argue, so long as skilled environmental mediators are in short supply, they must be very selective in the cases they handle. There may little gained by intervening in one case that will never be settled or in another where the parties can come to agreement on their own. Yet, until a case ripens, the intervenor may be unable to assess his or her potential contribution.

The second objection that some practitioners have to conflict anticipation is philosophical. Conflict, they note, is not necessarily a bad thing. As noted in the chapter 1, conflict may galvanize community organizations; it may put important issues on the public agenda; and ultimately, it may help produce more efficient and equitable outcomes. If conflict is nipped in the bud, it is argued, these important benefits may be lost. Conflict anticipation, they contend, can also be a mask behind which powerful corporate and governmental interests work to advance their own interests. Such groups may try to sponsor a mediator before a dispute actually emerges in order to "cool out" opposition. This danger, it is alleged, is greatest when conflict anticipation is practiced in site-specific disputes, where usually it will be the developer who initiates and who may underwrite the process. The problem is less pronounced in "policy dialogues" that deal with broad policy (such as mining, transportation, and consumption of coal), but even here the process can affect national goals.

The desirability of conflict anticipation is an issue that has sharply divided the small band of environmental mediators. It would be folly to pronounce judgment on which side is right, because so many of the arguments both pro and con are matters of assertion and fundamental principle. There is no way to prove, for example, that environmental disputes are ripe for settlement only when the parties have become polarized. The issue of ripeness is one of degree. Those who intervene late in a dispute cannot ignore its earlier history. Those who come in early may have to give greater attention to identifying parties and verifying facts, but these are matters that no mediator can ignore, no matter when he or she intervenes.

The philosophical debate pivots on the relative emphasis that is given to the costs of conflict and its benefits. No accountant or computer programmer can be expected to tell us whether the savings that can be realized from early intervention more than compensate for the benefits that can come from conflict. Though the question does not yield an easy answer, however, the debate has served the important purpose of bringing to the fore the larger social implications of intervention in environmental disputes. The issue of mediator responsibility, raised in chapter 10, really cannot be avoided.

Mandatory Negotiation

Introduction

Although judges have authority to promote negotiation when they see fit, present rules do not compel litigating parties to try to settle their differences before trial. The admonition about leading horses to water is often invoked to rebut proposals that people be legally required to negotiate. Yet, to state that people cannot be forced to volunteer is merely a truism and begs the question. The issue is not whether a horse can be forced to drink but whether having been led to a sparkling brook, it is more likely to do so of its own accord. (This distinction is more explicit in an earlier form of the proverb: "You may bring a horse to the river, but he will drink when and what he pleaseth.")

If negotiation of complex environmental issues can be beneficial not only to the parties immediately involved but to society at large, then should it not merely be encouraged but compelled? Parties who had been adamantly committed to litigation, if led to the bargaining table, might discover opportunities for mutual gain that will whet their appetite for settlement. A rule requiring negotiation spares either party from losing face by raising possible settlement. Are there any considerations that cut the other way?

Even if one endorses the principle of mandated negotiation, it may be difficult to implement. Practically speaking, it is meaningless to compel people to negotiate unless some sanction exists for failing to negotiate in good faith. But logically we cannot impose such a sanction unless we can recognize *bad* faith bargaining when we see it. What makes this such a difficult task is that posturing and intransigence are well-accepted tactics in the course of negotiation. Moreover, if the sanction for failing to reach agreement is simply a judgment imposed upon the parties, each side's willingness to reach agreement will be colored by their view of how they are likely to prevail if an order is made. Finally, if the supervisory authority is willing to impose sanctions on the parties or render a decision, many of the benefits of voluntary negotiation are lost. Specifically, the resulting "agreement" probably will not truly reflect the preferences of the parties. Many of these difficulties are illustrated by the following case and the note on collective bargaining that appears at the end of this section.

Case Study: Mandatory Negotiation under the Clean Air Act

Background

At present, there is only one federal environmental statute that compels negotiation, and it applies only to a narrow class of cases. Section 164(e) of the 1977 amendments to the Clean Air Act gives native American Indian tribes the right to request negotiations to resolve disputes over air quality on Indian lands. The Clean Air Act assigns air classifications to all regions in the country and establishes procedures for redesignating regions. In an area designated Class II, for example, moderate increases in the concentration of pollutants may be tolerated. If the same is redesignated Class I, however, a very small increase in sulfur dioxide (SO_2) or particulate matter may be regarded as a significant deterioration of the air quality and constitute a violation of the statute.

Section 164(c) of the 1977 amendments gives Indian tribes the right to redesignate regions which lie within the boundaries of reservations. The section 164(e) negotiations are intended to resolve controversies that arise over the redesignation of Indian lands:

> If any State affected by the redesignation of an area by an Indian tribe or any Indian tribe affected by the redesignation of an area by a State disagrees with such redesignation of any area, or if a permit is proposed to be issued for any new major emitting facility proposed for construction in any State which the Governor of an affected State or governing body of an affected Indian tribe determines will cause or contribute to a cumulative change in air quality in excess of that allowed in this part within the affected State or tribal reservation, the Governor or ruling body may request the Administrator to enter into negotiations with the parties involved to resolve such dispute. If requested by any State or Indian tribe involved, the Administrator shall make a recommendation to resolve the dispute and protect the air quality related values of the lands involved. If the parties involved do not reach agreement,

the Administrator shall resolve the dispute and his determination, or the results of agreements reached through other means, shall become part of the applicable plan and shall be enforceable as part of such plan. In resolving such disputes relating to area redesignation, the Administrator shall consider the extent to which the lands involved are of sufficient size to allow effective air quality management or have air quality related values of such an area.

The first, and apparently only, negotiation order under this provision took place in September 1979. It did not produce agreement, though progress was made on certain issues. That a settlement was ultimately reached in another setting reveals some shortcomings in this particular mandated process.

The dispute arose over the plan of Montana Power Company to add two 700 megawatt coal-fired electric power plants to its existing facilities at Colstrip, Montana, located about 20 miles from the northern Cheyenne Indian reservation. The Montana Power proposal divided the tribe. Some members opposed the project on the grounds that it would threaten the sanctity and tranquility of the Indian lands. Others viewed the project as inevitable and believed the tribe should use the leverage provided by sections 164(c) and (e) to ensure that the tribe got its share of the jobs that would be created.

Montana Power first applied for a state siting permit from the Montana Department of Natural Resources in 1973. By the time the case was finally settled in 1980, numerous state and federal agencies had become involved and several lawsuits had been filed. In 1977, the northern Cheyenne tribe exercised its rights under section 164(c) and changed the air quality designation of its tribal lands from Class II to Class I. The consequence of the redesignation was to impose much more stringent restrictions on allowable emissions from new sources of air pollution in the area. The redesignation also jeopardized Montana Power's pending application for a PSD (prevention of significant deterioration) permit that was pending with EPA. Without the PSD permit, the plant could not be built.

To resolve the dispute over the PSD permit, the northern Cheyenne invoked section 164(e), requesting that EPA initiate negotiations. Although many other parties and agencies had an interest in the outcome of the PSD issue, EPA invited only representatives of Montana Power and the northern Cheyenne to meet with the agency. On the eve of the first negotiation session, EPA announced that it would issue the PSD permit and that it would be subject only to conditions that would be the subject of the negotiations. There were, in fact, some important unresolved issues: The Indians, for example, were very concerned about particulate emissions that threatened the area's spectacular visibility. Yet, for those who opposed the plant, the announcement of EPA's decision was a great setback. The range of possible resolutions, at least within the context of 164(e) negotiations, had been narrowed considerably.

The participants in the 164(e) negotiations met for several days. Relations were cordial, but a review of the transcript of the proceedings shows that both sides regarded the meetings as an opportunity to make their particular case to EPA. Priorities were discussed, but there was little in the way of compromise or joint problem solving. A number of factors worked against the parties' reaching agreement. Once EPA announced its intentions to issue the PSD permit, the parties had little incentive to settle on their own. Uncertainty over EPA's actions could have led each side to be more flexible just as uncertainty over the actions of the god committee brought about a settlement in the Grayrocks Dam case discussed in chapter 3. The Indians were also represented by a lawyer who had relatively little authority to compromise; there were no Indians on the negotiating team.

The talks broke off without a settlement. EPA issued the PSD permit and imposed conditions that were intended to address the Indians' environmental concerns. Unsatisfied with EPA's decision, the tribe brought suit to overturn the permit. While these legal battles were being fought, other regulatory processes were going forward. In time, the Indians and the power company were negotiating again, but this time in the context of the issuance of the state siting permit. Although the terms of the PSD permit were limited to air pollution issues, the siting permit encompassed a much broader range of concerns. Moreover, the Montana State Siting Board had broad statutory authority to require compensation for the socioeconomic impacts of development. The negotiations thus covered not just pollution control, but jobs, employment training, and community relations. In 1980, the northern Cheyenne agreed to drop their suit in return for which Montana Power agreed to conditions imposed in the siting permit. These conditions provided job guarantees to the tribe, compensation for air pollution monitoring costs, a stipend to underwrite the cost of additional municipal services, scholarships, and other concessions.

The complete history of the dispute is described in Timothy Sullivan's "The Difficulties of Mandatory Negotiations," Chapter 3 in *Resolving Environmental Regulatory Disputes*, which was edited by Lawrence Susskind, Lawrence Bacow, and Michael Wheeler (Cambridge: Schenkman, 1983). Sullivan has identified several reasons for the failure of mandated negotiations. From the background described previously and the details Sullivan provides in the excerpt that follows, you should have enough information to evaluate his conclusions.

Analysis

Introduction. In the Colstrip power plant controversy, federally mandated negotiations failed to produce an agreement, whereas discussions required by the Montana Siting Board led to a negotiated settlement between Montana Power and the Northern Cheyenne. Although the EPA directed and supervised the 164(e) negotiations, little good-faith bargaining took place. On the other hand, unstructured bargaining between Montana Power and the Northern Cheyenne yielded a settlement. The EPA, with the assistance of a brace of technical and legal experts, failed to resolve the siting dispute, while representatives of Montana Power and the Northern Cheyenne produced a settlement that rested on nontechnical items, such as bus service and police protection. This section will examine the potential of mandated negotiations for resolving environmental disputes, the efficacy of quasi-judicial bargaining procedures, and the ability of the EPA to act effectively in local siting disputes.

Mandated Negotiations Rarely Produce Good-Faith Bargaining. Neither groups nor individuals will bargain in good faith unless they expect to realize gains or believe that they can avoid losses. In industrial relations, labor and management bargain most seriously when they believe that a settlement serves their interests more than a strike. A similar principle holds in bargaining between nations. Fred Ikle, in *How Nations Negotiate*, states, "Without common interests. there is nothing to negotiate for, and without conflicting interests there is nothing to negotiate about." (Harper & Row, 1976, p. 2)

In order for mandated negotiations to succeed they must offer gains to the negotiators. Section 164(e) of the Clean Air Act amendments expresses respect for the semi-sovereign status of Indians on their tribal lands by requiring that the EPA attempt to resolve disputes through negotiation rather than by unilateral administrative decision. Despite this statutory mandate for bargaining, however, little bargaining took place between the Northern Cheyenne and Montana Power during the 164(e) negotiations. This was partly the result of unrealistic expectations negotiators brought to the bargaining and partially due to the narrowness of the agenda. These factors made it impossible for the negotiators to bargain.

During the years of dispute, the lack of a shared understanding between Montana Power and the Northern Cheyenne of each other's rights and powers encouraged tests of strength rather than cooperative efforts. Montana Power had constructed two clean power plants at the Colstrip site with little opposition. It had no reason to expect that a proposed expansion of those facilities would produce a seven-year regulatory battle. Similarly, the Northern Cheyenne felt that the Clean Air Act Amendments had given them the power to stop the construction of the Colstrip power facility. As time passed, each side's hopes for a unilateral settlement were fueled by minor court and regulatory victories.

Even at the start of the 164(e) negotiations, six years after Montana Power's first permit application, each side still hoped for victory. Montana Power wanted to receive the PSD permit without further conditions or delays. Representatives for the Northern Cheyenne hoped to prevent issuance of the permit, to add new conditions to the permit that would force Montana Power to abandon the project, and to prepare the ground for future litigation. Thus, at the start of the 164(e) negotiations, each side hoped to achieve an outcome that was totally unacceptable to the other side.

Montana Power and representatives for the Northern Cheyenne not only had different views of what would constitute an acceptable settlement, but they also held different views of how the EPA would conduct the negotiations. Representatives for the Northern Cheyenne expected these meetings would determine only the ground rules for future negotiations. Thus, they came unprepared to move beyond their initial bargaining positions.

The announcement by [EPA Regional Administrator] Williams of his intention to issue a PSD permit to Montana Power forced the negotiators to assess the bargaining session more realistically. Since the EPA's power to issue permits was so great, Williams could limit the discussions to permit conditions and issues important to the EPA. This also forced the competing sides to reassess their expectations of the outcome.

Besides limiting the range of possible outcomes, Williams' announcement focused bargaining on a narrow set of environmental issues; negotiations sought to determine acceptable levels of emission of the three major pollutants: TSP [total suspended particulates], NO_x [nitrous oxides], and SO_2 [sulfur dioxide]. The interests of Montana Power and the Northern Cheyenne were directly opposed on these issues, and so the narrow focus polarized the negotiators. The Northern Cheyenne insisted on stricter emissions standards and Montana Power resisted them. Similarly, to conclude the bargaining, the EPA proposed additional permit conditions.

Although these environmental issues were highly relevant to the conflict, both parties had other major concerns. For example, the Northern Cheyenne feared the adverse socio-economic impacts that could accompany energy development. The tribe desired the jobs that construction could bring. Relaxation of pollution restrictions would not guarantee the Cheyenne an acceptable share of construction jobs. Similarly, Montana Power cared about its relations with the community, labor relations, and customer service. Focusing on technical issues obscured the parties' real concerns and reduced their ability to explore common ground. The EPA's interpretation of the Clean Air Act prohibited the participants in the 164(e) negotiations from addressing these other issues. The narrow agenda which framed the proceedings precluded opportunities for parties to discover areas of agreement and negotiate a settlement.

In contrast, the negotiations over the Montana siting permit provided the disputants with an agenda that allowed both sides to develop a common interest in settlement. The siting permit required that the Northern Cheyenne and Montana Power develop an employment program for the tribe and mitigate the impacts of the development of facilities. These requirements allowed the parties to link their positions on environmental issues to a larger agenda in which mutual concessions could produce a settlement. Through negotiations, the disputants produced an agreement that benefited both sides: Montana Power gained a withdrawal of the Northern Cheyenne's legal challenges to the Colstrip power plants and a better reputation in the community; the Northern Cheyenne received promises from Montana Power to control impacts of development and to give the Cheyenne a share of the construction jobs.

An expanded agenda that included the range of relevant issues allowed the negotiators to create a mutually beneficial settlement.

Quasi-Judicial Resolution Adversely Affects Negotiation. Under section 164(e) of the Clean Air Act, the EPA Administrator has the authority to resolve a dispute when negotiations fail. The focus of power in [regional administrator] Williams and [regional counsel] McClave caused the 164(e) negotiations to resemble a legal proceeding more than a negotiation session. Williams functioned as a judge and McClave, as a hearing officer. This assignment of powers gave a judicial character to the negotiations and affected the style of the proceedings, the choice of bargaining representatives, and the development of bargaining positions in ways that discouraged bargaining.

The style of the 164(e) negotiations was formal and legalistic. In the 164(e) negotiations, lawyers represented both the Northern Cheyenne and Montana Power. Furthermore, a lawyer for the EPA chaired the meetings. Negotiations were conducted formally, and the EPA even kept a verbatim transcript of the proceedings. Expert witnesses testified to support the position of each side. The final outcome was a legal document that carried the force of law. Finally, EPA officials decided the terms of the agreement.

This manner of conducting the negotiations discouraged direct interaction between the principals of the dispute. The negotiating lawyers tied their positions to the terms of the Clean Air Act and the powers that it gave them to appeal EPA decisions. Fundamentally, these were legal positions. Despite the importance of the final agreement to the Northern Cheyenne, [the] Indians did not participate in the negotiations. Representatives for the Northern Cheyenne had expected only preliminary meetings, and so tribal members did not join the sessions; the judicial and technical nature of the proceedings made it unlikely that members of the tribe could contribute significantly to the bargaining. The competing lawyers presented the positions of Montana Power and the Northern Cheyenne like legal cases. Thus, the 164(e) negotiations became a permit hearing rather than true bargaining sessions.

Although EPA officials realized that the rigid format and narrow agenda would inhibit negotiation, they believed that the Clean Air Act Amendments required such an approach. Unlike the Environmental Impact Statement of the National Environmental Policy Act, which requires consideration of the socioeconomic impacts of proposed federal actions, the PSD permit process is a more narrow administrative regulatory procedure. It is not clear how the EPA could have broadened negotiations to consider these other impacts at this late point in the conflict.

The Montana state siting process, on the other hand, required discussion and legitimized a broad framework of issues around which the disputants could successfully negotiate a settlement. This suggests that reliance on negotiations to resolve environmental disputes over development projects in the future may require statutory language that legitimizes local issues as considerations in a site review. The Environmental Impact Statement may prove to be a more appropriate vehicle than a regulatory permit process for resolving such disputes.

Local Disputes Require Local Resolution. One may question the ability of a federal environmental agency, often located far from the community in question, to settle a local siting dispute. In the Colstrip controversy, the EPA failed to settle a dispute between local and regional groups. It was within the state siting board's framework that members of the Northern Cheyenne tribal council and Montana Power resolved their dispute. The negotiated agreement included local and practical issues, such as bus service to the construction site, automatic paycheck deposits at the reservation bank, and hiring procedures to foster the employment of Indians. This dispute and the terms of its settlement suggests that environmental controversies can obscure more immediate disputes over economic development and local control of growth. In this case, the Northern Cheyenne were concerned with jobs and the boom-town effects of development as well as the environmental impacts of the power plants.

The EPA's mission, environmental protection, may further limit its ability to act in local disputes. As an institution, it is legally and philosophically unprepared to balance the values of local communities and developers. Nevertheless, federally mandated programs designed to ensure maintenance of environmental standards invariably draw the EPA into local conflicts. In the Colstrip controversy, the EPA sought refuge in long procedural and technical reviews. Although this showed the agency's commitment to giving each side a full hearing, it delayed construction of the plant for many years and increased the ultimate costs to all involved. In the end, the situation left many of the disputants unsatisfied.

The Northern Cheyenne, although they appreciated the power given to them by the Clean Air Act, felt that they should control development in the vicinity of the reservation. Montana Power, although grateful for receiving a permit, felt that it had designed a clean plant and that the delays stemmed from the EPA's bias against development. Thus reliance on procedures brought only criticism to the EPA.

The existence of opposing local values that are not balanced by statutes creates a complex dilemma that often prevents effective EPA action. Distributional issues that underlie a conflict often result in inaction. Since the United States Constitution gives the legislature the power to tax, the EPA lacks the authority to resolve satisfactorily disputes that distribute benefits and costs to a community. Without an act of either the legislature or the electorate, only a voluntary agreement between the disputants will end the conflict in a constitutionally acceptable way. Only local resolution of the dispute can produce such an agreement.

Study Questions on Colstrip

1. Suppose you are an EPA regional administrator. An Indian tribe has just requested negotiations pursuant to section 164(e) to resolve a controversy over a proposed power plant that is to be located adjacent to the Indian reservation. How would you conduct the 164(e) negotiations? Would you do anything differently from Williams and McClave?

2. Suppose that section 164(e) required that the negotiations be supervised not by EPA but by a neutral mediator appointed by EPA. Do you think the outcome would have been any different? Why or why not?

3. Suppose you could amend section 164(e) so that it created incentives for the parties to bargain in good faith. How would you amend it? Would you include sanctions for "bad faith" bargaining?

Cross-Reference Problem

Undoubtedly, there are some people who are so stubborn they will refuse to honor an order, even when doing so would clearly be in their best interests. All of us may have this tendency to some degree when we value the principle of independence more than the specific issue at stake. It is hard to believe, however, that people who can see significant potential gains in negotiation would nevertheless decline to engage in it simply because they had been required to do so. There may, of course, be specific aspects of the required process—such as the verbatim transcript requirement in Colstrip—that may discourage participation in some instances, but the mandate itself should not necessarily be counterproductive.

An examination of cases may identify more specifically the problems that may be generated by a mandate. Consider the following questions as they apply to the Brown Paper case (chapter 4) and the Holston River case (chapter 5) presented earlier.

1. Would a law requiring the parties to negotiate in those cases have had any effect for good or for ill?

2. If you were to draft a mandatory negotiation order applicable to such cases, at what point would you require the parties to sit down and negotiate?

3. What new issues or problems would a mandatory negotiation requirement raise in the Brown Paper and Holston River cases?

Good Faith Bargaining

Although section 164(e) of the Clean Air Act Amendments authorizes EPA officials to negotiate, the law specifies neither the negotiation procedure nor the duties of the parties. It does, however, acknowledge that the parties may fail to reach agreement, and it empowers EPA to break impasses when they arise.

Advocates of mandatory negotiation of environmental disputes have turned to collective bargaining to illustrate the mechanics of compulsory negotiation. Indeed, when private companies voluntarily come to the bargaining table to meet environmentalists and government regulators, they likely bring with them habits and expectations that have developed over decades of bargaining with unions within the provisions established by the National Labor Relations Act. This act requires bargaining in "good faith" and defines it as "the mutual obligation of the employer and the representative of the employees to meet at reasonable times and confer in good faith with respect to wages, hours, and other terms and conditions of employment." The definition, however, provides that "such obligation does not compel either party to agree to a proposal or require the making of a concession." Unlike section 164(e) of the Clean Air Act Amendments, the NLRA contains no provisions for breaking deadlocks.

Although the good faith provision has been part of the law since 1947 (and was predated by a similar agency requirement), it remains controversial. Although there is now little debate about the value of collective bargaining, some observers doubt that the process is advanced by the good faith requirement. The rule, in prescribing a subjective state of mind, is said to be inherently infirm. Not only is it hard to control people's attitudes, but courts may have difficulty in determining whether a bargainer has acted in good or bad faith. Because people ordinarily will not implicate themselves, judges necessarily must evaluate objective conduct in order to infer subjective motives. Inquiries that are troublesome for psychiatrists are no less so for judges.

It is not surprising that there has been considerable litigation over claims of bad faith. There is no landmark decision that resolves the issue; often the cases turn on specific facts. Some cases involve acts that are claimed to be proscribed, whereas in others, bad faith is alleged in an entire pattern of conduct. For example, it is clear that an outright refusal to meet would be illegal because the law requires labor and management to meet at reasonable times. The far more common case, and one that is harder to decide, involves a party who is willing to meet but appears to be practicing delay. What is a reasonable negotiation schedule? Bad faith has sometimes been found when a party has insisted that the other side change the composition of its bargaining team, has reneged by refusing to sign a document embodying terms previously agreed upon, or has refused to provide information that was necessary to the other side.

The line between proper and improper conduct can be elusive. Commentators sometimes state that the good faith provision requires that a party listen to the other side with an open mind; yet courts have consistently ruled that the law does not compel either party to agree to a proposal or to make a concession. In *Chevron Oil v. NLRB*, the Fifth Circuit Court of Appeals declared that "adamant insistence on a bargaining position . . . is not in itself a refusal to bargain in good faith." Yet such adamance may be unlawful if it is part of a larger pattern of intransigence.

Some critics argue that, at most, the law only requires ritual. They point to the well-known General Electric case in which a federal appeals court upheld the National Labor Relations Board's condemnation of "Boulwareism"—the tactic of presenting a firm package at the outset. If a company has calculated the best offer it can make, what purpose, they ask, is there in requiring it to waste time by putting forth less attractive proposals? The court, however, found the practice to violate the premise that collective bargaining represents a mutual search for desirable outcomes.

Even if good faith could be defined more precisely, there might still be difficulty in determining when the requirement applies. Although good faith bargaining requires the parties to meet and state their positions and responses, courts have held that the rule does not prohibit the parties from exerting economic pressure outside the bargaining room, no matter how disabling the consequences. Strikes and lockouts are permissible, the good faith requirement notwithstanding. For better or worse, the requirement to bargain in good faith does not equalize the bargaining power of the parties, nor does it control the way in which it is exerted. Finally, we note that the courts have struggled with developing an appropriate remedy for a breach of the duty to bargain in good faith. Typically, the NLRB will issue an order compelling the derelict party to bargain, and occasionally a fine will be imposed. However, given the difficulties inherent in determining whether a party is in compliance in the first place, it is open to question whether these remedies are effective.

This catalog of some of the difficulties that have arisen in mandating labor negotiation is not intended as a brief that is arguing for revision of collective bargaining statutes and regulations. Nevertheless, proponents of mandatory environmental negotiation must acknowledge the problems that have arisen in the labor context and try to mitigate them. There may be aspects of environmental disputes, however (most notably the difficulty in identifying the interested parties), that make it an even less promising field for this device.

Encouraging Negotiation through Incentives

Introduction

The federal law in the Colstrip case mandated negotiation but only as an adjunct to litigation. Section 164(e) of the 1977 amendments to the Clean Air Act provided that if the parties could not settle their differences among themselves, then the EPA administrator could break the impasse. If any of the parties opposed his decision, however, they could challenge it in court; indeed, the Northern Cheyenne did appeal the EPA's issuance of the PSD permit.

One of the lessons of the Colstrip case (and decades of experience under the National Labor Relations Act) is that simply ordering people to negotiate does not ensure that they will negotiate in good faith. There may be other ways, however, of getting people to bargain. Carrots may be more effective than sticks. The Massachusetts Hazardous Waste Facility Siting Act uses incentives and compensation to encourage negotiation between developers of hazardous waste facilities and potential facility opponents. In contrast to section 164(e), it limits legal appeals, for the most part, to procedural issues. It provides prospective host communities with an incentive to bargain in good faith by holding out the prospect of compensation for any community willing to accept a facility. The entire law is structured to encourage consensual agreement through face-to-face bargaining. As you read the following description of the act, consider how the law responds to the various negotiating issues we have identified in earlier chapters.

An Overview of the Massachusetts Hazardous Waste Facility Siting Act

Background

The Massachusetts Hazardous Waste Facility Siting Act was adopted in 1980 after the defeat of a number of proposals to site new hazardous waste facilities by local community opposition. Prior to adoption of the act, the legislature flirted briefly with a bill that would have preempted local authority to use zoning and police powers to exclude unwanted facilities. Preemption was rejected, however, after legislators from three communities under active consideration for hazardous waste facilities convinced their colleagues to exempt their communities statutorily from further consideration as possible sites. This political show of force prompted the legislature to turn to a siting process incorporating incentives as a strategy to overcome local opposition.

There are five important elements to the Massachusetts act. First, the act gives a developer the right to construct a hazardous waste facility on land zoned for industrial use if the developer obtains the required permits and completes a negotiated or arbitrated siting agreement with the host community. Second, the act limits the ability of local communities to exclude hazardous waste facilities without first showing that such facilities pose special risks. Third, the state provides potential host communities with technical assistance grants to promote local participation in the siting process and effective negotiation with developers. Fourth, the act requires that deadlocks between developers and host communities be submitted to arbitration. Finally, the act provides compensation to abutting communities that are likely to be affected by new hazardous waste facilities in adjacent jurisdictions. (A more complete description of the Massachusetts Hazardous Waste Facility Siting Act and its implementation appears in Lawrence Bacow's and James Milkey's "Overcoming Local Opposition to Hazardous Waste Facilities: The Massachusetts Approach," *Harvard Environmental Law Review* June 1982.

The State Role in the Siting Process

The developer and host community have the primary roles in the siting process; state agencies oversee the process but have no independent authority to site facilities or to override local decisions. Three state agencies share that oversight role. The Department of Environmental Management (DEM) is responsible for planning and is charged with assessing the state's requirements for hazardous waste storage, treatment, and disposal facilities, and for attracting developers to the state. The Department of Environmental Quality Engineering (DEQE) oversees facilities once they become operational and grants the necessary permits, licenses, and enforces the relevant environmental and safety regulations. Because of their mandates, these two agencies lack the neutrality necessary to referee negotiations between developers and communities. Consequently, the legislature created a new agency—the Hazardous Waste Facility Site Safety Council—to oversee the negotiation process. The council has 21 members and includes representatives of all parties involved in and affected by the siting of hazardous waste facilities.

Initiating the Siting Process

A prospective developer initiates the siting process by filing a notice of intent (NOI) with the council. The NOI describes the prior experience of the developer in the hazardous waste field, the general characteristics of the proposed facility, and how the developer intends to finance it. The developer need not have a specific site in mind when he or she submits the NOI; he or she can submit a nonsite-specific NOI and rely upon the siting process to identify potential sites.

Within 15 days of receiving a complete NOI, the council must decide whether the project is "feasible and deserving" of state assistance. If the council votes affirmatively, both the host and abutting communities become eligible to receive technical assistance grants to support their participation in the siting process. The feasible and deserving review is intended to be a rough screen used to eliminate projects that are technically unsound, projects that are unnecessary given existing in-state disposal and processing capacities, projects that are precluded by existing law, and projects proposed by developers who are either disreputable or financially insecure.

Within 30 days of the filing of a NOI, a local assessment committee (LAC) is formed to represent the interests of the host community in negotiations with the developer. The LAC is chaired by the community's chief executive officer, typically its mayor. The chief of the fire department and the chairmen of the local board of health, the planning board, and the conservation commission also serve on the LAC, as do four members of the area most immediately affected by the proposed facility. These latter members are elected by a vote of the ex officio members. In addition, up to four other members may be named to the LAC by the chief executive officer, provided they are confirmed by a majority of the community's local legislative body.

The Negotiating Process

After the developer has submitted environmental and socioeconomic data in the form of a preliminary project impact report, negotiations begin between the developer and the LAC. The negotiations are intended to result in a formal "siting agreement" that describes the measures that the developer will take to mitigate any adverse impacts associated with the facility as well as any compensation that might be paid. A facility cannot be constructed without an approved siting agreement. The statute states the "terms, conditions, and provisions" that the siting agreement must include, in addition to listing some optional provisions. However, because these "requirements" are general in scope and permissive in tone, they serve more to illustrate the range of potential negotiations rather than to constrain the final result.

Arbitration

If the developer and the host community fail to establish a siting agreement, the state council may declare an impasse and compel the parties to submit all unresolved issues to what the statute calls "final and binding arbitration." If the parties fail to agree on the choice of an arbitrator, the council may appoint one. The act itself contains no explicit criteria to be employed by the arbitrator in rendering a decision other than to say that the arbitrator shall "resolve the issues in dispute between the local assessment committee and the developer." Judicial review of the arbitration award is available only to show fraud or partiality by the arbitrator.

State and Local Permit Requirements

Before constructing a hazardous waste facility in Massachusetts, a developer must obtain a license from DEQE as well as certification of the site from the local board of health. By law, DEQE cannot issue a license unless it finds that the facility

> does not constitute a significant danger to public health, public safety, or the environment, does not seriously threaten injury to the inhabitants of the area or damage to their property, and does not result in the creation of noisome or unwholesome odors.

Communities cannot impose new permit requirements on hazardous waste facilities after the effective date of the act. As a result, in most jurisdictions the only local permit required of a developer is site certification by the local board of health. At the same time the legislature adopted the siting act, however, it also limited the circumstances under which a local board of health could refuse to certify a site. The local board must approve that a site must be certified if the proposed facility "imposes no significantly greater danger . . . than the dangers that currently exist in the conduct and operation of other industrial and commercial enterprises in the commonwealth not engaged in the treatment, processing or disposal of hazardous waste, but using processes that are comparable."

The statute also limits the power of localities to use zoning to exclude unwanted facilities. The zoning enabling act now permits hazardous waste facilities to be built as a matter of right on land zoned as industrial at the time a developer initiates the siting process by filing an NOI. Thus, a municipality cannot subvert the project by subsequently rezoning land. The state retains the power to seize a site through eminent domain, but it can only do so with the approval of the local city council, board of aldermen, or board of selectmen.

Declaration of an Operational Siting Agreement

After completion of the siting agreement either through negotiation or arbitration, the council reviews the agreement. Recently adopted regulations require the council to approve the agreement if it contains all provisions mandated by the regulations and complies with the terms of the act. If the council approves the agreement, the developer prepares a final project impact report, which is similar to the preliminary report but includes comments received by the developer, responses to these comments, a copy of the siting agreement, and relevant data derived from the negotiations. After the appropriate agencies approve the final report, the council decides whether to declare the agreement "operative and in full force and effect." This declaration establishes the siting agreement as a "nonassignable contract binding upon the developer and the host community, and enforceable against the parties in any court of competent jurisdiction."

Abutting Communities

Abutting communities are also directly involved in the siting process. They are invited to all briefing sessions conducted by DEM and are eligible for technical assistance grants from the council. Moreover, abutting communities may also petition the council for compensation to be paid by the developer for

> demonstrably adverse impacts . . . imposed upon said community by the construction, maintenance, and operation of a hazardous waste facility in a host community.

Unlike compensation for the host community that is determined through bilateral negotiations with the developer, the council fixes the compensation to be paid abutting communities after a public hearing. If the abutting community is unsatisfied with the council's award, it may request that the compensation issue be submitted to impartial arbitration. The developer has no comparable right of appeal.

STUDY QUESTIONS ON THE SITING ACT

1. What incentives does the law provide for host communities to abandon obstructionist tactics and negotiate in good faith? Does the law provide comparable incentives to developers?

2. Is the procedure for resolving impasses between the developer and the host community (i.e., binding arbitration) likely to encourage or inhibit reaching agreement through negotiation?

3. Why do you think the drafters of the act created a local assessment committee to represent the interests of local communities in the siting process? Why not simply let the mayor or chairman of the board of selectmen negotiate on behalf of the community?

4. The act does not require a developer and a host community to seek the services of a mediator. Is this an important omission? Should the Site Safety Council attempt to mediate their differences?

5. You are the mayor of a Massachusetts town. You have just been notified that a developer has filed the necessary papers with the state indicating interest in building a hazardous waste facility in your town. The press has asked for your reaction. Given your understanding of the negotiating incentives created by the

act, what response do you give? Can you think of any actions that a developer might take in announcing his or her intentions that might influence your response?

6. The Massachusetts statute was drafted to meet one very specific problem: the siting of a hazardous waste facility that could serve Massachusetts industry. As one person connected with its implementation has observed, "It only has to work once to be a success." Nevertheless, could a similar process be used to solve other types of problems? Would any of the cases we have studied have been appropriate for this type of process?

The Uses of Compensation

Many environmental disputes really are disputes over whether something should be built; the Grayrocks, Foothills, and Colstrip cases are all good examples. Powerful interests align themselves on each side asserting that their favored alternative is the only rational choice. Such disputes frequently are bitter because as long as the only options under consideration are "build" or "don't build," one side's losses are the other side's gains. In the lexicon of chapter 2, these are zero-sum disputes. The parties seeking to preserve the status quo (usually the environmental interests) will fight as long as their resources hold out.

As the Grayrocks case from chapter 3 illustrates, however, compensation can be a helpful tool for resolving such disputes. In effect, it introduces a third alternative into the discussion: Build but with compensation. This approach is especially attractive in disputes involving projects that, like hazardous waste facilities, provide benefits that are distributed regionally and social costs that are concentrated locally. In such cases, the local social costs—noise, pollution, latent threats to public health, and the like—are likely to outweigh any modest benefits that may accrue to the host community. As a result, proposals to build such facilities inevitably provoke intense local opposition. Because local opponents stand to lose much more if a facility is built in their neighborhood than do the potential beneficiaries of the facility, they will invest large amounts of time and money to thwart its construction. By contrast, the potential beneficiaries (who each have relatively little at stake) will do little if any lobbying for the site under consideration. The end result is that these facilities do not get built.

Compensation attempts to redistribute some of the benefits from the winners to the losers to offset the locally concentrated social costs. In theory, if the compensation is large enough, it should leave the recipients better off than with the status quo. Thus, compensation should provide an incentive to negotiate. For a more detailed discussion of the rationale for compensation in facility siting disputes see Michael O'Hare, Lawrence Bacow, and Debra Sanderson's *Facility Siting and Public Opposition* (New York: Van Nostrand Rheinhold, 1983).

For compensation to be effective, it must succeed in assuaging the concerns of those opposed to the proposed facility. In the Grayrocks Dam case from chapter 3, the power company obtained the acquiesence of the conversationists by guaranteeing the streamflow of the North Platte River, thus ensuring the continued well-being of the endangered whooping crane. But such guarantees are not always feasible. Had it been impossible simultaneously to build the dam and preserve the habitat of the crane, it is unlikely that the conservationistis would have been influenced by an offer of compensation. Often, environmental activists are motivated not out of self-interest but out of genuine concern for the long-term protection of parts of the environment that they may never enjoy personally. (People donate money to save trees and baby seals in parts of the world they will never see.) In such cases, direct offers of compensation not only will be rejected but will provoke righteous indignation. What is offered as compensation may be regarded as a clumsy attempt at a bribe. Indeed, in the Grayrocks case, the National Wildlife Federation rejected out of hand the power company's first cash offer of compensation. Only when the money was used to create a trust fund for the crane did it become acceptable. To the extent that compensation agreements can be restructured so that the benefits accrue directly to the environment, they stand a far greater likelihood of acceptance by conservationists who view their role as that of a steward for nature's bounty.

Environmentalists. however, are not the only ones who are uncomfortable about accepting offers of compensation. Although people implicitly make trades involving environmental amenities all the time (e.g., they live next to airports, stadiums, and power plants because they can obtain more housing for their dollar), they still sometimes balk at compensation when it is offered directly. Drawing on the work of O'Hare, Bacow, and Sanderson (1983), we have described below a number of reasons why.

First, not everyone is alike. Some people are comfortable exchanging amenities for compensation while others are not. Unfortunately, people who dislike these exchanges tend to congregate in high amenity locations. While people who already live next to airports might be willing to bargain over noise, people who live in rural areas usually do so because they enjoy their solitude.

Second, people are sometimes reluctant to give up environmental quality in return for compensation because of the irreversible nature of the decision. People know what their neighborhood is like right now. They can't be certain what it will be like after the developer constructs his 50,000-seat stadium. Moreover, much of the information about the future environment comes from the developer, a source not to be trusted. (The rationale for providing technical assistance grants to communities under the Massachusetts Hazardous Waste Facility Siting Act is to provide them with a source of expertise and information that is independent of the developer.) Given a choice between the status quo and compensation which is tied to a permanent and uncertain future environment, many people elect the status quo.

Third, where risks to human health are involved, offers to compensate are frequently rejected as morally repugnant. Most people view their continued good health as an entitlement that is not for sale. While they may at times voluntarily place it in jeopardy by smoking cigarettes or engaging in dangerous recreational activities like skydiving, these decisions generally do not involve careful calculations of benefit and cost. For example, relatively few cigarette smokers have consciously decided that the joys of smoking are worth five to seven years of foregone life expectancy. Instead, the decision to smoke is more frequently than not an emotional rather than an intellectual decision.

Finally, even when people are amenable to exchanging compensation for environmental amenity, the collective nature of the decision may frustrate the exchange. Usually a developer must negotiate a compensation agreement with representatives of a number of different groups simultaneously. In the Grayrocks case, the power company dealt with the Wildlife Federation as well as representatives of Nebraska and Wyoming. If each of these groups had had a different view of what constituted a fair exchange, the trust fund settlement might not have been forthcoming. What made the settlement possible is that it recreated the status quo, albeit in a slightly modified form, and this proved to be an outcome acceptable to all parties concerned.

13

Epilogue

Introduction

There have been a number of themes that have recurred throughout this book: the need to identify the incentives of the parties to settle a dispute, the problem of facilitating the participation of all affected parties, the means by which compliance to agreements can be secured, and so forth. The textual material and cross-reference problems that have followed the principal cases have been intended to emphasize the breadth of these themes. There have been several issues, however, that—though implicit in much of the preceding material—deserve separate treatment. These are considered next.

Small-Scale Disputes

All of the cases presented in this book represent major controversies. Because so much was at stake in every instance, each touched the lives of many people. Typically, the health, economic security, and values of thousands of individuals pivoted on whether the proposed development would go forward, and if so, under what terms.

The broad scope of these cases is also apparent in their geography. Whether the Brayton Point power station (see chapter 8) would be allowed to convert from oil to coal-fired generators potentially affected both air quality in large parts of Masschusetts and Rhode Island and utility rates throughout New England. Indeed, in the 301(h) rulemaking case (chapter 11), some of the parties lived on opposite sides of the globe.

The cases were also major ones in respect to the degree of government involvement. In each one the Environmental Protection Agency was a principal actor. In many, one arm of government was in conflict with another. In the Jackson Hole case (chapter 7), for example, the battle was between a town and a county, whereas the Foothills case (chapter 9) pitted the federal government against an array of state and local agencies. As several cases demonstrated, even within a single agency or organization there can be division and discord.

These various cases were selected for the book because complex disputes provide a fertile arena in which to study negotiation; many issues can be identified and explored in the context of a single case. It is fair to ask, however, whether the lessons that are illustrated by these major confrontations apply to disputes of a smaller scale? There are, after all, countless environmental controversies that are played out in local neighborhoods. They may involve a landowner who is seeking to have property rezoned from residential to commercial—over the vehement objections of neighbors. Another fight may be over a proposed extension of a sewer line: proponents may wish to improve groundwater quality by eliminating inadequate septic systems, whereas opponents may fear tax increases and new development. In a third instance, conservationists may be up in arms if a planned road widening threatens stately oak trees.

Such disputes are played out daily in cities and towns throughout the country—indeed, the world. Perhaps each one may involve just a handful of people and have consequences that are felt in only a small area. Yet, collectively these cases have a significant impact on community life. If it were possible to resolve some of these disputes more efficiently and equitably, the public and private gains could be substantial. How, then, is the negotiation of these small disputes similar to or different from the settlement of more complex disputes that have been described in this book?

The scale of disputes can significantly affect the pace and tone of negotiation—notwithstanding some important differences that are cataloged later. The basic structure of environmental disputes is similar whether they are big or small. As a result, the mode of analysis that has been developed in this book should be as relevant to neighborhood controversies as it is to cases that cross state and national boundaries.

Differences of scale do not necessarily make smaller disputes easier to settle than larger ones. For example, small cases, by definition, usually involve fewer parties than do large ones. In a neighborhood dispute, the parties often have dealt with one another in the past and are likely to cross paths again. As noted in chapters 4 and 7, a continuing relationship may enhance prospects for agreement; yet, in localized disputes, such closeness may complicate matters. For example, neighbors often bring to the negotiating table the baggage of past disagreements. Statements may be made in haste and anger, frequently within earshot of the entire neighborhood. Saving face and honor may prove to be the most important (and elusive) goal for each of the parties. When personality dominates substance, outcomes that are good for both sides are likely to be ignored. To put it another way, where the parties live in close proximity, it may be very difficult to separate the people from the problem. Indeed, the people may be the problem.

To be sure, character and temperament play important roles in large disputes as well. The bargaining positions and strategies of corporations, agencies, and special interest groups may reflect the personalities of their leaders. Yet, the fact that many people usually participate in formulating internal policy often moderates the influence of any one individual. By contrast, it is easier for the principals in small disputes to get sidetracked with ad hominem charges and countercharges that may have little relevance to the issue ostensibly in dispute. As a result, a mediator may prove particularly helpful. The intervenor need not be a professional mediator, of course, but rather a third party who is trusted and respected by both sides. For the same reason, it may behoove the parties to have agents negotiate on their behalf, particularly if their animosity runs so deep that the principals cannot stand to sit in the same room with each other.

Small-scale disputes also differ from their larger counterparts in that lines of communication usually are shorter; hence things can happen much faster. In the Colstrip, Holston River, and Foothills cases, negotiators often had to check back with their parent organizations for instructions. This sort of consultation sometimes serves as a device to gain time to evaluate new proposals or simply to stall. When parties negotiate for themselves, however, new offers may require immediate response; a party to a small dispute may have to be better prepared when entering into a bargaining negotiating session.

Small disputes differ from large ones in that they typically turn on somewhat less complex technical issues (though this difference is really just one of degree). Many of the cases described in this book involved sophisticated technical questions (e.g., the assimilative capacity of the Holston River in Tennessee; dispersion of air pollutants in Berlin, New Hampshire; the relationship between growth, water supply, and pollution in Denver, and so forth). By contrast, small disputes usually raise narrower questions, although these may not necessarily yield concrete answers. The issue may be one of engineering (will leachate from the proposed new town landfill contaminate the town well); law (does the smoke that blows into Smith's house from the Jones's new wood-burning stove constitute a nuisance); or policy (should the town permit development of the land adjacent to the town hall or should it maintain it as open space). As in larger cases, technical expertise may be helpful to an extent, but in either instance the parties may ultimately have to cope with uncertainty over important points. In small cases, moreover, the parties may lack the funds or sophistication to retain an army of experts. On one hand, this may make the mechanics of negotiating simpler because the number of participants at the bargaining table will be limited. Yet on the other, deferring to experts can be an effective way of saving face when an explicit concession is not in the offing.

Unlike the principal cases discussed in the book, smaller disputes are more likely to fall into familiar patterns. A New England town, for example, will receive several dozen requests for zoning variances in the course of a year. Such cases are more similar than they are different. In each one, a landowner seeks dispensation from a provision of the zoning code that allegedly imposes a personal burden. Neighbors worry about adverse impacts, and zoning boards of appeal are concerned with acting consistently with planning goals. Because these cases are routine, precedents exist for their resolution. Moreover, the resolution of current disputes can expect to influence the disposition of future cases. Thus, the parties negotiate while looking back and looking ahead. In contrast, just about all of the cases that we have analyzed and discussed are unique. Cities other than Denver may debate water needs and growth controls, but because the circumstances in the Foothills case are inevitably different from those elsewhere, the outcome in that case cannot be read as a precedent for other such disputes. In major cases, there thus may be a greater degree of uncertainty.

In short, differences in the number of parties, the relationships among them, the complexity of the issues, the role of experts and mediators, and the availability of precedents all give the negotiation of small-scale disputes a distinctive look. Nevertheless, the analytic principles described in the book are just as relevant to the small cases as they are to major controversies. Whether acting as a participant in a particular dispute or serving as a formulator of policy for the handling of a class of many small cases, one still must identify the incentives that the bargainers face, analyze the structure of the problem they confront (is it zero or nonzero-sum, two or multiparty, one time or repeated, and so forth), deal with problems of representation, evaluate mediation, invent solutions, fashion compensation, and ensure compliance.

It is hard to imagine a case in which the parties would not be better off if they did not engage in this sort of thinking. Indeed, the return on investment in such analysis may be even greater in small disputes where the initial temptation is to react instinctively, than in large complicated problems where the consequences of every move must be planned and considered. The two problems that follow are intended to test your ability to apply your analytic skills to localized environmental disputes. The first describes a problem of environmental negotiation that has been undertaken but not completed; the second involves an attempt to use mediation to avoid unnecessary confrontation.

PROBLEM 1

Instructions

The following problem is based on an actual case involving a proposed development in New York State. The information is drawn from a newsletter circulated by the Mohonk Trust, an organization that manages a 5,000-acre preserve that is close to the proposed development site. Because the organization depends on voluntary contributions for its support, it deliberately chose not to take a position on the project, though it has expressed concern about possible adverse environmental impacts. It also has provided technical information on local soil and water conditions to all who have requested it.

The background presented later describes the case as matters stood in mid-1980. You should analyze the case from the points of view of principal parties and as a hypothetical disinterested state official: What reasons might the parties have to settle their differences through negotiation? What obstacles might interfere with this goal? How might they be overcome? If you must make assumptions of fact to formulate an answer, be sure to consider the implications of other premises.

(As an alternative, the case may be used as the basis for a role-playing exercise, with people assuming the positions of the various parties. The problem that follows this one deals more explicitly with mediation, but it should be instructive to ask yourself how mediation might be initiated in this case as well.)

Background

At issue is the fate of the old Lake Minnewaska resort, which was originally developed in the late nineteenth century. The property lies at the geographical center of the 25,999-acre Shawangunk Mountain natural area, which is bounded on the north by the Mohonk Trust Preserve and on the south by the state park. Two hotels operated on the site for many years, but gradually the property went into decline. In 1970, a bankruptcy foreclosure sale was averted at the eleventh hour by the sale of a 7,000-acre parcel that became Minnewaska State Park. In 1977, another crisis led to the sale of an additional 1,350 acres to the state park. But the problems continued: the original building, Cliff House, was utterly consumed by fire in 1978, and the Wildmere building was judged beyond renovation. Minnewaska terminated its operations in 1979.

A plan by the Marriott Corporation represents the most serious proposal yet offered for resolving Minnewaska's long-standing financial problems. Marriott has proposed to build a $78 million resort conference center consisting of a 400-room hotel (on the Wildmere site), 300 condominium units (on the Cliff House site), an 18-hole golf course, 6 indoor and 5 outdoor tennis courts, 6 racket ball courts, a swimming pool, an indoor equestrian center, facilities for downhill and cross-country skiing, and parking lots for almost 1,000 cars.

Marriott's proposal falls within the scope of the New York State Environmental Quality Review Act that requires a written statement of the project's environmental, social, and economic impact. Marriott also sought a 10-year property tax reduction, an incentive that communities sometimes provide to developers. The company also needed to obtain a variety of local building, health, and safety permits. Marriott set the state environmental quality review process into motion by filing a draft impact statement, and the public hearings began some months ago.

On the evening of July 14, 1980, the auditorium of Rondout Valley High School was packed and sweltering. More than 100 individuals and groups—including the Mohonk Trust—had asked for time to speak. The speakers fell into three main groups. In the first group, some individuals were entirely in favor of the proposal, usually on economic grounds. Marriott anticipates that the new resort would employ 450 people full-time and have a first-year payroll of $6 million. There would also be large increases in school-tax and sales-tax revenues. Other speakers were inclined to endorse a new hotel, but they had questions about some aspect of the proposal—often related to its sheer size. The third group was opposed to any new resort and asked that the land be acquired instead for public recreation.

No one knows exactly where the people of Ulster County stood on the issue. In a newsletter poll conducted by State Assemblyman Maurice Hinchey, 55% of the respondents favored public ownership of the land, and 68% were opposed to granting Marriott a business-incentive tax abatement. Hinchey himself said,

> I think it makes sense to have a hotel up there, but if they push the level of development beyond the point of what it ought to be, it will not be a good development for them or for the community.

On the morning of July 15 the nuts-and-bolts work of state environmental quality review began with sworn testimony by expert witnesses before the hearing officer. Marriott came to court with a phalanx of consultants and lawyers in support of the draft impact statement. The state law, however, does not provide for an ombudsman to represent the general public. Although every citizen has the right to give testimony and to cross-examine witnesses, it was obvious to Marriott's critics (properly called "intervenors") that a thorough review of the inch-thick impact statement would require coordination and some professional help.

Friends of the Shawangunks, a 10-year-old citizen group, took the lead in raising money, hiring a respected attorney (Philip Gitlin), and doing the necessary research on the environmental issues. They were joined by Citizens to Save Minnewaska, a newly formed group whose primary focus was the social and economic issues. Many other organizations—including Minnewaska State Park, the Catskill Center, and the Department of Environmental Conservation—were represented at the hearings, and individual citizens did indeed appear to have their say.

The Mohonk Trust has played a special role by making its knowledge of the mountain environment available to all parties—the intervenors as well as Marriott's consultants. Nowhere has the trust's data file proved more important than

EPILOGUE

on the question of water supply, which emerged as the crucial issue of the hearings.

Briefly, Marriott would prefer to draw all their water from Lake Minnewaska—there is no proven alternate source. But the new resort's human population (a peak of 2,000 people) and water consumption (200,000 gallons per day in summer) would be much greater than that of the former hotels. In a moment of high courtroom drama, a grandson of the builder of the original resort produced water-level figures for the years 1938–1942. When these were correlated with Mohonk's weather station records, they showed that Marriott's proposed consumption would lower the lake by 16 feet. A calculation for the drought years of the mid-1960s showed that the new resort would have used up 80% of the lake water.

In August, these projections were basically confirmed by an expert hydrologist from the Department of Environmental Conservation, and Marriott asked for a recess to conduct new research. Because Marriott has no desire to build a resort overlooking a disappearing lake, they may have been saved by their critics from a multimillion-dollar mistake.

Although water supply is the central issue, it is by no means the only one, and every issue turns out to be multifaceted. It may seem self-evident, for example, that a $78 million investment would be good for Ulster County's economy, but the details cannot be overlooked, such as the fact that many of the jobs would be minimum-wage positions or that property-tax revenues would be reduced during the 10-year tax abatement period. The actual calculation of "benefit" and "loss" proved to be highly complicated, just as in a natural ecosystem.

Among the many other issues have been: soil erosion, air and noise pollution, the effect on vegetation and wildlife, the visual character of the new buildings, the plan for using treated sewage water on the golf course, the expansion of the golf course in the state-held conservation easement area, casino gambling, the potential breakup of the century-old trail network, the increase in highway traffic, and the cloudy future of other lands retained by the current owners. (The trust lands would be most directly affected by trail changes, reduced water flow in the Coxing Kill, and highway traffic.)

The final chapter of the impact statement is entitled "Reasonable Alternatives," and Marriott's proposal cannot be fully evaluated without a perspective on these options. Marriott has steadfastly maintained that a smaller resort complex would not be economically feasible. New York State, with equal consistency, has said that it does not have the money to acquire Minnewaska for public recreation. It is a fact that Minnewaska has substantial debts, and one possible outcome is sale of the property to satisfy the creditors. Even Marriott's critics admit that an auction could produce a worse result.

CHAPTER 13

Problem 2

Instructions

The following hypothetical case was developed by Lawrence Susskind. It presents a controversy at a early stage. What should the mediator attempt to accomplish at the first meeting? What should his or her long-term goals be? How might the presence of the mediator affect the bargaining strategies of the various parties.

This hypothetical case may be analyzed in the same way that the principal cases were in earlier chapters. Be sure to consider the ways in which the small scale of the dispute might alter the conduct and consequences of negotiation. The problem also provides a rich vehicle for role playing.

Background

Seaport is a small New England coastal town with 7,000 year-round residents and 15,000 summer residents. Seaport is one hour from the state's largest city by car or train. One-third of the land in Seaport is held by the federal, state, or local government (most of this land is on the coastline). Thirty-seven landowners control undeveloped acreage that comprises more than 20% of all the land resources in the town. Seven of these individuals own more than 40 acres each. The cost of a single family home in Seaport is about $100,000. An acre of residential land (in legal lots of suitable size for home building) sells for about $65,000 on the outskirts of town; an acre of oceanfront land costs between $150,000 and $200,000.

Seaport relies solely on its zoning by-laws, subdivision controls, and building codes to control the pace and quality of development. There are two people per acre in Seaport, or one home for every two acres.

The state Department of Community Affairs says that there are some 1,000 households in Seaport that ought to be receiving some form of financial assistance for housing (i.e., they spend more than 25% of the income for rent). The DCA says that an additional 250 household units require new or rehabilitated units.

Seaport can expect more than 750,000 visitors between May and October each year. Almost 80% of these will be "daytrippers" who spend less than a full day in town shopping, using the beaches, or sightseeing. Guests staying in commercial lodgings (hotels, motels, inns, or guest houses) account for another 15% of the influx. The remaining 5% or so are tourists who stay with relatives or friends. Tourists spend an estimated $10 million in Seaport each year. This money circulates through the local economy creating as much as $15 million

EPILOGUE

worth of business. There are over 5,000 business establishments in Seaport—about 75% of all business in town are tourist dependent.

The Seaport area has few rivers, ponds, or lakes, but it has an abundance of rocky ledges and drainage problems. Precipitation must be stored to protect against drought. Seaport's potable water supply cannot guarantee the town the water it needs in a dry year. Seaport is one of the few municipalities in the state with an approved sanitary landfill, although the landfill's capacity will be exceeded in another few years. The town has a sewage system that covers the downtown area and a few other outlying districts. The treatment plant, built about 10 years ago, will be adequate for only another 10 years at current growth rates of about 3% per year.

The Tremonts, one of the town's long-standing families, have decided to use part of a vacant 25-acre site (which it has owned for generations) near the center of town to develop a 100-unit low-income housing project, with funding from the state Housing Finance Agency. The site sits high atop a hill overlooking the ocean; its current market value is over $3 million, although it is currently assessed at $500 thousand and brings in only $11 thousand each year in property taxes. For decades the site has been used by local residents as a picnic/play area with the tacit permission of the Tremont family.

The Tremonts have decided that the town needs more low-income housing. The youngest member of the family, Emily Tremont, is a 30-year-old attorney interested in expanding her family's real estate business. She has a small development company that proposes to act as the developer for the Pigeon Hill low-income housing project. The units she proposes to build will be turned over to the local housing authority to manage, once they are constructed. She has submitted plans and designs to the state Housing Finance Agency showing 100 units in 5 two-storied clusters at the top of Pigeon Hill with a social service center in one of the buildings. As Figure 13 illustrates, the units will be 5 minutes walking distance from the railroad station and the shopping mall in the center of town.

Under state law, another agency, the state Housing Appeals Board, has the power to override local authorities should Seaport attempt to reject a request for the necessary rezoning of the land to permit multifamily housing. The site has been zoned as "single family residential" for more than two decades. The state Housing Appeals Board will not overturn the town's rejection of the site plan or the rezoning request if there are legitimate health and safety reasons for rejecting the project. The state law also requires Seaport to issue a consolidated (i.e.. water, sewer, zoning, conservation, etc.) permit within 90 days from the date that a plan is formally filed with the planning board.

Ms. Tremont has indicated to the head of the Seaport Town Council (a five-member council that administers all town funds and departments) that she plans to ask for a consolidated permit review and that if turned down by the town,

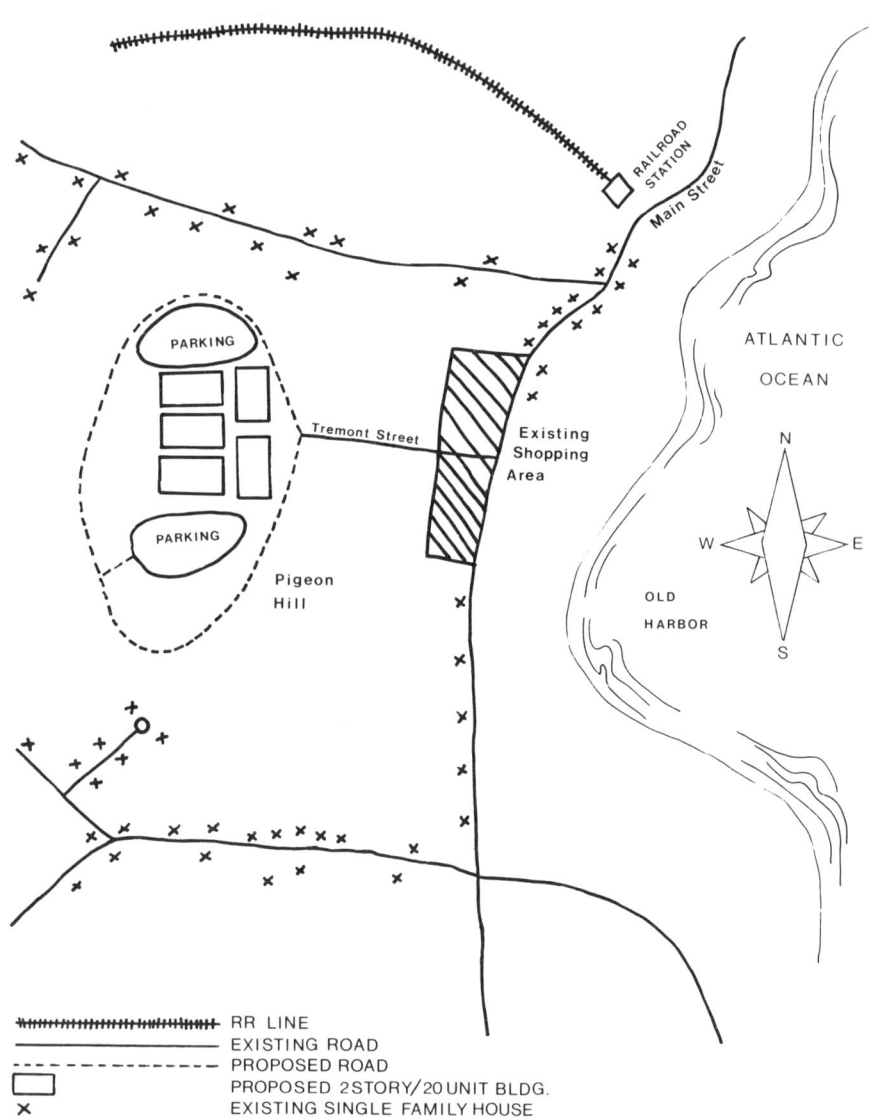

FIGURE 13. Map of Seaport.

EPILOGUE

she plans to appeal to the state. She has the money and the will to press her case. The family backs her completely. They will be donating the land at no cost, which is why the price of the units will low even if no federal or state rent supplements are available.

The proposed plan, prepared by a well-known architectural firm, indicates not only that 100 units and a social service center will be built but that additional commercial development along the edge of the site nearest the town center is also planned. Five thousand square feet of new tourist-oriented shops are proposed to be built on a commercially rezoned portion of the site. The new shopping area would be owned and managed by the Tremont Corporation. Ms. Tremont's limited dividend corporation has received a positive response from the state Housing Finance Agency regarding the plans for the project.

The local housing authority has indicated that it will accept responsibility for managing the new public housing units but only if the town doubles its annual administrative allocation from $50,000 to $100,000. At present, the housing authority manages only 25 units of public housing. The town water department claims that the existing pumping station in the central section of town is not adequate to provide sufficient water pressure to serve the 100 new units. (In addition, the water department claims that the water pressure throughout the entire central section of town would be diminished if the project proceeds.) A new pumping station would cost the town about $400 thousand, according to the water department. Even if the new station were built, there is the additional concern about the adequacy of the water supply. The town's fire chief claims that the proposed project will not be adequately protected from fire because of the pressure problem. He also claims that the layout of the project will make it impossible for fire trucks and ambulances to get in and out because of the single access road to the site. The state Housing Finance Agency has indicated that it will only fund the project if the units are made avaiable to any and all low-income applicants (not just the Seaport residents). The agency insists that the project must include at least 50 units for low-income families with children.

The town's Conservation Commission has asked the Tremont family to donate the land to a conservancy trust, arguing that the site is the best possible area for a park (because of the spectacular ocean view) and that the aged town sewage system cannot possibly handle the additional burden of 100 units in the center of town. The Conservation Commission claims that if the system fails, the town's limited water supply will be contaminated. A new sewage system would cost several million dollars.

The abutters to the site are mostly wealthy summer residents who have built expensive summer homes at the top of Pigeon Hill and along its slopes. They will all be in full view of the proposed low-income housing project. The senior citizens association in Seaport feels that the town really needs more housing for the elderly, and it opposes the state's requests that 50% of the units be family units and that the units be available to outsiders.

The Town Council has invited a well-known mediator from the nonprofit Center for Environmental Mediation to help organize an informal meeting of all the interested parties before Ms. Tremont submits her formal request for a 90-day consolidated permit review. The local newspaper, the *Seaport Eagle*, has just reported in an exclusive interview with the governor that he has agreed that the town should not be forced to accept a low-income housing project that it does not feel is in the best interests of the town—regardless of what the state housing law intends.

The meeting is about to begin. The positions presently taken by the principal parties are summarized next.

The Town Council (represented by the head) wants to avoid a state housing appeals review and has taken no posiion on the project. But the members do worry that the addition of new family units will increase the cost of providing services.

The Tremont family (represented by their local attorney) wants to make an important social contribution to the town and to the region.

Ms. Tremont (representing herself) wants to build the 100 units and the 5,000 square feet of commercial space.

The Seaport housing authority (represented by its elected head) favors the project and agrees to manage the units but only if the town increases the authority's annual operating budget.

The Conservation Commission (representing by its elected head) strongly opposes the project on the grounds that the site has always been used for recreation and is unsuitable for development because it would overburden the sewage and the water systems.

The Abutters Association (represented by its appointed head, a wealthy attorney and summer resident) strongly opposes the project on environmental grounds and because property values would be diminished. The association is willing to fight all the way to the Supreme Court. It argues that the state housing appeals system is a violation of home rule and that rezoning would be a breach of the public trust.

The state Housing Finance Agency (represented by its gubernatorially appointed head) favors the project if one-half of the units (at least 50) are for low-income families and if all the units are open to non-Seaport residents. It will provide necessary funds for construction.

The state Housing Appeals Board (representing by a staff member) will grant the zoning override if there is a local consensus on public health and safety issues; it will send an observer to the meeting.

The Seaport Planning Board (represented by the elected head) will grant the zoning change only if there is a consensus in the community regarding the desirability of the project; otherwise it will fight the state Housing Appeals Board.

The Board of Trade (represented by an appointed head) wants the new commercial development to proceed but opposes the housing project on the ground that it will create a bad image in the eyes of others.

The Senior Citizens Association (represented by its president) opposes the project as planned because it wants more units set aside for seaport's elderly.

The Seaport Fire Department (represented by the chief) strongly opposes the project on the grounds that the site and the site plan would not permit adequate fire protection.

The *Seaport Eagle* (represented by its editor in chief) has written editorials arguing that decisions about projects such as these should be made without state interference.

The mediator is very anxious to prove that mediation can "work" and also wants to enhance the reputation of the Boston center.

Politics and Alternative Dispute Resolution

This book is premised on the belief that negotiation is often a promising method for resolving environmental controversies. In some instances, it may yield results more quickly and economically than do competing alternatives like litigation, though as a number of the principal cases in this book illustrate it is not necessarily fast or cheap. The real virtue of environmental negotiation lies in its potential to produce decisions that are more acceptable to affected parties and that are consequently easier to implement than those that are imposed by courts.

Like any other tool, however, negotiation can be misused. In the early 1980s, there was growing concern that the EPA was trying to employ negotiation as a smoke screen to cover up lax prosecution of environmental standards. Ann Gorsuch Burford, the former administrator of the EPA, came under heavy attack from environmentalists, members of Congress, the press, and the public for eschewing a policy of vigorous enforcement of federal claims for cleanup of abandoned hazardous waste dumps in favor of a strategy of negotiated settlements. Stories of "sweetheart" deals and co-optation of the enforcement process have raised broader questions about the legitimacy of negotiation as a strategy for achieving regulatory compliance.

Such doubts about environmental negotiation really rest on two concerns: one is about outcome; the other, about processes. Specifically, it is feared that unless administrative decisions are subject to judicial review, an agency will be free to cut deals that favor particular interest groups at the expense of the public at large. It would be naive to say that this never happens. At the same time, however, it is erroneous to assume that litigation is necessarily superior in this regard. A negotiated enforcement process may be manipulated and exploited, but litigation also is vulnerable to similar tactics. Prosecuters of all types enjoy considerable discretion in what suits they bring, what allegations they make, and what penalties they seek. Moreover, any agency with enforcement responsibilities must allocate its scarce prosecutorial resources; inevitably, some cases are tried with more vigor than are others. Once a case is before a judge, of course, he or she can aggressively try to get to the heart of the matter, but, in much environmental litigation, the issues tend to be procedural rather than substantive. Moreover, it is difficult for any judge, unschooled in technical matters, to compensate for an agency's failure to produce relevant scientific evidence. In short, to the extent there can be several levels of judicial review of administrative actions, the focus usually is on whether the agency observed procedural requirements in reaching its decision—not on whether the decision itself was "correct."

Although judicial review does not really solve the problem of outcomes, it does address the second concern—that of process. In court, the parties are clearly identified; they may present their own witnesses and cross-examine those of others; and they participate in hearings that are open to the public. All these attributes of the legal process contribute to its legitimacy. To the extent that a particular negotiation is carried out differently, the process runs the risk of appearing illegitimate, no matter how benign the settlement actually is. In the case of the Superfund controversy (involving settlements with firms which had dumped toxic wastes), the real problem was not with attempting to settle out of court but with limiting the parties at the bargaining table. Administrator Burford's actions were called into question largely because these negotiations were conducted bilaterally between EPA and the firms alleged to have dumped illegally. Other parties with an interest in these negotiations (such as the affected communities, adjacent landowners, and environmental groups) did not participate. The desire of EPA to exclude these interests, coupled with other actions taken by EPA, led observers to question the administration's commitment to environmental objectives and to conclude that the negotiations really were a device to subvert enforcement.

As most of the cases in this book demonstrate, effective negotiation is usually promoted when there is full participation by all interested parties. Although a regulating agency and a company might initially wish to exclude others from the bargaining table, it usually should be in their interest to include anyone who could later challenge a settlement and thus prevent its implementation. Moreover, when all affected parties are at the table, there is a better chance that all the relevant issues will be raised and that the parties will be better situated to make efficient trades. As desirable as full participation may be, however, sometimes power and communication considerations require a different approach. In the Foothills case (chapter 10) Representative Tim Wirth found it necessary to

EPILOGUE

begin mediation by meeting first with a few powerful parties—and he was roundly criticized by those who were excluded. Ultimately, he quieted this outcry by drawing the others into later bargaining sessions. If EPA had similarly included other interested parties in its Superfund settlement talks, for example, the likelihood of scandal might well have been diminished considerably. In cases where there are just too many interested parties to include them all, proponents of negotiation must find ways to obtain representative participation. Where this is impossible, negotiation simply may not work.

The issue of participation by affected parties is closely tied to the question of public access to negotiation sessions. This is an era in which openess in government is highly valued. Many of the recent reforms of administrative process—the Freedom of Information Act, open meeting laws, limits on ex parte communication, and so forth—have made the operation of government much more visible to the public. Controversy over the Superfund negotiations was heightened by the fact that they were conducted in private.

It is often the press that is loudest in complaining about government involvement in closed-door bargaining sessions. As noted in chapter 10, even if the press cannot really be considered a party to a dispute, it most certainly has the capacity to shape and influence public perceptions of the conduct of negotiations. Moreover, the disputants themselves may try to exploit press coverage in order to develop support within their various constituencies and thus develop bargaining strength. In other cases, however, the parties may believe that it is in their interests to try to exclude journalists. They may be afraid that their actions, if reported, would alienate certain groups. Also, they may realize that it is hard to float possible compromises if outsiders are going to scrutinize and possible criticize them.

Certain kinds of agency proceedings must be conducted in the public eye, administrative rulemaking, for one. Public access or "sunshine" laws usually create exceptions, however, for negotiations over possible settlements of lawsuits. In the present political climate, perhaps agencies may find it prudent to be more open in these situations than the law actually requires. Whatever is gained by privacy may be more than outweighed by the suspicion that is generated when the government works behind closed doors. Some balance between these competing considerations may be possible. There are, after all, degrees of public access. Even if some sessions are limited only to the principals, others can be open.

In the end, the Superfund experience, for all its negative aspects, may provide an encouraging lesson about regulatory negotiation. It does demonstrate that when the government fails to include affected parties in negotiation and when it limits information about proceedings, it will come under harsh criticism. Perhaps, what has been repudiated is not negotiation in general but a style of negotiation that is vulnerable to misuse and exploitation.

Study Questions

1. Would the concerns that were raised by the Superfund cases be adequately met if all enforcement settlements negotiated by an agency had to be approved by a court? If so, should such agreements carry any presumption of validity? What should the standard of judicial review be? What should be the consequences of a judicial finding that an agreement was not in the public interest?

2. Would the problem of special deals be lessened if a mediator participated in the negotiation? In divorces and other domestic litigation, a guardian ad litem is sometimes appointed to represent the children. Should some sort of guardian of the public interest be given a role in regulatory negotiation? What role might he or she play? What sort of experience and qualifications should such a person have?

3. There are important environmental lawsuits in which only private parties participate, typically an environmental group and a corporation. Should we be any less concerned about the out-of-court settlements these litigants make when the terms clearly will affect the public interest?

Conclusion: Negotiation and Public Policymaking

This book has emphasized what might be termed *microanalysis of environmental* disputes. We have concerned ourselves with issues that confront parties actually embroiled in a specific dispute. Who should be at the bargaining table? What are their incentives to negotiate? How can the deal be made binding, and so forth? We have given less attention, however, to a broader set of questions that deal with the wisdom of making policy seriatim through a process of negotiation.

In chapter 1, we considered how disputants might view the relative advantages of settling their differences though negotiation or litigation. From the parties' perspective, negotiation in some instances might resolve disputes more efficiently, both in terms of providing a quicker and less expensive process, and, more important, by facilitating outcomes that maximized their collective welfare. But what about society at large? Over the years will we have better environmental policy if we resort to negotiation to resolve environmental disputes intead of litigation? Has our concern for process efficiency and the interests of particular parties obscured larger social objectives? Might the whole be less than the sum of the parts?

One danger in encouraging people to resolve disputes through negotiation, as opposed to litigation, is that they will strike deals that are good for themselves and bad for society as a whole. For example, might not the environmentalists in the Grayrocks Dam case have simply accepted the initial offer of compensation for the loss of the whooping crane and allowed the dam to be built and the creature's habitat to be destroyed? What is to prevent disputants from advancing their immediate interests at the expense of future generations? What is to prevent them simply from being wrong? In environmental cases, the consequences of an error in judgment may be irrevocable. The hope is that the same concerns that motivated the environmentalists to sue in the first place—the welfare of the crane—would also work to reject the initial offer of a direct compensation payment. Still, there is no guarantee that even the most sincere litigants will not sometimes be wrong. In the Grayrocks case, for example, the environmentalists might have mistakenly calculated that the whooping crane could probably survive the changes in waterflow and that the cash offered by the power company could be better put to other projects. The simple answer is that negotiation does not eliminate the possibility of serious mistakes—but neither do other processes. Well-intended judges, dealing with fragmentary evidence and uncertain science, may make decisions that they later regret, along with the rest of us. To the extent that negotiation allows a fuller examination of relevant data and encourages the participation of all affected parties, better decisions may be encouraged.

It is possible, of course, that even if a given settlement is efficient and equitable within the context of a particular dispute, it may be inconsistent with policies that have been adopted in other cases. Environmental planning requires extensive coordination. The ultimate effectiveness of an air pollution control system at one factory depends on the abatement strategies that are used elsewhere. Litigation allows greater consistency and coordination than does negotiation, it might be argued, because judges are bound to decide cases according to precedent. Like cases are supposed to be decided alike. Because of this respect for history, radical shifts in judge-made law are rare. The system operates predictably. Change in judge-made policy occurs incrementally, thus giving society time to reflect upon the wisdom of the change. By contrast, each negotiation stands alone. The parties are free to fashion any deal that accommodates their interests, without regard for whether their actions will be consistent with the resolution of prior disputes. But, whereas in theory, negotiating parties may be free to ignore history, in fact they rarely do. People bargain in the shadow of the law. To the extent that precedents exist (be they legal precedents or otherwise), they create expectations for future negotiators. This certainly has been true in collective bargaining over labor contracts. A wage gain obtained by one union creates expectations of comparable gains by another. Similarly, the fact that management acceded to a particular union request in one industry often influences the willingness of management to accede to the same request in another industry. In fairness, it must be noted that environmental disputes—at least major ones—are less apt to fall into familar patterns. As was noted in the previous section, the terms that were reached in the Foothills dispute (chapter 9) cannot really be expected to set a precedent for other cases. In situations where the issues are of first impression, that is, they have never been addressed before, the parties are free to fashion whatever agreement suits their needs. Yet, this is also true when such cases arise in litigation: Judges obviously are not bound by precedent where no precedent exists.

In short, where negotiation has shortcomings as a policymaking device, these same shortcomings are clear when other mechanisms—litigation and unilateral bureaucratic action—are substituted. Why is it, then, that many people still feel more comfortable relying on adjudication rather than negotiation in shaping social policy?

As a mechanism for making judgments about complex issues, adjudication is intuitively attractive. The plaintiff and the defendant each advance the strongest possible argument in defense of their positions, and a judge renders an impartial decision. To a first approximatiom, the judge's decision represents "the truth," or alternatively, "the right answer." Of course, given the uncertainties inherent in making judgments about complex scientific issues that lie at the heart of many environmental controversies, usually there is no unique right answer. Nonetheless, the judicial process creates the illusion of one that is often appealing.

By contrast, negotiating the resolution of the same controversy lays bare all of the uncertainties and complexities that render environmental policymaking so difficult in the first place. As the parties dissect their points of disagreement, they are forced to confront the elusive nature of facts. Inevitably, facts and data are negotiated. Moreover, the parties may be forced to compromise on matters of principle. The pulling and hauling that is the essence of bargaining forces the parties to recognize that there are no right answers—only compromises worked out on intermediate positions. Instead of creating the illusion of truth, bargaining embraces the accommodation of competing interests. Moreover, the process of compromise forces each side to acknowledge the legitimacy of the claims of the opposition.

To be sure, litigation also reveals differences of opinion over facts, theories, values, and principles. What distinguishes adjudication from negotiation is that the former does not require the parties to accept the legitimacy of competing claims or to acquiesce in the final outcome. In fact, the availability and frequency of appeals would suggest that the system expects people to take issue with the judge's decision. Thus, litigation preserves the illusion of a right answer and allows the parties to cling to their version of it.

Bibliography

"Agenda for Environmental Negotiation." *Environmental Impact Assessment Review* 3-1 (March 1980) and subsequent issues.
Alexander, Tom. "A Promising Try at Environmental Detente for Coal." *Fortune* 97-3 (1978):94–102.
Bacow, Lawrence S. *Bargaining for Job Safety and Health.* Cambridge: M.I.T. Press, 1980.
Bacow, Lawrence S., and Milkey, James R. "Overcoming Local Opposition to Hazardous Waste Facilities: The Massachusetts Approach." *Harvard Environmental Law Review* 6(1982):265–305.
Bellman, Howard S. "A Mediator Is a Mediator Is . . ." *Environmental Consensus* (now called *Resolve*) (Spring 1980):4–5.
Bellman, Howard S., Bingham, Gail, Brooks, Ronnie, Carpenter, Susan, Clark, Peter, and Craig, Roger. "Environmental Dispute Resolution: Practitioner's Perspective of an Emerging Field." *Environmental Consensus* (now called *Resolve*) (1981): 1–7.
Bellman, Howard S., and Sachs, Andy. "Parallels in Labor and Environmental Mediation." Working Paper of the Public Disputes Program, Program on Negotiation, Harvard Law School, 1984.
Bellman, Howard S., Sampson, Cynthia, and Cormick, Gerald W. *Using Mediation When Siting Hazardous Waste Management Facilities. A Handbook* (SW 944). Prepared for the U.S. Environmental Protection Agency, Office of Solid Waste. Available from the Government Printing Office, 1982.
Bingham, Gail. "Does Negotiation Hold Promise for Regulatory Reform?" *Resolve* (Fall 1981):1–8.
Bingham, Gail, Vaughn, Barbara, and Gleason, Wendy. *Environmental Conflict Resolution: Annotated Bibliography.* Washington, D.C.: The Conservation Foundation, 1981.
Bosselman, Fred P. "Buying Off the Neighbors: Negotiated Private Settlements of Development Disputes in Japan." *Environmental Comment* (May 1977):12–13.
Boulding, Kenneth E. *Conflict and Defense.* New York: Harper, 1962.
Boulding, Kenneth E. "Conflict Management as a Learning Process." In Anthony deReuck and Julie Knight (Eds.), *Conflict in Society.* Boston: Little, Brown, 1966.
Busterud, John. "Mediation: The State of the Art." *Environmental Professional* 2-1(1980):34–39.
Caldwell, Lynton K., Hayes, Lynton R., and McWhirter, Isabele M. *Citizens and the Environment: Case Studies in Popular Action.* Bloomington: Indiana University Press, 1976.
Carpenter, Susan L., and Kennedy, W. J. D. "Information Sharing and Conciliation: Tools for Environmental Conflict Management." *Environmental Comment* (May 1977):21–23.
Carpenter, Susan L., and Kennedy, W. J. D. "Environmental Conflict Management." *Environmental Professional* 2-1(1980):67–74.
Carpenter, Susan L., and Sachs, Andy. "The Decision to Intervene." Working Paper of the Public Disputes Program, Program on Negotiation, Harvard Law School, 1984.

Cifrino. Deborah. "Tearing Down the Wall through Environmental Mediation." *Conservation News* 43-19(1978):8–11.
Clark, Peter B. "Consensus Building Mediating Energy, Environmental and Economic Conflict." *Environmental Comment* (May 1977):9–12.
Clark, Peter B., and Straus, Donald B. "Computer Assisted Negotiations: Bigger Problems Need Better Tools." *Environmental Professional* 2-1(1980):75–87.
Colosi, Thomas. "Negotiation in the Public and Private Sectors." *American Behavioral Scientist* 27-2(1983):229–253.
Cormick, Gerald W. "Mediating Environmental Controversies: Perspectives and First Experience." *Earth Law Journal* 2(1976):215–224.
Cormick, Gerald W. "The 'Theory' and Practice of Environmental Mediation." *Environmental Professional* 2-1(1980):24–33.
Cormick, Gerald W., and Patton, Leah K. Environmental Mediation: Potentials and Limitation." *Environmental Comment* (May 1977):13–16.
Coser, Lewis A. *The Function of Social Conflict.* New York: Free Press, 1956.
Coser, Lewis A. "The Termination of Conflict." In Amitai and Eva Etzioni, (Eds.), *Social Change: Sources, Patterns, and Consequences.* New York: Basic Books, 1964.
Creighton, James L. "A Tutorial: Acting as a Conflict Conciliator." *Environmental Professional* 2-1(1980):119–127.
Deutsch, Morton. *The Resolution of Conflict: Constructive and Destructive Processes.* New Haven: Yale University Press, 1973.
Druckman, Daniel (Ed.). *Negotiations: Social-Psychological Perspectives.* Beverly Hills: Sage Publishers, 1977.
Ehrmann, John R., and Bidol, Patricia A. *A Bibliography on Natural Resources and Environmental Conflict. Management Strategies and Processes* 84. Chicago: Council of Planning Librarians, 1982.
Fisher, Roger. *International Mediation: A Practitioner's Guide.* New York: International Peace Academy, 1978.
Fisher, Roger, and Ury, William. *Getting to Yes: Negotiating Agreement Without Giving In.* Boston: Houghton Mifflin 1981.
The Ford Foundation. *New Approaches to Conflict Resolution.* New York: Ford Foundation, 1978.
Golten, Robert J. "Mediation: A 'Sellout' for Conservation Advocates, or a Bargain?" *Environmental Professional* 2-1(1980):62–66.
Golten, Robert J., and Sachs, Andy. "The Interplay of Environmental Litigation and Negotiation." Working Paper of the Public Disputes Program, Program on Negotiation, Harvard Law School, 1984.
Greenberg, Michael R., and Straus, Donald B. "Up-Front Resolution of Environmental and Economic Disputes." *Environmental Comment* (May 1977):16–18.
Gulliver, P. H. *Disputes and Negotiations: A Cross Cultural Perspective.* New York: Academic Press. 1979.
Gusman, Sam. "Selecting Participants for a Regulatory Negotiation." *Environmental Impact Assessment Review* 4-2(June 1983):195–202.
Gusman, Sam, and Sachs, Andy. "Developing a Model for Policy Dialogues." Working Paper of the Public Disputes Program, Program on Negotiation, Harvard Law School, 1984.
Kriesberg, Louis. *The Sociology of Social Conflict,.* Englewood Cliffs, N.J.: Prentice-Hall, 1973.
Lake, Laura. *Environmental Mediation: The Search for Consensus.* Boulder, Colorado: Westview Press, 1980.
Laue, James, and Cormick, Gerald W. "The Ethics of Intervention in Community Disputes." In Gordon Bermant, Herbert Kelman, and Donald Warwick, (Eds.), *The Ethics of Social Intervention.* Washington, D.C.: Hemisphere Publishing, 1974.

Lee, Kai N. "Defining Success in Environmental Dispute Resolution." *Resolve* (Spring 1982).
Lesnick, Michael, and Crowfoot, James. *Bibliography for the Study of Natural Resource and Environmental Conflict*, 64. Chicago: Council of Planning Librarians. 1981.
Lord, William B. "Water Resources Planning: Conflict Management." *Water Spectrum* (Summer 1980):1–10.
Mernitz, Scott. *Mediation of Environmental Disputes: A Sourcebook*. New York: Praeger, 1980. (Available in paperback from the Conservation Foundation.)
Murray, Francis X. (Ed). *Where We Agree: Summary and Synthesis*. Report of the National Coal Policy Project. Boulder, Colorado: Westview Press. 1978.
O'Connor, David. "Environmental Mediation: The State of the Art." *Environmental Impact Assessment Review* 2(1978):9–17.
O'Hare, Michael, Bacow, Lawrence, and Sanderson, Debra. *Facility Siting and Public Opposition*. New York: Van Nostrand, 1983.
Raiffa, Howard. *The Art and Science of Negotiation*. Cambridge: Belknap Press/Harvard University Press, 1982.
Rapoport, Anatol. *Fights, Games and Debates*. Ann Arbor: University of Michigan Press, 1960.
Richman, Roger, and Gibson, William. "Environmental Conflict Resolution in Virginia." *Environmental Consensus* (September 1979):3–6.
Rivkin, Malcolm D. "Negotiated Development: A Breakthrough in Environmental Controversies." *Environmental Comment* (May 1977):3–6.
Rivkin, Malcolm D. *An Issue Report: Negotiated Development: A Breakthrough in Environmental Controversies*. Washington, D.C.: The Conservation Foundation, 1977.
Rubin, Jeffrey A., and Brown, Bert R. *The Social Psychology of Bargaining and Negotiation*. New York: Academic Press, 1975.
Sander, Frank E. A. "Varieties of Dispute Processing." *Federal Rules Decision* 70-79(1976):111–134.
Sander, Frank E. A., and Snyder, Frederick E. *Alternative Methods of Dispute Settlement—A Selected Bibliography*. Prepared for the American Bar Association Special Committee on Resolution of Minor Disputes. Washington, D.C., 1979.
Schelling, Thomas C. "An Essay on Bargaining." In *The Strategy of Conflict*. Cambridge: Oxford University Press, 1960.
Shorett, Alice J. "The Role of the Mediator in Environmental Disputes." *Environmental Professional* 2-1(1980):58–61.
Simkin, William E. *Mediation and the Dynamics of Collective Bargaining*. Washington, D.C.: Bureau of National Affairs, 1971.
Stockholm, Nan. "Environmental Mediation: An Alternative to the Courtroom." *Stanford Lawyer* 15-1(1980):21–25.
Straus, Ansel M. *Negotiation: Varieties, Contexts and Social Order*. San Francisco: Jossey-Bass, 1978.
Straus, Donald B. "Managing Complexity: A New Look at Environmental Mediation." *Environmental Science and Technology* 13-6(1979):661–665.
Sullivan, Timothy J. *Resolving Development Disputes through Negotiation*. New York: Plenum, 1984.
Susskind, Lawrence E. "Environmental Mediation and the Accountability Problem." *Vermont Law Review* 6(Spring 1981):1–47.
Susskind, Lawrence E. and Keefe, Frank. "The Negotiated Investment Strategy in Columbus," In *Report of the Negotiated Investment Strategy Project*. unpublished report on process and agreement for Columbus, Ohio, 1980.
Susskind, Lawrence E., and Weinstein, Alan. "Toward a Theory of Environmental Dispute Resolution." *Boston College Environmental Affairs Law Review* 9-2(1980–1981):311–351.

Susskind, Lawrence E., Bacow. Lawrence S., and Wheeler, Michael (Eds.). *Resolving Environmental Regulatory Disputes.* Cambridge: Schenckman, 1984.

Talbot, Allan R. *Settling Things: Six Case Studies in Environmental Mediation.* Washington, D.C.: The Conservation Foundation, 1983.

Thomas, Kenneth W. "Conflict and Conflict Management," In Marvine D. Dunnette (Ed.), The *Handbook of Industrial and Organizational Psychology.* Chicago: Rand McNally, 1976.

Tribe, Lawrence H., Shelling, Corrine, S., and Voss, John (Eds.), *When Values Conflict: Essays on Environmental Analysis, Discourse and Decision.* Cambridge: Ballinger, 1976.

Vaughn, Barbara J. "Environmental Mediation: Fighting Fair." *Planning* 46-8(1980):16–18.

Wall, James A. "Mediation: An Analysis, Review and Proposed Research." *Journal of Conflict Resolution* 25-1(March 1981):157–180.

Wehr, Paul. *Conflict Resolution.* Boulder, Colorado: Westview Press. 1979.

Young, Oran. *Bargaining: Formal Theories of Negotiation.* Urbana: University of Illinois Press, 1975.

Index*

Administrative law: enforcement (Brown Paper case), 56–73
    grantmaking (Jackson Hole case), 127–145
    mandatory negotiation, 328–338
    permitting (Holston River case), 77–91
    regulatory reform, 305–322
    rulemaking, 279–322
Administrative Procedure Act, 290
Air pollution: enforcement of Clean Air Act (Brown Paper case), 56–73
    mandatory negotiation under Clean Air Act (Colstrip case), 329–337
    power plant conversion (Brayton Point case), 158–187
Arbitration: contrasted with mediation, 157–158
    defined, 157
    in hazardous waste siting, 341
    of unresolved issues, 152

Bacow, Lawrence, 99–103, 306–307, 339–346
Bargaining power, 38–41
Bargaining range, 34–38
Brayton Point case, 158–187
Brown Paper case, 56–73

Chevron Oil v. NLRB, 338
Clean Air Act. *See* Air Pollution
Coalitions: strategy, 116–122, 304. *See also* Foothills case and Water Treatment Rulemaking case
Coase Theorem, 31–33

Collective bargaining: contrast with environmental negotiation, 74–75
Colstrip case (mandatory negotiation under the Clean Air Act), 329–337
Commons problem, 28–31
Compensation: to settle disputes, 54, 74, 344–346
Compliance with agreements: causes of non-compliance, 149–151
    enforcement mechanisms, 151–54
    and type of dispute, 148–149. *See also* Foothills Dam case and Jackson Hole case
Conflict anticipation/management, 326–328
Consent decrees, 154. *See also* Brown Paper case and Foothills Dam case
Cormick, Gerald, 191–194, 266–269
Courts. *See* Litigation

Data negotiation: elusive facts, 99–103
    Holston River case, 77–91
    judicial review, 91–96
    role of courts, 96–99. *See also* Foothills Dam case
Decision analysis: and negotiation theory, 23–26
Dunlop, John, 12

Endangered species. *See* Grayrocks Dam case
Energy conversion. *See* Brayton Point case
Energy Supply and Environmental Coordination Act, 157–187

---

*The major case studies are indexed to the extent that they illustrate general themes, but the detailed information in them has not been cited. The Brayton Point case, for example, can be found under air pollution, enforcement actions, mediation, and so forth, but not under the names of the people who were involved.

Enforcement of agreements. *See* Compliance with agreements
Enforcement of environmental regulations. *See* Brown Paper case
Environmental conflict: disputes over facts, 99–103
growth of, 1–4
origins, 5–10
Environmental litigation. *See* Litigation
Ethics: for mediators, 248–278. *See* Mediators

Fact-finding: contrasted with mediation, 156–157. For joint fact-finding, *See* Data negotiation
Federal lands disputes. *See* Foothills case and Gospel-Hump case
Fisher, Roger 26, 35, 45, 56
Flood control disputes. *See* Snoqualmie Dam case
Foothills Dam case, 195–247
Forestry dispute (Gospel-Hump case), 272–274
Foster, Charles, 190

Gladwin, Thomas, 3
Goldbeck, Willis, 105–109
Gospel-Hump case, 272–274
Grantmaking power of agencies (Jackson Hole case), 127–145
Grayrocks Dam case, 46–52
Growth control disputes. *See* Foothills case and Jackson Hole case

Hardin, Garrett, 28–29
Hazardous waste; Massachusetts siting law, 339–346
South Carolina regulations, 54–55
Highway construction dispute (West Side Highway case), 105–109
Holston River case, 77–91
Home Box Office v. FCC, 290, 305
Horowitz, Donald, 14–17

Incentives to negotiate, 26–28, 42–55
Institutionalizing negotiation: conflict anticipation, 326–328
creating incentives, 339–346
credentialing mediators, 254–256, 257, 258, 259
mandatory negotiation, 328–338

Institutionalizing negotiation (*cont.*)
rulemaking reform, 279–322
providing compensation, 344–346
removing procedural obstacles, 324–326
Intergovernmental disputes. *See* Brayton Point case, Foothills Dam case, Jackson Hole case, Water Treatment Rulemaking case, and West Side Highway case
Intervention, *See* Mediation

Jackson Hole case, 127–145
Joint fact-finding. *See* Data negotiation
Joint problem solving: Brown Paper case, 56–73
versus competition, 304
Judicial Review: scope of, 17–18, 91–96
of technical issues, 96–99. *See also* Litigation

Knaster, Alana, 191–194

Lake, Laura, 105–109, 113–115
Lawsuits. *See* Litigation
Litigation: advantages of, 12–17
costs, 10–12
incentives to settle, 43–44
simultaneous with negotiation, *See also* Brown Paper case, Foothills Dam case, Grayrocks Dam case, and Judicial Review
Lovi, Peter, 275–278
Lynn, Laurence E., Jr., 31–32

McCrory, John, 156–157, 252–253, 258–262
Malpractice: mediator's, 260–262
Mandatory negotiation under the Clean Air Act (Colstrip case), 329–337
Marcus, William, 31–32
Media. *See* Press
Mediation: defined, 156–158
initiation of, *See* Brayton Point case and Foothills Dam case
institutionalizing, 254–256, 257, 258, 259
of large disputes (Foothills Dam case), 195–247
phases of, 190
techniques, 156–194
See also Mediator, and Brayton Point

INDEX

Mediation (cont.)
    case, Foothills Dam case, Gospel-Hump case, Snoqualmie Dam case, and West Side Highway case.
Mediator: accountability of, 248–253
    with clout, 270–275
    credentialing or licensing of, 254–255
    duties to bystanders, 248–253
    ethics for, 248–278
    links to courts, 255
    malpractice, 260–262
    motives, 262–265
    payment for, 256, 262–265
    role of, 188–189, 191–194
    skills of, 187–194
    See also Mediation.
Milkey, James, 339–346
Models: of environmental impacts. See Brown Paper case, Brayton Point case, and Holston River case
Multiparty negotiation, 104–125. See Also Foothills Dam case, Snoqualmie Dam case, and Water Treatment Rulemaking case.

Negotiated rulemaking: agency's role, 310–312
    existing constraints, 289–291
    legal issues, 310–316
    proposed reforms, 305–322
    Regulatory Negotiation Act of 1980, 318–321
Negotiation: as alternative to litigation, 18–20
    analytical methods, 22–26
    bargaining power, 38–41
    bargaining range, 34–38
    among coalitions, 116–122
    commitment strategy, 38–41
    costs of, 27
    good faith bargaining, 332–334, 337–338
    incentives for, 26–28, 42–55
    mandatory, 328–338
    obstacles to, 28–33, 53–54
    participants in, 109–116
    and policy making, 359–364
    of small-scale disputes, 347–359. See also Compensation, Compliance with agreements, Multiparty negotiation, and Negotiated rulemaking.
Neutral. See Mediation

Newspapers. See Press
Nonzero-sum disputes, 33–38

O'Connor, David, 190

Permitting power of agencies (Holston River case), 77–91
Policymaking, 362–364
Politics of dispute resolution, 359–362
Pollution. See Air pollution, Hazardous waste, Sewage treatment disputes, Water pollution
Power plant: construction (Grayrocks Dam case), 46–52
    energy conversion (Brayton Point case), 158–187
Press: its role in environmental disputes, 244–247
Prisoners' dilemma, 29–31, 304
Problem solving. See Joint problem solving
Public participation: in negotiation, 109–116

Raiffa, Howard, 118–122
Referendums: to choose bargaining representatives, 147–148
Regulation. See Administrative law
Regulatory Negotiation Act of 1980, 318–321
Regulatory reform: 305–322
"Rethinking Regulation," (Harvard L. Rev. Note), 310–316
Rivkin, Malcolm, 74
Roosevelt, Theodore, 73–74
Rubin, Alvin, 44
Rulemaking power. See Negotiated rulemaking

Scenic Hudson I & II, 9, 10–12
Schelling, Thomas, 38–40
Science court, 99
Scientific disputes, 91–103
Settlement. See Compensation, Compliance, and Negotiation
Sewage treatment disputes. See Jackson Hole case and Waste Water Treatment case
Small-scale disputes, 347–359
Snoqualmie Dam case, 113–116
South Terminal Corp. v. EPA, 92–96
Stewart, Richard, 305–306, 307–310

Storm King, 9, 10–12
Stulberg, Josh, 188–189, 250–251, 274–275
Sullivan, Timothy, 109–116, 147–148, 272–274, 331–336
Susskind, Lawrence, 246–247, 249–262, 274–275, 325–326

Technical disputes, 91–103
Third parties. See Mediation
Thurow, Lester, 34
Tragedy: of the commons, 28–31

Uncertainty: as source of conflict, 7–8
Ury, William, 26, 35, 45, 56

Waste, Hazardous. See Hazardous Waste
Water: conservation and distribution (Foothills case), 195–247
diversion (Grayrocks dam case), 46–52
Water pollution. See Holston River case, Jackson Hole case, Water Treatment Rulemaking case
Water Treatment Rulemaking case, 280–305
Weinstein, Alan, 325–326
West Side Highway case, 105–109, 113
White Flint Mall case, 74
Will, Hubert, 44

Zero-sum disputes, 33–38